INOCULATED

HOW SCIENCE LOST ITS SOUL IN AUTISM

KENT HECKENLIVELY, JD

Skyhorse Publishing

Children's Health Defense

Skyhorse Publishing books may be purchased in bulk at special discounts for sales promotion, corporate gifts, fund-raising, or educational purposes. Special editions can also be created to specifications. For details, contact the Special Sales Department, Skyhorse Publishing, 307 West 36th Street, 11th Floor, New York, NY 10018 or info@skyhorsepublishing.com.

Visit our website at www.skyhorsepublishing.com.

10 9 8 7 6 5 4 3 2

Library of Congress Cataloging-in-Publication Data is available on file.

Print ISBN: 978-1-5107-6518-4
Ebook ISBN: 978-1-5107-6519-1

Printed in the United States of America

"It's troubling to me that in a recent Senate hearing on childhood vaccinations, it was never mentioned that our government has paid out over three billion dollars through a Vaccine Injury Compensation Program for children who have been injured by vaccinations."

—Congressman William Posey on the floor
of the US House of Representatives, July 29, 2015.

"There comes a time when silence is betrayal."

—Martin Luther King, Jr.

Table of Contents

FOREWORD

As we approach the beginning of 2021, the autism epidemic rages on unhindered and unnoticed by the federal government. In the early 1980s, the incidence of autism was about one in ten thousand. Now, it is one in forty-five, and continuing to grow at an exponential pace. I was first exposed to the autism epidemic in the late 2000s and heard recurring accounts of children regressing into the condition following one or more of their infant vaccines. To respond to this tragedy that so many families were and still are facing, I formed Focus Autism in 2009.

The controversies surrounding the cause of the autism epidemic are many, and the US government's response is woeful at best. The Centers for Disease Control and Prevention (CDC) continues to deny any relationship between autism and infant vaccines, despite the mountain of evidence supporting the theory. The revelations of Dr. William Thompson, epidemiologist at the CDC, add substantially to this evidence. Early in 2014, Brian Hooker revealed to me that he was in discussions with a CDC whistleblower and that the whistleblower was revealing a consistent pattern of corruption in the CDC's Immunization Safety Office.

What we suspected all along was now being confirmed by an insider. Covering up an autism epidemic and its causes is no trivial thing. The ramifications are astounding: the more than one million children in the United States alone now affected by the epidemic, the hundreds of billions of dollars it will require to take care of these children, not to mention five thousand vaccine injury cases in the U S government's "Vaccine Court" that were denied justice due to the cover-up.

And then there are the scientific careers that were laid waste for those brave souls who dared to enter this research area with an open mind. The term "Wakefielded" is now synonymous with being scientifically discredited, fileted by the press, and ostracized by one's own countrymen. Dr. Judy Mikovits was actually thrown in jail over her important work on the presence of retrovirus particles in vaccines. Likewise, Brandy Vaughn is constantly haunted by surveillance from the pharmaceutical industry, where she was once employed, simply because she has spoken publicly regarding the lack of appropriate testing for vaccines.

The impact that autism has on families is tremendous. Having a child with autism will clean out your bank account, ruin your marriage, and put a huge strain on any other children in the family. Adding insult to injury, the autism epidemic and genetic susceptibility to vaccine injury have led to many families having more than one child on the autism spectrum. And autism isn't *Rain Man*. This misconception is laughable to any parent of an affected child. *Rain Man* portrayed a very high-functioning young man with autism—this is nowhere close to the norm.

Children with low-functioning autism may or may not have the ability to speak, many are not toilet trained, and more are haunted by constant pain, head pain, gut pain, and joint and muscle pain. Many of these children have seizures, some of which require hospitalization. In the face of ever-increasing pressure to mandate all vaccines nationwide, it is imperative that word get out regarding the causes of autism including the identification of the children most vulnerable for this and other types of damage caused by infant vaccines.

It is far past time to peel back the layers of this controversy, expose the criminal actions by CDC officials and others within the government and the pharmaceutical industry, and help children recover. The fact that this crisis could have been prevented means that it can still be stopped dead in its tracks.

Kent Heckenlively's fine book *Inoculated* is the first systematic effort to tell the entire story of vaccines and autism, starting with the 1986 National Childhood Vaccine Injury Act, a law that fundamentally changed the arrangement of legal checks and balances regarding vaccine safety. Vaccines are the only products in our society that are not covered by our traditional civil justice system. I want to stress that there are still many things we do not know about autism. But the one thing that everybody should agree upon is that no subject can be off-limits when it comes to the health of our children.

—Barry Segal, Founder of Focus For Health

CHAPTER ONE

The Call

November 7, 2013

Dr. Brian Hooker was a fifty-year-old associate professor of biology and chair of the Math and Science Division at Simpson University. A Christian liberal arts college in Redding, California, it perched quietly on the coast near the Oregon border. Dr. Hooker sat in his faculty office preparing his class lecture notes when the phone rang.[1]

Hooker looked up from his papers and saw the caller was from the 404 area code. He knew from long experience that it was probably from the Centers for Disease Control and Prevention (CDC) in Atlanta, Georgia. *How were they going to harass him now?* he wondered. For more than a decade, Hooker had been battling the CDC as part of a large group of parents who believed that their children had developed autism and other neurological problems as a result of their vaccines and that the CDC was not conducting an honest investigation into their concerns. Most Americans were unaware that, in 1986, Congress had passed—and President Reagan had signed—the National Childhood Vaccine Injury Act, establishing a separate court to adjudicate claims of vaccine injury. The *New York Times* reported that after signing,

> Mr. Reagan said he had approved the bill "with mixed feelings" because he had "serious reservations" about the vaccine compensation program . . . The program would "be administered not by the executive branch, but by the Federal judiciary," Mr. Reagan said, calling it an "unprecedented arrangement" that

> was inconsistent with the constitutional arrangement for separation of powers among the branches of the Federal Government.[2]

The bill had been drafted in large part by Congressman Henry A. Waxman, a California Democrat. The Justice Department had urged a veto of the bill. However, the measure was strongly supported by Vice President George H. W. Bush, Commerce Secretary Malcom Baldrige, Secretary of Health and Human Services Dr. Otis R. Bowen, and Secretary of the Treasury and former White House Chief of Staff James A. Baker. Reagan expressed hope that later changes would address his constitutional concerns.[3] The changes were never made.

In the opinion of many parents, the so-called "Vaccine Court" was an affront to the concept of justice. In the first place, it gave pharmaceutical companies complete immunity from being sued for damages from vaccines. The fund to compensate children who suffered vaccine injuries would come from a seventy-five-cent tax that would be added to the cost of every vaccine. In essence, the public was self-insuring for vaccine injuries.

The law also eliminated many of the cornerstones of a traditional civil court, such as the requirement that defendants (in this instance, the pharmaceutical companies that manufactured the vaccines) had to produce relevant documents, or that the scientists employed by these companies could be compelled to testify. For Hooker and many of the parents, it seemed the pharmaceutical companies had convinced government officials, though a combination of financial contributions and apocalyptic claims of what would happen if vaccines were subjected to the same type of review as other consumer products, that such an approach would have devastating consequences for public health.

The pharmaceutical companies were removed from the equation, and in its place, the United States government was on the hook for any injuries or deaths caused by vaccines. The government also licensed the vaccines and promoted their use through public education programs. One could say the United States had become certifier, promoter, and purchaser (through low-cost or free immunization programs), while the Vaccine Court was expected to determine the truth about vaccine injuries and provide adequate compensation. Many saw it as an inherent conflict of interest or, like Reagan, wondered if such a setup was even constitutional.

This "unprecedented arrangement" left parents like Hooker relying on approaches such as making requests under the Freedom of Information Act (FOIA) for the relevant data. Hooker had made many FOIA requests over the years. He had also received several calls from people who identified themselves as officials for the CDC and who questioned why he was making so many requests, and then when he asked for their names, they would refuse to divulge their identity. He was aware that, to many people outside of the autism world, such claims would sound vaguely conspiratorial, all part of the "antiscience" and "kooky" labels the media loved to pin on them, but in reality it was more annoying than frightening. Hooker didn't fear that anybody would come after him, just that the bureaucrats in the CDC would do everything in their power to avoid taking an honest look at vaccines and autism. The answers given by the CDC in response to his FOIA requests were normally provided months or years after he'd made the request and were generally not responsive to the questions he'd asked.

Hooker waited for the message to go to voicemail.

But the caller didn't leave a message.

Hooker went back to his lecture notes, tried to concentrate on his upcoming class, but couldn't stop wondering who had called. He got the number from his phone and returned the call. Nobody answered, but it went to voicemail. Hooker was surprised to discover it was the voicemail for Dr. William Thompson, a senior scientist at the Immunization Safety Division of the CDC.

Hooker remembered Thompson. They had talked a good deal between 2002 and 2003, when Hooker had contacted the CDC with concerns about the research they were doing on thimerosal, a mercury derivative that was being used in vaccines, which many parents suspected might be a factor in the development of their child's autism. Mercury was well known as being one of the most dangerous substances on the face of the Earth, and its use in vaccines was a reasonable cause for concern. Because of Hooker's scientific background and training, he became something of a leader among the parents, and William Thompson was designated by the CDC to be Hooker's point of contact with the agency. Hooker was not impressed with Thompson at this stage of their relationship. In one of their initial conversations, Hooker recalled Thompson talking about his daughter, who was of a similar age to Hooker's autistic son, and saying, "Well, my daughter got all the same vaccines as your son, and she's fine."

Hooker was stunned by the hubris of such a statement. It was a bit like somebody claiming he'd smoked for forty years and not come down with lung cancer, so smoking must be safe for everybody. Thompson struck Hooker as a run-of-the-mill bureaucrat with very little interest in doing the right thing. Still, it had been many years since he'd last spoken to Bill

Thompson. He still had Thompson's email address, so at 11:22 a.m. on November 7, 2013, he sent Thompson the first email of what would prove to be one of the most unusual relationships in science and would reveal the greatest medical scandal in American history:

> Bill:
>
> Did you just call me? I have a meeting that is starting but will be available after 3:00 p.m. EST.
>
> Brian.[4]

November 8, 2013

Thompson replied the following day:

> Brian,
>
> Believe it or not, that was a mistake on my part. I had come across this number and it was written next to the name Senator Patty Murray. I apologize for making this call to you. And I won't do it again.
>
> Thanks,
>
> Bill Thompson[5]

November 9, 2013

Thompson's innocent explanation didn't convince Hooker, and he suspected something else was going on. After all, his relationship with Thompson had ended when Hooker joined the Autism Omnibus group of more than five thousand parents in the Vaccine Court in 2003, and he'd been told that since he was now an "adversary" of the CDC, Thompson could no longer communicate with him. It was an absurd claim to make, as the National Childhood Vaccine Injury Act clearly indicated that the Vaccine Court would be a no-fault, nonadversarial system. The assumption of the law was that all parties would be interested in the safety of childhood vaccines. There wasn't supposed to be an "us" and "them." Hooker wrote back:

> Bill,
>
> Your account of the call makes no sense. A seasoned government scientist like yourself would know that DC numbers for Congress start with a 202 area code (224 prefix for the Senate and 225 prefix for the House). Also, if you would want to call Senator Murray's office, why wouldn't you simply look up her number at murray.senate.gov?

> Could you please tell me the "real" reason you were trying to get in touch with me by phone? I don't have time for more CDC lies.
> Brian.[6]

On that same day, Thompson responded to Hooker.

> Brian,
> Seriously, this wasn't a lie. I was reviewing notes from a call you and I had back in 2003. I am going to be providing study related notes as part of the most recent congressional request so I have to review study notes that go back to 2000. This is no small task and I was curious whether Senator Murray's staff would pick up from this number because I wasn't sure whether she was still in office. I apologize because I know it's probably difficult to discern the purpose of such a call from me.
> Bill[7]

Hooker read the email and decided to respond the following morning. He found himself troubled by Thompson's email for reasons he couldn't quite put into words. There was something different in the tone. Maybe he was just imagining that things had changed. Hooker figured it was time to bring the conversation to a close but couldn't help adding a parting shot:

> The congressional request was not initiated by me and I no longer live in Washington State. Don't worry—you answer to someone other than me and that's fine. I just wish for once you would do a clean cohort study that wasn't "overmatched" to the hilt to absolve yet another vaccine and vaccine-component from causing neurodevelopmental disorders in children.
> Brian[8]

Thompson was well aware of Hooker's interest in thimerosal, the mercury derivative used as a preservative in vaccines that Hooker believed to be at least a contributing factor in his son's autism. Hooker was convinced that the email would scare Thompson off for good and any positive feelings that had developed in their brief exchange would quickly vanish and be replaced by the more familiar mutual loathing between the two sides. But Thompson continued to surprise Hooker:

> I am in complete agreement with you. My recent paper with Jack Barile which reanalyzed the 2007 [data in] the NEJM [*New England Journal of Medicine*]

> article is a good summary of where I stand on that paper. The thimerosal-autism study was absolutely a bust because we found a protective effect of thimerosal which we all agree doesn't make sense. So it was probably a sampling issue. The matching was agreed upon up front by many different folks including Safeminds so we published what we found and tested.
>
> Just so you know, there will be new documents that will be shared in this next congressional request.
>
> Bill.[9]

None of this was making any sense to Hooker. This wasn't the way CDC scientists spoke to members of the parent community. It actually seemed like they were having a civil discussion about vaccines and autism and some of the various scientific challenges in determining the truth.

Had Thompson actually said the CDC's thimerosal/autism study was "a bust"?

Hooker looked at the email again. There it was: "The thimerosal-autism study was absolutely a bust because we found a protective effect of thimerosal which we all agree doesn't make sense." As a scientist, Hooker knew what Thompson's words meant. The CDC study was unreliable. One of the study's own authors didn't believe the results, and if what he was saying was true, neither did the other authors.

Hooker tore himself away from the email to look at the article Thompson had coauthored with Jack Barile and found it quickly online.[10] The article found a small, but statistically significant, association between thimerosal and tics. Hooker knew this was significant because many children with autism had tics as a comorbid condition. The last line of the article was even more striking: "Given that the association between thimerosal and tics has been replicated across several different studies, it may be informative to consider additional studies examining the associations using more reliable and valid measures of tics."

They exchanged a few more emails, continuing the friendly tone. Thompson continued to make hints that he was really on Hooker's side, but after years of double-talk from government scientists, Hooker wasn't interested in wasting time. They all knew the battle lines of the controversy. If something was breaking, that was fine. If not, he wasn't going to waste any more time on it. Hooker knew the players and who was likely to have had control over the information about thimerosal. He was going to go for broke.

The Health and Human Services (HHS) website states, "From 2008 through 2011, Dr. Orenstein was Deputy Director for Immunization

Programs in the Vaccine Delivery Department of the Global Health Program at the Bill and Melinda Gates Foundation. His primary focus at the foundation has been on polio eradication, measles control, and improving routine immunization programs." Orenstein's biography at the HHS website covers his time as a former assistant surgeon general of the United States and director of the National Immunization Program, where he "successfully developed, promoted, facilitated, and expanded new vaccination strategies to enhance disease prevention. Dr. Orenstein has co-authored numerous books, journals, and reviews. Along with Stanley Plotkin, MD, and Paul Offit, MD, he co-edited Vaccines, 6th edition in 2012—the leading textbook in the field."[11]

On Monday, November 11, 2013, at 8:23 a.m., Thompson replied in an email, "I will call you in 30 days. I will tell you why then."[12] Thirty days from that date would have been December 11, 2013.

But Dr. William Thompson couldn't wait a month to talk to Dr. Brian Hooker. On Wednesday, November 13, 2013, Thompson called Hooker from his car, and they began a series of dramatic conversations that would lay bare the extent to which a cabal of leading scientists at the CDC actively concealed research findings of great importance, damaged an entire generation of children, and poisoned the debate about vaccines and neurodevelopmental disorders for more than a decade. When the conversation was finished, Hooker wrote down the following:

> Notes from phone conversation with Bill Thompson, 11/13/13. The phone call was brief as he was traveling in his car to teach class at 1:30 p.m. EST.
>
> Bill was very friendly—teaches in the medical school at Morehouse University (Atlanta, GA) a historically black university, with a diverse (40% African American) med school population. He is now with the National Center for Birth Defects and Developmental Disabilities at the CDC. He listed his tenure at CDC which is consistent with his CV. [curriculum vitae—resume]
>
> Regarding the Barile paper—structured equation modeling is a matrix-based technique where eigenvalues (i.e., averages) for each outcome are assessed. Bill indicated that a weakness of the model was that it did not include interaction terms for pre and postnatal thimerosal effects.
>
> Bill indicated that he was talking to me in fulfillment of my FOIA requests (which was odd and not according to CDC policy at all) and seemed willing to talk further in the future. He also indicated that he is gathering

> information for a Congressional request (most likely via Issa) as well and he acted like he knew that I was involved in the Congressional request.
>
> Bill wanted to talk only on his cell phone and not while he was in CDC property. He did not indicate what happens in 30 days (as he stated something would permit him to talk in 30 days in his earlier email). He has 4 years until his 20-year anniversary at the CDC. He would like to retire and teach psychology at a small university. He quipped that we could collaborate on papers.
>
> We also joked that we were twins separated from birth. He turns 50 in December and I just turned 50. He has two children (14 and 13) and I have a similarly aged son (15). I told him that Steven was doing well and described what his life was like. Bill sounded concerned and truly grateful to get an update.[13]

When Hooker finished up the brief phone call, he felt he and Thompson had started the first steps of a dangerous dance. There were things that had given Hooker pause as to whether to continue the relationship. The initial phone call that Thompson claimed was an attempt to get in touch with US Senator Patty Murray's office didn't make sense. The assertion by Thompson that the CDC's own thimerosal research was "absolutely a bust" made Hooker believe he wasn't talking to just another uncaring bureaucrat, but a man who had undergone some significant personal change. And if Thompson was talking to Hooker in response to Hooker's various Freedom of Information (FOIA) requests and a Congressional request (presumably from Congressman Darrel Issa's House Oversight committee), why did he only want to talk to Hooker "on his cell phone and not while he was on CDC property"? It all sounded very cloak-and-dagger for a college biology professor. And what event was happening in thirty days that would allow Thompson to speak more freely?

In addition, it was reasonable to consider that this might be some sort of strange CDC entrapment strategy to punish him for his repeated requests for information about the vaccine program. Hooker resolved to see how it would all play out.

***

Dr. Brian Hooker was an unlikely figure to be in the middle of the greatest battle in modern science. He grew up in the small town of Redlands in Southern California, a community located roughly midway between Los Angeles and Palm Springs.[14] His father was a banker and his mother worked in public health, fostering his interest in science. Brian recalls his mother always being interested in science and technological advances, as well as what was going on in medicine, and she passed that enthusiasm onto her son. Brian's older sister became an accountant, following the interest of their father in business matters.

The Hookers were Baptist, a Christian faith that emphasizes the love and sacrifice of Jesus Christ for humanity, the ability of all people to have a personal relationship with God, the validity of each individual's interpretation of scripture, the importance of a local church, and the need to be a witness for justice in society. Since their founding in the seventeenth century, Baptists have taken leading positions in the fight against slavery, the Civil Rights movement, the promotion of women in church and society, and efforts for ecological responsibility. The expectation that a Baptist should be a positive force for good in the world was part of the legacy with which Brian Hooker grew up.

After high school, Brian went to the California Polytechnic State University (Cal Poly) in San Luis Obispo and graduated with a Bachelor of Science degree in Chemical Engineering in 1985. After college, Hooker tried to get a job in the chemical engineering field, but jobs were scarce, and he ended up working as a fire inspector for an insurance company. He did not find the job intellectually challenging and applied to and was accepted into the graduate school at Washington State University in Pullman, which had a prolific research program. Brian met his wife, Marcia, in 1989 when he walked into a Baptist church and saw her on the front altar as a worship leader. "She was a reason to keep coming around!" he later recalled. Brian and Marcia got engaged about a week before he had to defend his dissertation. Brian received his PhD in Biochemical Engineering in 1990, and the following year he and Marcia got married.

From 1993 to 2009, Dr. Hooker worked for the Pacific Northwest National Laboratory, located on the Hanford Nuclear Reservation, established in 1943 as part of the Manhattan Project during World War II. The site was home to the first full-scale plutonium reactor, and plutonium manufactured at the site was used in the first nuclear bomb exploded in the New Mexico desert in 1945. As the great mushroom cloud rose into the sky, project leader Dr. Robert Oppenheimer uttered his chilling words from the Hindu religious text, the Bhagavad-Gita, demonstrating science's amazing power and terror, "Now, I am become Death, the destroyer of worlds."

The plutonium from the Hanford site was also used in the atomic bomb detonated over Nagasaki, Japan, killing more than a hundred thousand people and ending the war in the Pacific. Hooker's research dealt mainly with environmental cleanup, focusing on bioremediation of toxic wastes in soil. In addition to bioremediation, Hooker developed a specialty in plant molecular biology with the goal of using different genetic modification techniques to grow therapeutic proteins in plants. One of the highlights of this time was being part of the first research group to clone a blood-clotting factor in a genetically-modified plant.

The pace of pure research was grueling, and, in 2010, Simpson University offered Hooker the opportunity to go back into teaching, something he had always intended to do at some point in his career. Hooker looked forward to university life in a small, Northern California town, not far from the Pacific Ocean. Another factor in his decision to change careers was that Brian and Marcia had a son, Steven, born in 1998, who developed autism. His son's autism had propelled Hooker into a different world, and he knew that the academic freedom of Simpson would also allow him to devote significant time to this personal quest to determine what had happened to his son.

By 2013, Brian Hooker had been pounding on the door of the CDC for eleven years, trying to get his questions answered. He did not have any idea that Bill Thompson would soon provide answers, which would confirm his worst fears.

***

On the day after their first phone call, Bill Thompson forwarded twenty-five separate emails to Brian Hooker.[15] Even more alarming, Thompson was sending them from his CDC computer, not his home computer. Why was this man, who only wanted to talk on his cell phone and away from CDC property, now sending all these emails from his work computer?

In the email messages, Thompson told Hooker that something big was going to happen in the National Center for Birth Defects and Disabilities in the CDC in the next thirty days and that he was forwarding many documents to Congress, only letting the Office of the Director of the CDC review the documents and make a decision. Hooker suspected that if the CDC had been withholding documents, it would have been done by employees at a level lower than the Office of the Director. Were documents being concealed from the CDC director? The possibility boggled Hooker's mind.

Hooker was aware that there was one attorney in the Office of the Director with whom he'd been at odds with over several requests, and he asked Thompson about the attorney. Thompson replied he never went near that attorney and considered him a "scumbag." Thompson went on in an email to note that was the first time he'd ever said anything about that attorney to anybody, although he'd long held the opinion. Thompson said he'd been collecting this information for the past ten years and it would be good for Hooker's "book."

Hooker replied he didn't have time to write a book on these developments, but it seemed cathartic to Thompson to be sharing it all.

The next day would bring even more surprises.

***

On Friday, November 15, 2013, Bill Thompson called Brian Hooker for their second conversation. After a few minutes of social chitchat, Thompson got down to the purpose of his call: "You're going at it all wrong with the Geiers [fellow collaborators] and trying to get into the Vaccine Safety Datalink. Why are you doing that? That's just the wrong way to go. You need to be requesting the public use dataset. They're publicly available and they're available to you by law. You need to go through a particular procedure, and I can give you the email of the individual to contact. They have to give you these data-sets if you request them, but you have to do it the right way."[16]

Hooker was stunned by the sudden revelation and the urgency in Thomson's voice. "Okay," he replied, "I had no idea you could do that."

***

Something must have broken in William Thompson, because in that moment it seemed to Hooker that this CDC employee had decided to change from being a bureaucrat to operating as a scientist whose sole obligation was to tell the truth. "If you follow my lead, I will guide you through this," said Thompson. "You will have more data than you know what to do with. And I will show you where the issues are with the CDC results." As Thompson continued to talk, Hooker got the feeling this was going to be a wild ride.

Thompson went onto speak about why Hooker's efforts with the father-and-son team of Mark and David Geier was unlikely to yield any useful information. "It's very difficult and very expensive to get into the Vaccine-Safety Datalink," Thompson told Hooker. "And you're always in danger of being kicked out because if one of the HMOs doesn't like what you're looking at or publishing, they can kick you out."[17]

Hooker knew that the Geiers had spent hundreds of thousands of dollars to get into the Vaccine Safety Datalink and had been kicked out twice for just the type of searching that Thompson had mentioned. Hooker also knew that there were supposed to be public datasets available to researchers, but nobody had shown him how to access that information. But here was Thompson, offering to be Hooker's guide. Unbeknownst to Hooker at the time, Thompson would actually be the one answering the email requests for information.

Thompson told Hooker whom to email at the CDC, how to structure the emails, and what to request. Hooker outlined five different datasets, got the emails ready, and showed them to Thompson. The first dataset that Hooker wanted regarded the CDC's investigation of thimerosal and, specifically, the Verstraeten study.[18] The actual conclusions have been a matter of heated discussion, with parents claiming that little was actually shown, one way or the other. This is the conclusion from their article in the journal *Pediatrics*: "No consistent significant associations were found between TCVs [thimerosal-containing vaccines] and neurodevelopmental outcomes. Conflicting results were found at HMOs for certain outcomes. For resolving the conflicting findings, studies with uniform neurodevelopmental assessments of children with a range of cumulative thimerosal exposures are needed."[19] That hardly sounds like a ringing endorsement of their safety.

The Verstraeten study of the Vaccine Safety Datalink on thimerosal had come in for sharp criticism from the parent community, specifically in the way in which it had systematically reduced or eliminated associations between thimerosal exposure and neurodevelopmental disorders. The actions of the Verstraeten team were the subject of an analysis titled "A Brief Review of Verstraeten's 'Generation Zero' VSD Study Results" by the group Safe Minds (Sensible Action for Ending Mercury-Induced Neurological Disorders), and part of their review is reproduced below:

> Between February 2000 and November 2003 Thomas Verstraeten and his supervisors at the National Immunization Program produced four separate generations of an analysis designed to assess the impact of vaccine mercury exposures on neuro-developmental disorders in children. . . . With each generation, elevated and statistically significant risks were reduced and/or eliminated.
>
> But before these four generations of reports were produced, Verstraeten conducted an earlier analysis of these issues in November and December of 1999. He never prepared a formal report on this work, but statistical tables

> obtained by Safe Minds in a FOIA request (and not previously analyzed) demonstrate large and statistically significant mercury exposure effects that in many cases exceeded the findings of the later reports . . .
>
> The results of the Generation Zero analyses are striking and more supportive of a causal relationship between vaccine mercury exposure and childhood developmental disorders (especially autism) than any of the results reported later:
>
> - Relative risks of autism, ADD, sleep disorders and speech/language delay were consistently elevated relative to other disorders and frequently significant. Disease risk for the high exposure groups ranged from lows of 1.5X-2 times to as high as 11 times the disease risk of the zero exposure group.
> - Many other outcomes showed no consistent effect, while a few appeared to show a protective effect from vaccine mercury exposure (most likely children with these diagnoses were immunized later).
> - The strongest effect was for the highest levels of mercury exposure at the earliest time of exposure, consistent with the idea that infant brain development is most sensitive to the earliest exposures.
> - The elevated risk of autism for the highest exposure levels at one month ranged from 7.6 to 11.4 times the zero exposure level. This increased risk level corresponds to the tenfold increase in autism rates seen since vaccine mercury exposures increase starting in 1990.[20]

Based on these findings from Safe Minds, as well as mercury's well-known neurotoxic properties, it was understandable why Hooker would first want to look at the original datasets for the Verstraeten study. When Hooker had earlier asked for this information from the CDC, he'd been told that the original datasets had been destroyed but he might be able to reassemble it from the Vaccine Safety Datalink, which he had been trying to do for the past several years. For those familiar with scientific research, failure to provide the raw data on which a conclusion is based is highly suspicious.

"You can try to get the thimerosal data," Thompson told Hooker, "but the first thing you want to do is get the dataset from the DeStefano study, regarding the MMR vaccine."

Hooker felt some of his earlier excitement begin to dissipate. He was aware of the Frank DeStefano paper of 2004,[21] which had been the death knell among most of the scientific community of British researcher Dr. Andrew Wakefield's suggestion that the MMR vaccine should be studied for its connection to autism and other neurodevelopmental disabilities. "Bill, I don't know anything about the MMR vaccine and I don't even think my son was injured by it, so I'm hesitant to start on it."

"Just trust me," said Thompson. "Go ahead and get it. I will show you some things about that particular dataset. It's very straight forward and easy to analyze."

Hooker went ahead and composed an email requesting the dataset from the DeStefano study on the MMR vaccine. Just as Thompson had predicted, the study was provided, and Hooker started to examine it. He saw it was from school districts in five counties in metropolitan Atlanta and contained about 625 children with autism and about 1,800 matched controls. Hooker had often been disappointed in previous CDC studies with the way the government scientists matched cases to controls, but as he examined the dataset, it became clear to him that this time they had done a good job. The children were matched by gender, and not on race, and there were also data on several other vaccines.

The first data analysis Hooker ran was on children who received the MMR vaccine before thirty-six months and those who received the MMR vaccine after thirty-six months. The odds ratio for the earlier group was a 1.49 increase over those who received the MMR shot after thirty-six months.[22] The CDC had actually published that number in the DeStefano study. This was a statistically significant number, but the CDC explained it away by saying that children diagnosed with autism were receiving special education services that required them to get the MMR shot. Hooker knew that explanation was preposterous. In the early 2000s, an autism diagnosis was rarely made before a child was three years old.

Hooker then ran the analysis on just boys, as the rate of autism is known to be higher in males than females, and found that the odds ratio went up to 1.67.[23] The CDC had also reported this number in the DeStefano study. When he ran the analysis on the girls, he was not able to find an increased odds ratio. Hooker wryly noted that the DeStefano study authors had not commented on how their earlier explanation for the increase due to inclusion in special education services failed to explain the negative finding in the girls.

The next analysis Hooker utilized was for African American boys and girls. The odds ratio was a 2.6fold increase.[24] In science, an odds ratio above 2.0 is considered a clear and convincing signal and is often considered proof of causation in a legal case. Hooker then analyzed the effect for the African American boys and found the odds ratio jumped to 3.36.

This was a smoking gun.

He went back to the DeStefano paper and noticed they had done something unusual with the African American cohort: They had run the analysis only on those African American group members who had a valid State of Georgia birth certificate. As Hooker calculated it, the CDC scientists had thrown out about ninety of the 220 African American children with autism, lowering the pool by 40 percent and dramatically skewing the odds ratio.

Hooker compiled the information, checked and double-checked his numbers, typed up the results, and then scheduled a conference call with Thompson.

When Hooker got Thompson on the phone, he told him what his analysis had revealed.

"Oh, you found it?" replied Thompson.[25]

"Yeah, I found it. Tell me what I found?" Hooker would later come to believe he had passed some kind of test in Thompson's mind. If he quickly found the association, Thompson would tell him more. If he had failed, Thompson would have decided Hooker wasn't worth his time.

Thompson replied that when he first analyzed the data, he quickly saw the effects in the African American population but did not see it in any other racial groups.

"Did you do the analysis on just African American males?"

"No, I didn't, but I figured it would be an even stronger association because I assumed you wouldn't see the association in girls."

Hooker shared with him the analysis he had done on the African American males.

"How long did it take you to find it?" Thompson asked.

"About thirty minutes after I started programming."

"Yeah, it just jumps off the page, doesn't it?"

Hooker asked him why the DeStefano paper only showed the numbers from the birth certificate cohort.

"Yeah, you're absolutely right. We shouldn't have done that," Thompson admitted. He went on to tell Hooker about being present at a meeting with his coauthors on the paper and what one of them had said about the need to bury the effect. Thompson said the coauthor's comment was filled with such unimaginable hubris that he would never forget it in his entire career. But he didn't want to share it with Hooker because he didn't have it documented. Hooker thought the comment had to be something along the lines of "We're doing this for the greater good," a sentiment he'd heard in various forms from people involved in the issue. Hooker wondered how science today would be different if Galileo or Darwin had ever decided to lie about the motion of the planets or the facts about evolution for the "greater good." And, arguably, the truth about whether vaccines were causing devastating, lifelong disabilities was of more immediate importance to the public than the movement of distant celestial bodies or the change in life-forms over millions of years.

Even though he did not want to share the comment, Thompson said it was at that meeting that a decision was made to bury the effect of the earlier MMR shot on African American children by looking only at individuals with a valid State of Georgia birth certificate. They had said this was the only way they could determine the race of a child, but this information was also available from school records and had not been a reason to remove children of any other race.

Hooker was stunned by these revelations. He had believed the CDC was trying to downplay the risk of vaccines and neurodevelopmental disorders, but this was a completely different order of magnitude. They had identified a clear signal from earlier administration of the MMR vaccine among African American males, one of the most vulnerable groups in the country, and they had decided to conceal this information. Science had demonstrated the parents were right. The scientists at the CDC had betrayed their profession and the public's health.

***

It seemed to Hooker that Thompson's revelations lifted a great weight off of Thompson's shoulders. At one point, Thompson apologized to Hooker for his son's autism.

Hooker replied that his son was born in 1998 and the events Thompson were describing happened from 2001 to 2004, well after Hooker's son had developed autism. "You're not responsible for that. You don't have to bear that burden. There are others who have done things, but not you."

Thompson started to talk about how the entire series of events had affected him emotionally. He shared how devastating it was to be in this CDC culture of intimidation and fear, the profound disgust he had with many of his coworkers, and how they had let this situation go on for so long. Thompson also strongly believed that there were people in the Immunization Safety Office who were trying to make a difference, but they were being systematically targeted and transferred to different divisions, presumably by people like CDC Director Dr. Julie Gerberding or the head of the National Immunization Program, Dr. Walter Orenstein.[26]

Hooker was surprised to learn that one of the people William Thompson admired was Robert Chen, director of the Immunization Safety Office from the late 1990s to the early 2000s. Hooker considered Chen to be a "bad actor" because at one time when he was on a conference call with parent petitioners under the National Childhood Vaccine Compensation program who were requesting documents from Chen's office that he did not want to produce, Chen had said, "If I had a gun I would shoot you all."[27]

But to Thompson there was a different side to Robert Chen, a more honorable one. In the early 2000s, Chen had been very vocal that the Immunization Safety Office did not belong in the same division as the National Immunization Program. Chen even advocated that the Immunization Safety Office be moved out of the Centers for Disease Control because of the inherent conflict of interest. The CDC and the National Immunization Program were promoting vaccines; indeed, the CDC even had copatents on some vaccines. This conflict of interest did not exist for any other consumer or medical product.

Chen had actually received a letter of reprimand from the head of the National Immunization Program, Walter Orenstein, for suggesting that the Immunization Safety Office was ill-equipped to protect the public due to this conflict of interest and should be moved to an office where its independence would be unquestioned. Thompson was convinced that the reprimand of Dr. Chen was unwarranted and wrote an email on October 16, 2002, to Dr. Orenstein, which is reproduced below:

> Dear Dr. Orenstein:
>
> I respectfully request that you withdraw the reprimand of Dr. Robert Chen. I believe the reprimand contains misleading and false information. I am also concerned regarding the impact of the reprimand on Dr. Chen's staff.
>
> Sincerely,
>
> William W. Thompson, PhD
>
> National Immunization Program[28]

Orenstein replied to Thompson's email on October 18, 2002:

> I am responding to your October 16 email concerning a matter related to Bob Chen. While I am not at liberty to discuss the substance of confidential personnel matters, such as disciplinary actions, suffice it to say that no such action would have been taken without much forethought and discussion. This

> is an internal management matter on which I fully support the actions of Bob's Division Director, Melinda Wharton.
>
> Without speaking to the particulars of the personnel issues, I can assure you that NIP management continues to strongly support vaccine safety-related activities and research. Furthermore, I can assure you that the NIP vaccine safety budget and vaccine safety datalink project have received funding increases for three consecutive years (FY's 2002002). The quality of vaccine safety activities and research performed by Bob and his group continues to be superlative and supported by all throughout the organization. We anticipate being able to resolve our management issues and continuing this productive relationship.
>
> Walt.[29]

It is difficult to read these emails and come to any other conclusion than that there were strong differences of opinion at the National Immunization Program. Thompson and Chen seemed to be on the losing side, while Walter Orenstein and Melinda Wharton held the upper hand. Chen was placed on probation from 2002 to 2004 as head of the Immunization Safety Office and was then assigned to a division dealing with the global HIV/AIDS crisis. As Hooker continued to talk with him, Thompson revealed more information, such as how the stress of concealing the MMR data had led him to try to commit suicide in 2004 and 2005. Thompson told Hooker that his second attempt at suicide took place on April 12, 2005, when he had taken an overdose of pills. The amount he'd taken was not enough to kill him, but he got into his car, hit a parked vehicle in De Kalb County, Georgia, and then fled the scene of the accident. Thompson was picked up by police and spent April 12 and 13 in jail, only to be released on April 14, 2004, after paying a six-hundred-dollar fine. Thompson confessed to Hooker that he was concerned about the suicide attempts and the arrest because the CDC might use it to try and discredit or marginalize him if he ever became a whistleblower.[30]

As Thompson continued to share information, Hooker felt he was beginning to put together a time line of the actions of this cabal of CDC scientists who were determined to never find an association between vaccines and autism or other neurodevelopmental disorders. It seemed that Thompson had first found the association in November 2001, reported it to his coauthors (DeStefano, Karapurkar Bhasin, Yeargin-Allsopp, and Boyle), and they bandied it around until July 2002, when they made the decision that they were not going to publish the results. Between July and September of 2002, the cabal decided to get together on a Saturday afternoon and throw all of their results in a trash can. In this way, the results just sat around at the CDC for a good fifteen months until January 2004, when it became imperative for a team from the CDC to discuss the research into autism and vaccines because the Institute of Medicine (IOM) was preparing a meeting on February 9, 2004, to address the topic.

Hooker would later discover that in addition to the finding regarding African American males, similar effects were observed for what was termed "isolated autism," meaning there were no previously existing conditions that might have contributed to the development of autism (often referred to as "regression autism," in which the child was normally developing and after a vaccination suffered a severe decline) nor was there an effect at twenty-four months. The MMR vaccine/autism study had revealed not one, but three different groups who were affected by earlier administration of the MMR shot.

Thompson was slated to present the information on the MMR vaccine and autism from the Atlanta school districts at the IOM meeting, and his conscience was bothering him. It was one thing to conceal evidence. One could say they were just taking their time.

But it was another thing to go out and affirmatively lie about the issue. On February 2, 2004, Thompson decided to take matters into his own hands and write directly to Dr. Julie Gerberding, who at the time was the director of the CDC. Thompson was also concerned because a congressman, Dave Weldon (R, Florida), who was also a medical doctor, had spoken to autism parent groups with their concerns about the integrity of the science being performed on this issue. Weldon had asked very direct questions to the CDC, which they had not answered. Thompson's actions were outside the chain of command, and he would be reprimanded for directly contacting Dr. Gerberding. The letter is reproduced in full below:

> February 2nd, 2004
>
> Dear Dr. Gerberding:
>
> We've not met yet to discuss these matters, but I'm sure you're aware of the Institute of Medicine Meeting regarding immunizations and autism that will take place on February 9th. I will be presenting the summary of our results from the Metropolitan Atlanta Autism Case-Control Study and *I will have to present several problematic results relating to statistical associations between the receipt of the MMR vaccine and autism* [italics added].

It is my understanding that you are aware of several news articles published over the past two weeks suggesting that Representative Dave Weldon is still waiting for a response from you regarding two letters he sent you regarding issues surrounding the integrity of your scientists in the National Immunization Program. I've repeatedly asked individuals in the NIP Office of the Director's Office why you haven't responded directly to the issues raised in those letters and I'm very disappointed with the answers I've received to date. In addition, I've repeatedly told individuals in the NIP OD [National Immunization Program, Office of the Director] over the last several years that they're doing a very poor job representing immunization safety issues and that we're losing the public relations war.

On Friday afternoon, January 30th, 2004, I presented the draft slides for my IOM presentation to Dr. Steve Cochi and Dr. Melinda Wharton. The first thing I stated to both of them was my sincere concern regarding presenting this work to the Institute of Medicine if you have not replied to representative Weldon's letters. I have attached the draft slides for your review. I have been told that you have suggested that the science speaks for itself. In general I agree with the statement, but as you know, the science also needs advocates who can get the real scientific message out to the public.

In contrast to NIP's failure to be proactive in addressing immunization safety issues, you have done an amazingly effective job addressing the press on a wide range of controversial public health issues including SARS, Monkey Pox and Influenza. The CDC needs your leadership with respect to the IOM meeting because I may very well be presenting data before a hostile crowd of parents with autistic children who have been told not to trust the CDC. I believe it is your responsibility and duty to respond in writing to Representative Weldon's letters before the Institute of Medicine meeting and make those letters public. Otherwise, you give the appearance of agreeing with what has been suggested in those correspondences and you're putting one of your own scientists in harm's way. This is not the time for leadership to act politically. It is a time for our leadership to stand by their scientists and do the right thing. Please assist me in this matter and respond to Representative Weldon's concerns in writing prior to my presentation on February 9th.

Sincerely,

William W. Thompson, PhD

Epidemiologist[31]

When Brian Hooker later read the letter, he was struck with a number of conflicting emotions. In reading between the lines, it seemed as if Thompson were trying to get the CDC director to directly engage with the issue of immunizations and autism, rather than letting lower-level officials obscure the issue. Hooker suspected this was part of an all-too-familiar pattern of behavior in which those in charge could later claim ignorance of any illegal or unethical actions of their subordinates if such actions caught the attention of any investigative bodies.

The Weldon letters, although respectful, had been scathing in their accusations. The CDC director was at a moral crossroad with the Weldon inquiry about misbehavior of CDC scientists in the thimerosal/autism investigation and the request by one of her own scientists, Dr. William Thompson, to discuss problematic results in the work of some of those same scientists in the MMR/autism investigation. The same group of scientists was involved in assuring the public that both mercury-containing vaccines and the MMR vaccine were safe for their children.

If she accepted Thompson's invitation to discuss the issue, he planned to tell her about the birth certificate issue regarding African American boys, as well as the difference in the other two groups. He expected that she would find some diplomatic solution to quietly discipline the scientists involved, as well as release the troubling information. As public health scientists, it was their job to protect the citizens of the United States, not cover up for the pharmaceutical companies or government programs that supported vaccines. Even questions from a congressman did not seem to move the CDC director to take action.

The letter Congressman Dave Weldon sent to Dr. Julie Gerberding on October 31, 2003, and that formed part of Thompson's concerns in early 2004 is reprinted below in its entirety:

> Dear Dr. Gerberding:
>
> I am writing to follow up our conversation about the article (Verstraeten et al.) that will be published in the November 2003 issue of *Pediatrics*. I have reviewed the article and have serious reservations about the four-year evolution and conclusions of this study.
>
> Much of what I observed transpired prior to your appointment a year ago as the Director of the Centers for Disease Control and Prevention [CDC]. I am very concerned about activities that have taken place in the National Immunization Program [NIP] in the development of this study, and I believe the issues raised need your personal attention.
>
> I am a strong supporter of childhood vaccinations and know they have saved us from considerable death and suffering. A key part of our vaccination

program is to ensure that we do everything possible to ensure that these vaccines, which are mandatory, are as safe as possible. We must fully disclose adverse events. Anything less than this undermines public confidence.

I have read the upcoming *Pediatrics* study and several earlier versions of this study dating back to February 2000. I have read various emails from Dr. Verstraeten and coauthors. I have reviewed the transcripts of a discussion at Simpsonwood, GA between the author, various CDC employees, and vaccine industry representatives. I found a disturbing pattern which merits a thorough, open, timely, and independent review by researchers outside of the CDC, HHS, the vaccine industry, and others with a conflict of interest in vaccine related issues (including many in University settings who may have conflicts).

A review of these documents leaves me very concerned that rather than seeking to understand whether or not some children were exposed to harmful levels of mercury in the 1990s, there may have been a selective use of the data to make the associations in the earliest study disappear. While most childhood vaccines now only have trace amounts of mercury from thimerosal containing vaccines [TCVs], it is critical that we know with certainty if children were injured in the 1990s.

Furthermore, the lead author of the article, Dr. Thomas Verstraeten, worked for the CDC until he left over two years ago to work in Belgium for GlaxoSmithKline {GSK], a vaccine manufacturer facing liability over TCVs. In violation of their own standards of conduct, *Pediatrics* failed to disclose that Dr. Verstraeten is employed by GSK and incorrectly identified him as an employee of the CDC. This revelation undermines this study further.

The first version of the study, produced in February 2000, found a significant association between exposure to thimerosal containing vaccines [TCVs] and autism and neurological developmental delays [NDDs]. When comparing children exposed to 62.5 mg of mercury by 3 months of age to those exposed to less than 37.5 ug, the study found a relative risk for autism of 2.48 for those with a higher exposure level. (While not significant in the 95% confidence interval for autism, this meets the legal standard of proof exceeding 2.0.) For NDDs the study found a relative risk of 1.59 and a definite upward trend as exposure levels increased.

A June 2000 version of the study applied various data manipulations to reduce the autism association to 1.69 and the authors went outside of the VSD database [Vaccine Safety Datalink] to secure data from a Massachusetts HMO (Harvard Pilgrim, HP) in order to counter the association found between TCVs [thimerosal containing vaccines] and speech delay. At the time that HP's data was brought in, HP was in receivership by the state of Mass., its computer records had been in shambles for years, it had multiple computer systems that could not communicate with one another (*Journal of Law, Ethics, and Medicine* Sept. 22, 2000), and it used a health care coding system totally different from the one used across the VSD [Vaccine Safety Datalink]. There are questions relating to a significant underreporting of autism in Mass. The HP [Harvard Pilgrim] dataset is only about 15% of the HMO dataset used in the February 2000 study. There may also be significant problems with the statistical power of the HP dataset.

In June of 2000 a meeting was held in Simpsonwood, GA, involving the authors of the study, representatives of the CDC, and the vaccine industry. I have reviewed a transcript of the meeting that was obtained through the Freedom of Information Act [FOIA]. Comments from Simpsonwood meeting include (summary form, not direct quotes):

- We found a statistically significant relationship between exposures and outcomes. There is certainly an under ascertainment of adverse outcomes because some children are just simply not old enough to be diagnosed, the current incidence rates are much lower than we would expect to see (Verstraeten);
- We could exclude the lowest exposure children from our database. Also suggested was removing the children that got the highest exposure levels since they represented an unusually high percentage of the outcomes (Rhodes);
- The significant association with language delay is quite large (Verstraeten);
- This information should be kept confidential and considered embargoed;
- We can push and pull this data any way we want to get the results we want;
- We can alter the exclusion criteria any way we want, give reasonable justifications for doing so, and get any result we want;
- There was really no need to do this study. We could have predicted the outcomes;
- I will not give any TCVs [thimerosal containing vaccines] to my grandson until I found out what is going on here.

Another version of the study—after further manipulation—finds no association between TCVs and autism, and no consistency across HMOs between TCVs and NDDs [neurodevelopmental disorders] and speech delay.

The final version of the study concludes that "No consistent significant associations were found between TCVs and neurodevelopmental outcomes," and that lack of consistency argues against an association. In reviewing the study there are data points where children with higher exposures to the neurotoxin mercury had fewer developmental disorders. This demonstrates to me how excessive manipulation of data can lead to absurd results. Such a conclusion is not unexpected from an author with a serious, though undisclosed, conflicts of interest.

This study increases speculation of an association between TCVs and neurodevelopmental outcomes. I cannot say it was the author's intent to eliminate the earlier findings of an association. Nonetheless, the elimination of this association is exactly what happened and the manner in which this was achieved raises speculation. The dialogue at the Simpsonwood meeting clearly indicates how easily the authors could manipulate the data and have reasonable sounding justifications for many of their decisions.

The only way these issues are going to be resolved—and I have only mentioned a few of them—is by making this particular dataset and the entire VSD database open for independent analysis. One such independent researcher, Dr. Mark Geier, has already been approved by the CDC and the various IRBs to access this dataset. They have requested the CDC allow them to access this dataset and your staff indicated to my office that they would make this particular dataset available after the *Pediatrics* study is published.

Earlier this month the CDC had prepared three similar datasets for this researcher to review to allow him to reanalyze CDC study datasets. However when they accessed the datasets—which the researchers paid the CDC to assemble—the datasets were found to have no usable data in them. I request that you personally intervene with those in the CDC who are assembling this dataset to ensure that they provide the complete dataset, in a usable format, to these researchers within two weeks. The treatment that these wellpublished researchers have received from the CDC thus far has been abysmal and embarrassing. I would also be curious to know whether Dr. Verstraeten, an outside researcher for more than two years now, was required to go through the same process as Dr. Geier in order to continue accessing the VSD [Vaccine Safety Datalink].

You have not been a part of creating this current situation, but you do have an opportunity to help resolve this issue and ensure that confidence and trustworthiness in the CDC and our national vaccination program is fully restored. I would ask that you work with me to ensure that a full, fair, and independent review is made of the VSD database to fully examine this

> matter. I would like to meet with you at your earliest convenience to move this process forward.
>
> Thank you for your consideration. I look forward to working with you on this urgent matter of great importance to our nation's most precious resource, our children.
>
> Sincerely,
>
> Dave Weldon, M.D.
>
> Member of Congress[32]

Imagine you are head of the CDC and you receive such a letter in October of 2003. Four months later, in February 2004, you have still not responded to the letter, and one of your own employees wants to talk to you about similar concerns. How can you not call everybody in and have a discussion? These are not petty concerns. This is about the active harming of children under the guise of a program that purports to protect them from disease.

Even though Dr. Gerberding was informed by Congressman Weldon of his suspicions regarding the behavior of scientists in the National Immunization Program during their investigation of thimerosal, she did not invite William Thompson to her office for a discussion of his concerns about the MMR vaccine and autism rates among African American males. Thompson was removed from the February 9, 2004, presentation at the Institute of Medicine on Immunizations and Autism and later placed on administrative leave. Frank DeStefano replaced Thompson and did not report the findings, providing instead an altered dataset that showed there was no reason to be concerned about the MMR vaccine.

Dr. Gerberding left the CDC in 2009 and in January 2010, like Thomas Verstraeten before her, accepted a position working for a pharmaceutical company. Gerberding became president of the Vaccine Division at Merck Pharmaceuticals. Gerberding's biography on the website MyBio states:

> She was responsible for Merck's portfolio of vaccines, planning for the introduction of vaccines from the company's pipeline, and accelerating efforts to broaden access to Merck's vaccines around the world. Under her leadership, Merck's vaccines are now reaching more people than ever, and Merck became the global leader in the vaccine market.[33]

Thompson continued to press his concerns about vaccine safety at every opportunity, leading to his being placed on paid administrative leave a month later on March 9, 2004, a duty that fell to Robert Chen, whom Thompson had unsuccessfully tried to defend two years earlier.

In a three-page memorandum signed by Robert T. Chen, chief, Immunization Safety Branch, Thompson's conditions were laid out. The first page of the memorandum read in part:

> This memorandum is to notify you that you are being placed on paid administrative leave effective immediately.
>
> This action is being taken in order to provide adequate time for you to obtain and provide management the documentation requested in our March 9, 2004 Request for Documentation memorandum. Once the documentation has been assessed by management, further decisions will be made regarding your assignments and possible special accommodations. In the meantime, you will be placed on administrative leave and you will be notified when it is appropriate to return to duty after management has completed its assessment of the documentation you provide.
>
> This is not a disciplinary or adverse action. You will receive full pay and benefits while you are on administrative leave.[34]

The second page of the memorandum detailed two counseling sessions Thompson had with Robert Chen on February 4, 2004, and March 9, 2004. The second page of the memorandum prepared by Dr. Chen read in part:

> This memorandum will document the counseling sessions I had with you on February 4, 2004 and March 9, 2004 regarding the extremely stressful environment facing vaccine safety research, and the challenges you've faced in coping with those stresses; especially several documented instances of inappropriate and unacceptable behavior in the workplace (Annex 1).
>
> It is extremely unfortunate that this stressful environment and incidents occurred. CDC is working to reduce the stress and expects that you will work with your clinician to ensure that incidents of this nature do not occur again. Please understand that this memorandum is meant to advise you of our expectations and is <u>not</u> a disciplinary action. A copy of this memorandum will <u>not</u> be placed in your Official Personnel Folder [OPF]. *However, any recurrence of such behavior on your part will cause* ***a more severe penalty to be proposed.*** [italics and bold added].[35]

The third page of the memorandum detailed what Robert Chen believed were inappropriate behaviors by Thompson, and that entire account is reproduced in full:

Annex to March 9, 2004 Memorandum of Counseling

Specific instances of inappropriate and unacceptable behavior in the workplace

On February 2, 2004 you sent an email to Dr. Gerberding and various supervisors within the National Immunization Program [NIP] regarding your upcoming Institute of Medicine [IOM] presentation. In your email you criticized the NIP/OD for doing a very poor job of representing vaccine safety issues, and you requested that Dr. Gerberding reply to a letter from a congressional representative before you made your presentation to the IOM.

On or about February 26, 2004 Dr. Gina Mootrey approached you and asked questions about slides you had prepared for a previous influenza presentation. Dr. Mootrey was attempting to clarify a few points from your slides in order that Dr. Walter Orenstein could modify some of the slides for a different presentation to be made by Dr. Orenstein. You did not agree to assist.

On February 27, 2004 you approached Dr. Orenstein in the Building 12 parking lot, at which time you demonstrated inappropriate anger towards Dr. Orenstein, his request, and your perception that Dr. Orenstein was responsible for permitting a hostile environment within your organizational unit.

On February 27 and 29, 2004 you sent emails to Dr. Orenstein in which you alleged that Dr. Orenstein had not properly addressed various issues related to vaccine safety and expressed your opinion that he should apologize.

On February 29, 2004 you wrote additional emails to senior NIP staff stating that you had serious concerns regarding Orenstein's behavior and felt that it [was] harassment.

The general tone and content of your emails were inappropriate and gave the appearance that senior management had not fulfilled their public health obligations as they pertain to vaccine safety. Your actions had the effect of eroding the employment relationship between supervisor and subordinate, and appear to make a mockery of management's authority to direct the activities of this office. Furthermore, your interaction with Dr. Orenstein created concern about your level of anger being out of proportion to the facts [italics and bold added].[36]

When Hooker finished reading the document, he couldn't help but feel astonishment at how a government agency had gone completely crazy.

Thompson was complaining that the CDC was "not fulfilling their public health obligations as they pertain to vaccines," and the only thing his superiors could say in response was that his "level of anger" was "out of proportion to the facts"? No wonder Thompson had tried to commit suicide twice. Who could possibly bear the weight of knowing that every day in America responsible parents were taking their children in for their pediatric checkups and subjecting them to the possibility of a lifelong disability? And that your silence and the silence of other government scientists were allowing these injuries to continue to occur?

Although Thompson initially seemed to feel better about the disclosures to Brian Hooker, sometime on November 20, 2013, the weight of what he was disclosing seemed to close in on him. Thompson did not tell Hooker exactly what had happened on that day, but on November 21, 2013, they exchanged a series of texts in which Thompson suggested he had turned the corner for the better.

Thompson texted Hooker: "I got my shit back together this morning. I just [have] to tell you how much I admire you. You will be vindicated in the end. And I apologize for participating in the cover-up."

Hooker texted back: "You are my hero. Thank you so much for all you are doing—stay safe and family first."

Thompson responded: "From my perspective there are no heroes on this end. I will show you how to access lots of documents over the next several months in a legal manner."

"Again I appreciate it."

"I appreciate your kindness."

Hooker replied, "I believe the vast majority of the CDC scientists have done the best they possibly could at any given time."

Thompson answered: "I agree and we can discuss more as we go down this interesting journey together."[37]

* * *

Bill Thompson was very insistent that Brian Hooker had to publish the information and analyses he was getting on the MMR vaccine issue and African American boys. In addition to getting information the CDC had tried to keep hidden, Hooker continued to press Thompson on how this cover-up could have taken place and lasted so long. From what Thompson told him, Hooker came to believe the CDC scientists were so conditioned to believe there was no association between vaccines and autism that whenever they did see an association, they immediately concluded it had to be wrong.

Hooker asked about the data analysis runs that Thompson had performed to hide the race effect of the MMR vaccine and whether Thompson had any emails that might show definitively that the scientists were attempting to skew the data.

"We don't email that kind of stuff," Thompson replied, then proceeded to tell Hooker how those requests would be made verbally during closed-door meetings.[38]

Hooker was also troubled by the potential threat to Thompson if his identity were revealed. And yet on the other hand, if his identity were not revealed, it would be so much easier to dismiss the findings. At some points, Hooker thought Thompson imagined he could remain safely hidden for decades like the legendary "Deep Throat," the source used by *Washington Post* reporters Bob Woodward and Carl Bernstein in the Watergate scandal that brought down President Nixon in 1974. Hooker advised Thompson to get a whistleblower attorney, which he did. Hooker also encouraged him to contact the office of Congressman William Posey, who had taken over from Congressman Dave Weldon and had been so aggressive in questioning CDC Director Dr. Julie Gerberding. Thompson contacted Congressman Posey's office and eventually turned over thousands of pages of documents detailing the decades-long cover-up.

Thompson told Hooker on several occasions that he hoped to be subpoenaed by a Congressional committee so that he would be sworn to tell the truth and would reveal this information in the manner intended by law. Thompson seemed to believe that applying for whistleblower status would give him all sorts of legal protections. But Hooker told him that in his opinion, such protections were more illusory than real.

Even without the legal protections he had expected would come with whistleblower status, Thompson seemed to be resigned to his fate. "I am basically done lying," he told Hooker in one conversation.

The relationship between the two men became close, despite the fact that they had been on opposite sides of the greatest scandal in medicine. The two men were both scientists, of about the same age, and married with children of a similar age. And as they talked about their personal histories as well, a strong friendship developed. "I love you, man," Thompson told Hooker at the end of one conversation in which they had shared a great deal of personal information.

Hooker paused for a moment before replying. "I love you too, man."

Hooker couldn't help but imagine what the mainstream press would make of their friendship. In their view, scientists like Thompson were the vanguard of truth and progress, while parents like Hooker were supposed to be antiscience, knuckle-dragging cavemen. This was a war that could only be won by the annihilation of the other side.

Would the actual truth ever be told?

By March 2014, Hooker had a paper together on the MMR vaccine effect on autism in African American males. Hooker would send drafts to Thompson, they'd discuss how the analyses was going, and they'd make adjustments. In April 2014, Hooker submitted the paper to the journal *Translational Neurodegeneration*, in which he'd previously published.

The initial plan of the editors was to have two peer reviews, but after those two reviews came back, they added a third reviewer. Thompson was anxious to know how the peer review process was going, and when Hooker shared with him that he'd answered the questions of the reviewers to their satisfaction, Thompson was ecstatic.

The article, titled "Measles-Mumps-Rubella Vaccination Timing among African-American Boys: A Reanalysis of CDC Data," was published on August 8, 2014.[39] The conclusions section stated:

> The present study provides new evidence of a statistically significant relationship between the timing of the first MMR vaccine and autism incidence in African American males. Using a straight-forward, Pearson's chi-squared analysis on the cohort used in the DeStefano et al. study, timing of the first MMR vaccine before and after 24 months of age and 36 months of age showed relative risks for autism diagnoses of 1.73 and 3.36, respectively.

Hooker and Thompson talked on the phone that day. "Today I am vindicated," said Thompson. "The truth is out."[40]

CHAPTER TWO

The Insanely Good Soul of Dr. Andrew J. Wakefield

The great nineteenth-century Russian author, Leo Tolstoy, wrote in *The Kingdom of God is Within You*, "The most difficult subjects can be explained to the most slow-witted man if he has not formed any idea of them already; but the simplest thing cannot be made clear to the most intelligent man if he is firmly persuaded that he knows already, without a shadow of doubt, what is laid before him." In considering the nearly twenty-year controversy regarding the work of Dr. Andrew Wakefield, we might do well to consider Tolstoy's words. Is Dr. Wakefield the villain and enemy of science, as he is portrayed in the popular and scientific press? Maybe it is worth our time to review the story of Dr. Wakefield, especially in light of the Thompson allegations.

On a personal note, I must say that I have never encountered a controversy in which people have such strong opinions and yet at the same time have such a tenuous grasp of the facts. In the summer of 2011, I had the opportunity to work as a summer research associate at Lawrence Livermore Labs because I was a science teacher. Lawrence Livermore is one of our great national labs, and I was fortunate to be placed on a wonderful team that was studying viruses.

I underwent the training to become certified to work in a biosafety level 2 lab and will never forget that exciting day when I was allowed to enter the lab, dressed in my bioprotective gear and follow one of the researchers as he worked on one of his experiments. (I vividly remember when I met this researcher that he said, "My name is Maher El-Sheikh. Do you know what El-Sheikh means?" I told him I did not. He replied, "It means, 'the king.' So you should always call me 'the king,' even though everybody here considers me a lowly lab tech because I only have a master's degree!" He later got his PhD, and I told him I would always refer to him in the future as "Dr. King.") So there I was in a biosafety level 2 lab, watching "the king" perform his experiments as I hovered over his shoulder and asked what he was studying.

"We're working with Ebola today," he told me in a nonchalant voice.

"EBOLA!!!" I shouted, imagining blood pouring out of all my bodily orifices as I slowly succumbed to a painful and horrible death.

"Do not worry, my friend," he replied after a long moment, finally, cracking a smile. "It has been deactivated. If it was not, we could not work with it in a biosafety level 2 lab."

I guess that was the scientific equivalent of a fraternity prank on the new guy. Just for the record, I never got Ebola.

But I did get an earful when I mentioned to the senior scientist of our group that my daughter had autism and he started going on about "that bastard Wakefield"! It was the only time in my three months at the lab that I heard this scientist use a curse word.

I listened to him rant and rave about Wakefield for a few minutes, waited for the storm to pass, and since, like many autism parents, I'd read a good deal about Dr. Wakefield, I couldn't help myself and finally asked this senior scientist, "What exactly did he do wrong?"

The volcano suddenly went silent, and he stared at me. "He faked his results!" he said finally.

"How?" I asked, thinking that surely a top scientist at one of our leading national labs could clearly explain to me the sinister methods used by this supervillain. Strong opinions are certainly backed up by deep understanding, right?

I got nothing. After a few moments of awkward silence, the researcher said he had to get ready for a meeting, and I exited his office.

If you're one of those people who believe Dr. Wakefield is a scientist who faked his lab results but can't articulate why you think this, you're not alone. Even scientists at major government labs can't tell you what he actually did wrong. Maybe it's time to get some facts.

***

Standing well over six foot two, with movie star good looks and a voice that would be at home in the Royal Shakespeare Company, as well as being a competitive rugby player well into his forties, Andrew Wakefield did not seem destined to become the most controversial scientific figure of the late twentieth and early twenty-first century.

All of his family went to the same medical school, University of London, Saint Mary's Hospital Medical School. His great-grandfather attended the school, as did his grandfather, who spent his career as a general practitioner. Wakefield's mother and father met at Saint Mary's, where his father became a neurologist and his mother a general practitioner. His brother became a colorectal surgeon. He also had nephews who went to the school, as well as assorted aunts and cousins. "It's a wonderful medical school, and so far I'm the only one to have been struck off the medical register and lost my license!" he said with a laugh. "But it's been in a good cause."[1]

"Andy," as he is known in the autism community, graduated from Saint Mary's Hospital Medical School in 1981 and five years later qualified as a fellow of the Royal College of Surgeons. He spent several years studying small intestinal transplant surgery at the University of Toronto, finally returning to England at the end of the 1980s to take up a job at the Royal Free Hospital on a Wellcome Trust fellowship. He worked briefly as an academic surgeon before becoming a senior academic in gastroenterology and running a research team at the Royal Free until he was terminated in 2001 because of his research into the MMR vaccine and the development of autism.

In the mid-1990s, Andy and his team looked at the question of the measles virus and Crohn's disease, a bowel disorder that had been increasing in children. The disease had been virtually unheard of in children until the mid-1960s, raising the question of what had caused this change. In a remarkable irony, considering all that came later, Andy's work was even funded by the pharmaceutical giant Merck, which produced the MMR vaccine. Andy had personally presented the proposal at Merck's headquarters, suggesting the MMR vaccine might be linked to the development of Crohn's Disease, based perhaps on the age when the child received the vaccine. Merck agreed it was an important question and funded the research.

The MMR vaccine and Crohn's Disease paper was published in *The Lancet* in 1995 and showed a threefold increase in Crohn's Disease among those children who received the MMR vaccine as compared to children who got measles. Andy thanked Merck in the article for funding the paper. This caused one of the Merck executives to call Andy's boss, Roy Pander, and shout at him that they should never have put Merck's name on the paper. Andy believed he had to put Merck's name on the paper, since they had funded the study, and failing to do so would have been a violation of research guidelines. Still, he realized he was unlikely to get any further funding from Merck.

After the publication of the Crohn's Disease paper, Andy started getting calls from parents who reported that their children had been developing fine, received the MMR vaccine, then regressed into autism and developed terrible gastrointestinal problems and seemed to be in extreme pain. The parents complained that the medical community was not taking them seriously and asked if Andy would examine their children for gastrointestinal problems. Andy put together a team of specialists to examine the children and biopsied several gut tissue samples. Around the same time, Andy was introduced to Dr. John O'Leary, who was leading the world in a technique called TaqMan PCR, which at the time was a cutting-edge method of looking for very low copies of genetic material. When Andy first met with O'Leary, he was on a fellowship with his team at Cornell. The two agreed on a collaboration plan, and O'Leary used this technique on Andy's samples when O'Leary returned to Dublin, Ireland, to take up a post as professor of pathology.

Andy's team would code the samples, O'Leary's lab would perform the tests, and then the code would be broken in front of the independent trustees of the charities that funded the study. The data showed there was a high prevalence of the measles virus in the children whose parents reported a change after the MMR shot, and a very low prevalence in controls. O'Leary was able to further sequence the measles virus isolated, using a process called allelic discrimination. It showed the virus came from the vaccine, rather than a wild strain the child had somehow contracted. The paper was published in 1998 and caused a firestorm of controversy that continues to this day.

As Andy recounted in his book *Callous Disregard—Autism and Vaccines: The Truth Behind a Tragedy*, published in 2010:

> On February 28, 1998, twelve colleagues and I published a case series paper in *The Lancet*, a respected medical journal as an "Early Report." The paper described clinical findings in 12 children with an autistic spectrum disorder (ASD) occurring in association with a mild-to-moderate inflammation of the large intestine (colitis). This was accompanied by swelling of the lymph glands in the intestinal lining (lymphoid nodular hyperplasia), predominantly in the

> last part of the small intestine (terminal ileum). Contemporaneously, parents of 9 children associated onset of symptoms with measles, mumps, and rubella (MMR) vaccine exposure, 8 of whom were reported on in the original paper. The significance of these findings has been overshadowed by misunderstanding, misrepresentation, and a concerted, systematic effort to discredit the work. This effort, and specifically the complaint of a freelance journalist and an intense desire to subvert enquiry into issues of vaccine safety and legal redress for vaccine damage, culminated in the longest running and most expensive fitness to practice case ever to come before the United Kingdom's medical regulator, the General Medical Council.

The decision of the General Medical Council on January 28, 2010, removed Dr. Wakefield from the register of physicians allowed to practice medicine. It was viewed by parent groups around the world as a great injustice. Perhaps nowhere was this more evident than in the decision of the General Medical Council to call none of the parents. What doctor has ever been subjected to such a tribunal in which no representatives of his patients were ever called to testify? Wakefield was not punished for misdeeds done to his patients or their complaints. He was persecuted because his research struck at the very heart of the financial and ideological underpinnings of modern public health, the belief in the vaccine as an ultimate good, of which no criticism could be tolerated.

In an article from the *Daily Mail* on May 24, 2010, the day after he was struck off the roll of physicians by the General Medical Council, they reported:

> The doctor at the center of the MMR vaccine controversy has been struck off after being accused of 'callously disregarding' vulnerable children.
>
> Andrew Wakefield, 53, whose research claimed there was a link between autism and the measles, mumps and rubella jab, was yesterday branded dishonest, misleading and irresponsible by the General Medical Council.
>
> He has been banned from practicing in Britain after being found guilty of more than 30 charges of serious professional misconduct.
>
> In the longest, most expensive hearing in its 148-year history, the GMC accused the doctor of 'bringing the medical profession into disrepute.'[3]

It's clear that the article throws a great deal of accusations at Dr. Wakefield, but there is one fundamental thing missing: the evidence that the research is flawed or faked in any way. As an attorney, I'm trained to listen to what people say, knowing they will eventually get to what they believe to be the most important point. The fourth paragraph of the article seems to reveal Wakefield's true crime, "bringing the medical profession into disrepute."

Consider the following paragraphs in the same article and ask yourself what seems to be the overriding concern of the General Medical Council:

> The panel said he behaved unethically and showed 'callous disregard for any distress or pain the children might suffer.'
>
> It also heard how he ordered some children at the Royal Free Hospital to undergo unpleasant and often painful procedures such as colonoscopies, urine tests, lumbar punctures (injections into their spines) and barium meals—where they were force-fed gas pellets and acid to expand their stomach.
>
> The panel ruled that many of the children should never have been included in the research. It also found that Dr. Wakefield and colleagues had not been granted ethical approval to use the children in their research.[3]

In all of my investigations, I keep looking for some evidence that Andy Wakefield's results were the result of fraud or were somehow mistaken. Even the most vehement critic of Wakefield must admit that he was removed from the roll of physicians in England because of the conclusion he performed unnecessary tests, even though these were the standard tests administered at the time if there was the suspicion of gastrointestinal problems. Is this enough to make him the world's greatest scientific villain? These nonverbal children presented with signs of gastrointestinal distress, so he investigated the claims and published a case series report, a common practice in medicine. He suggested further research and, in an abundance of caution, suggested parents might want to have their children get the single shots for measles, mumps, and rubella, an option that was available in England at that time. Wakefield has always stated that these tests were medically permissible and that he had ethical permission to perform them, a claim that has been substantiated by other investigators, such as Dr. David Lewis of the National Whistleblower Center.

Despite the claims of my volcanic supervising scientist at Lawrence Livermore National Laboratories, no honest review will support the claim that Wakefield faked his results. A more accurate description might be that he was like Columbus, sailing into the western ocean and discovering a new land, but when he returned to Spain and announced his discovery he was told that he had exceeded his mission and in the process had committed heresy. No longer would ships be allowed to sail beyond a certain point in the western ocean.

The long trial and harsh punishment of Wakefield would serve as a stark warning to all who ventured into the gastrointestinal system that the authorities did not want them looking into autism.

***

One of the facets of the Wakefield persecution that has been overlooked is that Wakefield did not find himself alone in the docket at the General Medical Council (GMC) in Britain. He was joined by two of his fellow researchers, Professor John Walker-Smith and Professor Simon Harry Murch. Murch was found not guilty by the panel. (The *Lancet* paper had thirteen authors.) Although Walker-Smith had retired from practice in 2001, the GMC charged him, and he had to defend himself in the hearings from 2007 to 2010. He was found guilty, along with Wakefield, and stricken from the medical record. The prosecution of Walker-Smith was curious, not only because he had retired several years earlier, but also because he is considered one of the founding fathers of pediatric gastroenterology in Britain.

Even though he was more than a decade into his retirement, Walker-Smith petitioned his insurance carrier to pay for an appeal of the GMC ruling. Wakefield made a similar appeal to his insurance carrier but was denied. From February 13 to February 17, 2012, Walker-Smith presented his appeal in the High Court of Justice, Queen's Bench Division, Administrative Court. In March 2012, the High Court under Justice Mitting handed down an exhaustive fifty-nine-page decision, reviewing each of the children examined, as well as the General Medical Council's investigation. The decision was harshly critical of the GMC investigations and cleared Walker-Smith of all charges.[4] Justice Mitting wrote in the conclusion:

> For the reasons given above, both on general issues and the *Lancet* paper and in relation to individual children, the panel's overall conclusion that Professor Walker-Smith was guilty of serious professional misconduct was flawed, in two respects: inadequate and superficial reasoning and, in a number of instances, a wrong conclusion.[5]

For those unfamiliar with the language of a legal appeal, it's difficult to deliver a sterner rebuke than to say one's reasoning was "inadequate and superficial" and that in many instances the "wrong conclusion" was drawn. According to an article from the BBC:

> Mr. Justice Mitting called for changes in the way General Medical Council fitness to practice hearings are conducted in the future, saying: "It would be a misfortune if this were to happen again."
>
> Professor Walker-Smith, who retired in 2001, said, "I am extremely pleased with the outcome of my appeal. There has been a great burden on me and my family since the allegations were first made in 2004 and throughout the hearing that ran from 2007 to 2010. I am relieved that the matter is now over."[6]

Although Wakefield published his article with twelve other scientists, only three faced disciplinary actions and two were found guilty, but one of those was overturned on appeal. Wakefield now stands alone as the only person disciplined for his role in the *Lancet* article. Is it simply because his insurance company did not fund his appeal, as Walker-Smith's did? If Walker-Smith's initial decision in the General Medical Council was based on "inadequate and superficial reasoning" as well as in many instances the drawing of a "wrong conclusion," are we to believe these defects are not present in the case against Dr. Wakefield?

Attorney Mary Holland, a research scholar at New York University School of Law and director of the Elizabeth Birt Center for Autism Law and Advocacy, wrote of the exoneration of Professor Walker-Smith:

> Justice Mitting's impartial judicial decision marks a turning point in a long campaign to discredit the 1998 *Lancet* article and Dr. Andrew Wakefield in particular. To date, international media have failed to probe the GMC's ruling or to explore the many connections between Brian Deer, the Rupert Murdoch media empire, Glaxo Smith Kline, the *British Medical* Journal and numerous other bodies . . . "This victory for Dr. WalkerSmith is a triumph for all those who care about people with autism and bowel disease. I hope this decision leads to investigating the true causes of this global epidemic."[7]

While the exoneration of Professor Walker-Smith was a victory for those who believed in the work of Dr. Wakefield, it did not lead to any reexamination of his case. Media coverage of the story was minimal. Dr. Walker-Smith's name had been cleared, but the name of Andy Wakefield would still be spoken with the scorn and derision reserved for history's greatest villains.

***

Maybe Wakefield should have known from the very beginning that persecution would be at least part of the journey he would endure. After his research was published in 1998, he gave several talks at the CDC, and given what came later it should not be surprising that among those government scientists were many who would later come to exert great influence over Dr. William Thompson. Frank DeStefano was an important member of these meetings, as well as Robert Chen, who was Thompson's direct superior.

Another prominent member of these discussions was Marshalyn Yeargin-Allsopp, a CDC scientist who was herself African American and would later play a prominent role in covering up the increase in autism in African American boys from earlier administration of the MMR shot. Wakefield remembers being at a meeting at the Cold Spring Harbor Laboratory to discuss his work when Yeargin-Allsopp asked if he wanted to predict what proportion of children with autism was damaged by the MMR vaccine. Wakefield considered it a trick question. His research had only gone so far as to reveal evidence for twelve children whose autism could be linked to administration of the MMR vaccine. To give an answer to such a question would be mere speculation given what he knew at the time. Wakefield's paper in *The Lancet* had suggested further research to answer the question. He did not in any way suggest he knew the answer, even though his suspicions were that that number could be quite high.

One of the individuals Andy interacted with a great deal was Sallie Bernard, an autism parent and cofounder of the group Safe Minds, who had written extensively about the danger of mercury and how its presence in the vaccine schedule might be affecting the immune response of children. As Andy recalled, "We thought that looking at one vaccine or one component in isolation was naive. Because these things could clearly interact, potentiate the side effects by deviating the immune system, or injuring the brain, and making it more vulnerable."[8]

As a general rule, in these discussions, Andy found academic scientists to be more open to discussion of these issues than government scientists, who seemed to take personal offense to his research:

> The people I met in public health were deeply unimpressive. People in the American public health system were reflective of those in the English public health system, career bureaucrats who were desperate to exonerate vaccines. Their whole *raison d'être* was to promote vaccination policy and to protect the

> perception of a CDC with high-level integrity and scientific accountability, and open-mindedness, but it was quite the opposite. They were angry and dismayed that one should even question whether vaccines were causing these problems. I have to say I found them deeply unimpressive.[9]

I asked Andy about the work of Dr. Ian Lipkin, the Columbia University professor who had performed what was touted at the time as a "replication study" of his work, but which has come under harsh criticism from many corners. An example is a critique published on the website, *Fourteen Studies*, which among other things notes that the Lipkin study failed to look at unvaccinated children and of the twenty-five children with autism, only five reported problems after the MMR vaccine, when that was the critical factor in Wakefield's investigation. Nine of the twelve children in Wakefield's original work were reported by their parents to have developed autism after their MMR vaccination. Other criticisms revolved around the fact that the study design was created by the CDC, the very government agency that was charged with promoting vaccines, which is akin to a murder suspect being allowed to direct the detectives investigating the crime.

As I reviewed the study, I couldn't help but feel that the public discussion of it veered far away from the actual findings. For example, the findings of Wakefield's genetic expert, John O'Leary, were identical to those of the CDC laboratories and Lipkin's lab at Columbia University. And they did find the measles virus in the gut of one of the children with autism, while none of the control children had it. But they dismissed this difference as unimportant. The measles virus should not be persisting in the gut of any child. That alone called out for further investigation.

Andy had these concerns as well, but one of his was even more critical:

> The greatest concern is that when we looked for the virus throughout the bowel, the large intestine, and the terminal ileum with its lymphoid tissue, it was only in this hugely swollen lymphoid tissue that we found the virus. We did not find it anywhere in the colon. Now, Lipkin got his biopsies from Tim Buie, and Tim, bless him, was not very good at getting into the ileum, which is technically challenging to many gastroenterologists. I worked with guys who were extremely good at it. It wasn't me. I didn't do it. But they were very good. So we were always able to get samples of this lymphoid tissue where we found the virus. When Buie did his studies with Lipkin he provided tissues from the secum (the large intestine), and the ileum, and he didn't say how many came from the respective sites. Well, all of them that came from the

> secum would have been negative based on our earlier studies. We made this point clear. It was most definitely not a replication of our study. And what slayed me was Lipkin's subsequent reporting in the news media that this was the final word and it ruled out any possibility that MMR vaccine was causing autism. It didn't even go close to doing that. That really made me question the role Lipkin was actually playing in all of this and to what extent he understood the science and wanted the issue resolved in a scientific way.[10]

***

An extensive investigation of Wakefield's work was conducted by Dr. David Lewis, a thirty-year Environmental Protection Agency (EPA) scientist, recipient of a Science Achievement Award from EPA head Carol Browner, and the only EPA scientist to be lead author on papers published in *Nature* and *The Lancet*. In his book *Science for Sale*, Lewis devotes two chapters to the Wakefield case, looking first at the investigative reporter Brian Deer, as well as the working of the UK General Medical Council, which stripped Wakefield of his license to practice medicine. Lewis paints Deer as a puppet of powerful financial, scientific, and governmental interests who were intent on defending the MMR vaccine, regardless of the actual facts. Lewis wrote in his chapter on Deer:

> The plan carried out by Parliament member Evan Harris, who escorted Brian Deer over to *The Lancet* and accused Dr. Wakefield and his coauthors of scientific fraud, worked. They succeeded in getting the scientific community and the world media to blame Wakefield for a global public health disaster created by government officials in league with the vaccine industry in the United States, Canada, Great Britain, and elsewhere throughout the world. It worked because everybody had their hands in the pie: top government officials, leading universities funded by the vaccine industry, and even prestigious scientific journals on the payroll of Merck and GSG [Glaxo-Smith-Kline], all searching for a scapegoat.[11]

Lewis had even less respect for the work of the United Kingdom General Medical Council, whose hearings against Wakefield started on July 16, 2007, and lasted until April 14, 2010, a period of nearly three years. Lewis argues that many of the accusations that the General Medical Council leveled at Wakefield were demonstrably false and that when presented with evidence of their falsity, they simply refused to answer.

Science for Sale opens with the warning of President Eisenhower in his farewell address of the dominance of the nation's science by the federal government and how easily intellectual inquiry could be captured by a small minority of powerful interests. The Wakefield case was, to Dr. Lewis, a confirmation of Eisenhower's greatest fear:

> The problem is that, when government and industry are geared up to cover up, we have no way of knowing what the real toll is on public health or the environment. But one thing is safe to assume. Government agencies, big corporations, and the universities they fund will not confine the use of these tactics to attacking only scientists who represent small areas of interest. Nor will they hesitate to go after areas of science where they lack a broad consensus of support. Protecting government policies and industry practices knows no bounds. Whatever tools prove effective in reaching those goals will, sooner than later, be used with little, if any restraint. It is something, as President Eisenhower warned, to be gravely regarded. Our silence now will eventually bring an unbearable price for all to pay.[12]

Wakefield definitely had his defenders. Among the parent community, he was regarded by many as a scientist of unshakeable integrity, and certainly the writings of well-respected figures like Dr. David Lewis were a great consolation to him. But among the great mass of scientists and the general public, Wakefield was seen as a dangerous, if not a downright evil, figure. In their view, he was a man who wanted children to die.

And yet, in May 2014, Andy Wakefield would be given an enormous opportunity to right the injustice done not simply to him, but to an entire community of parents and their suffering children. He was determined that in this fight, he would grab the upper hand, and take science back from those who had corrupted it.

***

Brian Hooker had been communicating with William Thompson for approximately six months when he decided to bring Andy Wakefield into the conversation. The two men had known each other over the years, meeting at conferences. However, neither of them recalled any conversations of significant depth. Hooker had been interested in the mercury (thimerosal) component of vaccines, while Wakefield was more focused on the MMR shot.

As Wakefield recalled:

> It was absolutely fascinating that someone from the inside had clearly suffered a pang of conscience over a protracted period. And shared this information with Brian and given him access to it. At the time I don't think any of us realized the depth of the fraud that had taken place. We hadn't done a thorough analysis of the documents and gone into the background. But Brian had clearly picked up on the African-American issue and in a fairly straight-forward analysis had confirmed what Thompson himself had found and taken it further. Then the question became, had they committed fraud? And going back and looking at the analysis plan that Thompson had provided, they made it absolutely clear that race was going to come from the entire group. And when you get to the paper they'd reduced it to the birth certificate cohort.[13]

In addition to the change in the analysis plan on the reporting of African American children, Wakefield had noted the admission in the DeStefano paper that there was a small increase in autism among normally developing children if they received an MMR shot at an earlier time. Wakefield remembers he'd even written a rebuttal to the DeStefano article, which was not published. He said, "In that paper I noted that they'd made this observation in children without mental retardation, that there was an increased risk. They'd hidden a lot of the data, but what they revealed was instructive. My point at the time in the rebuttal was that's exactly the group in which you'd expect to find the effect. If you do not have mental retardation, then you're more likely to have had a period of normal cerebral development. You're not born with an injury. You acquire it."[14]

When asked his emotional state upon learning of this information, Andy recalled:

> I was delighted and shocked at the same time. I had always thought that someone might come forward. I am eternally optimistic about this. But I also felt that in the CDC or in the regulatory structure there's got to be someone who knows what's going on. I had always hoped and believed that someone might come forward. It was still astonishing and pleasant when they did come forward, though. That's the point when I said to Brian, 'this is absolute gold dust. You have to record this.' Because my experience with whistleblowers is that these people can slip through your fingers and you're left with no evidence. You miss gems that they disclose to you because you don't document them and I encouraged Brian to do that.[15]

Andy's suggestion to Brian that they tape-record his conversations with Thompson didn't sit well with Brian. Over the past months he had developed a close friendship with Thompson, and recording the CDC scientist felt like something of a betrayal. And yet, he couldn't discount Andy's argument that this was about more than one scientist, it was about millions of injured children, and their families, who still didn't have the truth.

"I have to say that Brian is a much more moral person than I am," said Wakefield. "He was against that [recording Thompson without his knowledge]. I had no scruples whatsoever when it involves the damage to children. Recording Thompson was a stroll in the park. That was not the issue. The issue was establishing the extent and the depth of the crime that had been committed. That was all that was important."[16]

Andy and Brian talked to attorney Jim Moody about how they might legally tape a phone conversation with Thompson. Moody investigated the laws of California and Oregon and found that the law in those two states required both parties to give consent to a phone recording. However, Washington State allowed a phone conversation to be recorded with the approval of just one of the parties. Hooker made plans to drive to Washington State, where he would call Thompson and eventually record four long telephone conversations with him. Even though he agreed to the plan, Hooker didn't like doing it but considered it a necessary evil.

In his review of the documents, Wakefield found himself wishing he could talk directly to Thompson. Wakefield had some specific questions about what and how certain things had been done. He brought this issue up with Brian Hooker, who thought it wouldn't hurt to ask Thompson if he'd speak directly to Wakefield. Amazingly, Thompson agreed to talk to Wakefield.

Wakefield recalls those first conversations as being extremely tentative. He said:

> I didn't want to scare him off. Here he was talking to the dreadful Andy Wakefield. But of course he didn't think that because he knew he'd been part of a cover-up of a hypothesis we shared with them. A hypothesis he tested and found to be positive. And if he hadn't committed this fraud, Andy Wakefield might still have a job. So I guess we both felt like we were on our first date. Just dancing around each other. But it was good. Very useful. We spoke to each other as scientists. I asked him scientific questions that he was able to answer. I didn't go into 'why did you do this?' or 'what drove you to do it?' or 'how could you have done this?' It was much more of a fact-finding inquiry

> because I had to be absolutely certain of the ground we were on. He was very forthcoming, and very frank, and very honest. There was an immense sense of relief that he was finally able to talk about this. It had clearly been vexing him for a long, long time. And indeed, since it first began, [it had bothered him] and that's evident from his meeting notes and his exchanges with colleagues at the time. He really suffered quite badly as a consequence of being coerced into this fraud.[17]

While Wakefield was excited by these developments, he was also filled with a feeling of foreboding. Brian had submitted one paper to the journal, *Translational Neurodegeneration*, and another to *Nature Neuroscience*, co-written with Dr. David Lewis. In the article submitted to *Nature Neuroscience*, Hooker had gone as far as describing his source as a whistleblower.

Wakefield said:

> At that point, Thompson was in real danger. Because in my experience, whistleblowers are in danger as long as the only people who know their identities are their enemies. And when Brian's paper went to *Nature Neuroscience*, a journalist from Nature Neuroscience contacted Walter Orenstein, who contacted Frank DeStefano. DeStefano then sent out an email to his colleagues on the paper, saying this may be coming your way. They then knew that something was breaking, that something was happening. As Brian says in the interview for my documentary, he thought we'd be dredging a river for Bill Thompson's body. Now paradoxically, the way to prevent that is to make the person a public figure.[18]

Wakefield then went on to describe the way in which he and Brian Hooker sought to protect Thompson against the forces they believed were gathering to silence him:

> And so I said to Brian, we need to do three things. We need to make sure he's got a whistleblower lawyer, which Brian had been talking to him about. But he was very reluctant to do that. 'No, I don't need that. I'm fine. I don't want my wife to know.' He really didn't understand the precarious nature of what he was involved in. He needed to get all of his documents to Congressman Posey, an official person, and then ideally to be deposed by Posey. And we needed to make him public for his own protection. Because then it's out there. These people are utterly ruthless. It may not be his coauthors. But once it is known throughout the CDC hierarchy and their friends in industry, then

> anything can happen. And the way to prevent that is to make him public and force him to file under the whistleblower act. And that's what he did. And Thompson is still at the CDC and as best as one can ascertain he's safe. All his documents are with Posey and he has given Posey a full and detailed statement only to be released upon his death. So he's got that added insurance.[19]

Wakefield's actions were not fully understood by many members of the autism community:

> It was the right thing to do, but people I've known for years in autism were furious. They thought it was some kind of stunt, which it was not. I've been at this far too long to pull stunts. I come from the bitter experience of having whistleblowers and lost them because they got frightened off and they were intimidated and they disappeared. And that wasn't going to happen to Bill Thompson if I could help it. Everybody had an opinion. Some people I've known for years and had been very friendly with were very upset and really quite abusive. And I'm afraid that's just the way it is. Hopefully, in the fullness of time, they will understand why it was necessary to do what we did.[20]

The public revelation of a high-level whistleblower in the CDC generated such controversy in the autism community that on August 27, 2014, Thompson's whistleblower attorney, Frederick Morgan Jr. of Morgan Verkamp, LLC, released a statement from Thompson. It is reproduced below in its entirety:

> **FOR IMMEDIATE RELEASE—AUGUST 27, 2014**
>
> **STATEMENT OF WILLIAM W. THOMPSON, Ph.D., REGARDING THE 2004 ARTICLE EXAMINING THE POSSIBILITY OF A RELATIONSHIP BETWEEN MMR VACCINE AND AUTISM**
>
> My name is William Thompson. I am a senior scientist with the Centers for Disease Control and Prevention, where I have worked since 1998.
>
> I regret that my coauthors and I omitted statistically significant information in our 2004 article published in the journal *Pediatrics*. The omitted date suggested that African American males who received the MMR vaccine before age 36 months were at increased risk for autism. Decisions were made regarding which findings to report after data were collected, and I believe that the final study protocol was not followed.

> I want to be absolutely clear that I believe vaccines have saved and continue to save countless lives. I would never suggest that any parent avoid vaccinating children of any race. Vaccines prevent serious diseases, and the risks associated with their administration are vastly outweighed by their individual and societal benefits.
>
> My concern has been the decision to omit relevant findings in a particular study for a particular sub group for a particular vaccine. There have always been recognized risks for vaccination and I believe it is the responsibility of the CDC to properly convey the risks associated with receipt of those vaccines.
>
> I have had many discussions with Dr. Brian Hooker over the last 10 months regarding studies the CDC has carried out regarding vaccines and neurodevelopmental outcomes including autism spectrum disorders. I share his belief that CDC decision-making and analyses should be transparent. I was not, however, aware that he was recording any of our conversations, nor was I given any choice regarding whether my name would be made public or my voice would be put on the internet.
>
> I am grateful for the many supportive emails that I have received over the last several days. I will not be answering further questions at this time. I am providing information to Congressman William Posey, and of course will continue to cooperate with Congress. I have also offered to assist with reanalysis of the study data or development of further studies. For the time being, however, I am focused on my job and family.
>
> Reasonable scientists can and do differ in their interpretation of information. I will do everything I can to assist any unbiased and objective scientists inside or outside of the CDC to analyze data collected by the CDC or other public organizations for the purpose of understanding whether vaccines are associated with an increased risk of autism. There are still more questions than answers, and I appreciate that so many families are looking for answers from the scientific community.
>
> My colleagues and supervisors at the CDC have been entirely professional since this matter became public. In fact, I received a performance-based award after this story came out. I have experienced no pressure or retaliation and certainly was not escorted from the building as some have stated.[21]

Dr. Wakefield and his wife, Carmel, were driving to Fayetteville and the University of Arkansas to visit their daughter, who was starting her final semester when news of this statement began to break. A few people contacted Andy and asked if he knew about it. They told him what they knew, and Andy's head was spinning. Since Andy was driving, he asked Carmel if she would text for him.

"Is the Press Release real?" Carmel texted to Thompson.

"Yes," came the reply.

Andy had Carmel text back, "Thank you. It was the right and honorable thing to do."

"I agree. I apologize for the price you paid for my dishonesty," Thompson texted.

"I forgive you completely and without any bitterness."

"I know you mean it and I am grateful to know you more personally," Thompson texted in ending the conversation.

As Andy Wakefield drove with his wife to their daughter's college, he allowed himself to believe for a brief moment that his sixteen-year odyssey might soon be over.

But the truth was that there was still a lot of road left to travel.

* * *

Although it was a great personal vindication to receive the apology from Dr. Thompson, as well as the press release, Andrew Wakefield was still aware of the terrible damage that had been done to his scientific and personal reputation over the years since he'd published his case study of twelve children in *The Lancet.* He conferred with Dr. Brian Hooker about the issue, and the two of them decided to file a formal complaint with the CDC and the Office of Research Integrity at the US Department of Health and Human Services. The complaint was drawn up by James Moody, an attorney with long experience in vaccine issues, and sent to these two agencies by Federal Express on October 14, 2014.

The thirty-four-page complaint opened with the claim of research misconduct, reviewed the allegations, and then included the press release from Thompson's whistleblower attorney on August 27, 2004. The complaint quoted directly from the press release:

> I regret that my coauthors and I omitted statistically significant information in our 2004 article published in the journal Pediatrics. The omitted data suggested that African American males who received the MMR vaccine before age 36 months were at increased risk for autism. Decisions were made regarding which findings to report after the data were collected, and I believe that the final study protocol was not followed.[22]

The complaint named those who had engaged in this research misconduct as Dr. Frank DeStefano, Dr. William Thompson, Dr. Marshalyn Yeargin-Allsopp, Dr. Tanya Karapurkar Bashin, and Dr. Colleen Boyle. The misconduct consisted of covering up "statistically significant associations between the age of first MMR and autism in (a) the entire autism cohort, (b) African-American children, and (c) children with 'isolated' autism, a subset defined by The Group as those children with autism and without co-morbid developmental disabilities."[23] The complaint also contained direct quotes from the four legally recorded telephone conversations between Hooker and Thompson.

The conversation of May 24, 2014, contained some of the most damning information regarding the development of autism among those children without any other health conditions, as Thompson explained:

> You see that the strongest association is with those [autistic cases] without mental retardation. The non-isolated, the non-MR [mental retardation] . . . the effect is where you would think it would happen. It is with the kids without other conditions, without the comorbid conditions . . . I'm just looking at this and it's like, "Oh my God!" . . . I cannot believe we did what we did, but we did . . . It's all there . . . It's all there. I have handwritten notes.[24]

The handwritten notes showed that on January 28, 2004, Thompson was extremely concerned about the upcoming Institute of Medicine meeting in February and wrote, "What should we do about the race effect??—shows large effect for blacks and no effect for whites."[25]

Below this note, Thompson had written himself some advice:

> Stay calm.
> Don't over react.
> We all have good intentions.
> Parents of autistic children have very difficult lives.[26]

As was detailed in the last chapter, Thompson's concerns led him to break the chain of command and, on February 2, 2004, to write directly to the director of the CDC, Dr. Julie Gerberding, an action for which, among others, he was placed on administrative leave. The complaint by Hooker and Wakefield detailed the great psychological burden this placed on William Thompson:

> Certainly, from this point forward, and likely for several months prior, there can be no doubt that The Group and Dr. DeStefano in particular were aware of Dr. Thompson's concerns about the study findings and the imminent public distribution of false and misleading research results in the midst of the growing vaccine-autism controversy. This is highly relevant to Dr. DeStefano's statements made in light of the current media coverage (see below).
>
> In the end, Dr. Thompson signed off on The Paper that was published in *Pediatrics* [Exhibit 11]. However, his name was withdrawn from the roster of those due to present at the IOM [Institute of Medicine] on February 9, 2004. In reporting a discussion that he had with his whistleblower lawyer Thompson stated:
>
>> Ya know, I'm not proud of that and uh, it's probably the lowest point in my career that I went along with that paper and I also paid a huge price for it because I became delusional.
>
> In his recorded call with Dr. Thompson of 5.8.14, Dr. Hooker pressed Dr. Thompson on whether he raised his concerns about the omission of significant data with The Group in the days leading up to the IOM meeting.
>
> **Dr. Hooker:** Did you raise that . . . did you raise the issue at the time?
>
> **Dr. Thompson:** I will say I raised the issue . . . I will say I raised this issue, the uh . . . two days before I became delusional.
>
> This reference is important: three days before the IOM presentation Thompson—faced with either presenting false data or taking responsibility for the vaccine-autism link in front of potentially hostile parents of autistic children—stopped sleeping and became profoundly depressed and "delusional." Crucially, he reports no prior history of mental disorder.
>
> Dr. Thompson went on to confirm, to Dr. Hooker, that the DeStefano 2004 paper was the reason for these acute psychological problems.[27]

The complaint makes several important allegations. The first is that Thompson made it abundantly clear to his fellow CDC scientists, including Frank DeStefano, that there was a strong signal from the data showing an increase in autism in those who received an earlier MMR shot, particularly among African American males as well as children who had no other medical conditions. It is this later group that Dr. Wakefield finds of critical importance because a period of normal development followed by a sudden change is more indicative of an injury than a genetic condition. The second is that Thompson genuinely suffered as a result of the stress of concealing a link between earlier administration of the MMR vaccine and autism. He stopped sleeping and "became profoundly depressed and delusional."

One can certainly sympathize with the pressure of this explosive information on Thompson, but as an autism parent and citizen of this country who wants to believe our public health authorities, this can only be described as a betrayal of our citizens on the most fundamental level. In my opinion, it is a lesser crime to steal military secrets and give them to our enemies than to conceal evidence of harm to our children and to allow this carnage to continue.

The complaint next moved to the issue of why the information concealed in this study is so critical to understanding the poisoned atmosphere that exists today on the question of vaccines and autism:

> Dr. DeStefano made the presentation to the IOM on February 9, 2004. His slide presentation is attached as Exhibit 19. In slide 17 of 40—and in direct contradiction to the Study Analysis Plan of May 11, 2001 [Exhibit 2]—Dr. DeStefano gave the source of the "race" data as the Georgia birth certificates.
>
> Dr. DeStefano's subsequent "race" slide based upon the Georgia birth certificate cohort (GBCC) analysis, claimed "no statistically significant associations [between age at first MMR and autism risk]." Slide 33 of 40 [Exhibit 20] Dr. DeStefano omitted and concealed from the IOM statistically significant associations between MMR and both race and "isolated" autism found by the Group.
>
> Dr. DeStefano's presentation to the IOM and in particular his omission of significant risks of autism in African American children vaccinated under 36 months of age, and those with "isolated" autism, were major factors in the IOM's recommendation for "no further epidemiology." The IOM's report states:
>
> > Of interest: "The committee wishes to comment on several of the other recommendations it made in its 2001 report on MMR and autism. First, the committee recommended exploring whether exposure to MMR vaccine is a risk factor for ASD in a small number of children. To date, <u>no convincing evidence of a clearly defined subgroup with susceptibility to MMR-induced ASD has been identified</u>.
> >
> > While the committee strongly supports targeted research that focuses on better understanding the disease of autism, from a public health perspective the <u>committee does not consider a significant investment in studies of the theoretical vaccine-autism connection to be useful at this time</u>.
> >
> > The nature of the debate about vaccine safety now includes a theory that genetic susceptibility makes vaccinations risky for some people, which calls into question the appropriateness of a public health, or universal,

> vaccination strategy. However, the benefits of vaccination are proven and the hypothesis of susceptible populations is presently speculative. Using an unsubstantiated hypothesis to question the safety of vaccination and the ethical behavior of those governmental agencies and scientists who advocate for vaccination could lead to widespread rejection of vaccines and inevitable increases in incidences of serious infections like measles, whooping cough, and HiB bacterial meningitis." [Underline added for emphasis.][28]

In a follow-up submission on February 16, 2015, Wakefield and Hooker detailed the risk for the subgroup titled "isolated autism," a term cooked up by the wizards at the CDC to mean normally developing children. They wrote:

> . . . [T]he valid result for the "Isolated" subgroup is the significant risk of 2.45 (1.205.00) described in the unadjusted analysis in 1.a. above. The authors knew this, and they also knew that this effect was being driven specifically by those vaccinated by 18 months—data they chose to conceal. Those vaccinated on schedule (i.e., 12–18 months) were the only group that, from repeated analysis of the data over time, consistently showed a significant increased autism risk.[29]

There it was, African American boys had an increased autism risk of 3.36 from earlier administration of the MMR shot. The rest of the population had an increased autism risk of 2.45 from earlier administration of the MMR shot.

Just to remind the reader, any finding of an increased risk above 2.0 is generally considered to be strong evidence of causation in a legal proceeding. Questions would have been raised. Those in Congress would have demanded answers.

If Thompson and his cohorts from the CDC had honestly presented their results, there would have been an enormous public outcry about the vaccine program, Dr. Wakefield's work would likely have been vindicated, and, more important, millions of children would not have been put at risk for a devastating lifetime disease. And presumably, the scientific and medical establishment in 2004 would have had a target for their massive resources at developing treatments or even a cure for autism. Who knows how much progress they could have made in the past twelve years?

As damning as these results were, let's talk about how incomplete the picture is, even with this startling information.

The scientists did NOT compare rates of autism in children who received the MMR vaccine with those who received NO MMR vaccine. Later does not mean safe. It just means not as dangerous. It's a bit like comparing lung cancer rates in adults who started smoking at fourteen years old, rather than eighteen. The more complete scientific picture would be comparing children who received the MMR shot with those who never did. For those who were also concerned with the larger question of whether the vaccine schedule as a whole was also a problem, the most complete picture would be those children who followed the schedule, and those who received no vaccines. We do not know the answer to this question because the CDC has refused to perform such a study, despite the fact that there are several well-identified populations in the United States.

The simple fact of the matter is that the Institute of Medicine, as well as the CDC, was probably hoping for the phony findings presented by Frank DeStefano, if not actively encouraging it. What else would explain their immediate conclusion that they "not consider a significant investment in studies of the theoretical vaccine-autism connection at this time"? In addition, the IOM goes on to conclude that the hypothesis should not be pursued any further, since questioning "the safety of vaccines and the ethical behavior of those governmental agencies and scientists who advocate for vaccination could lead to widespread rejection of vaccines . . ." If a product is unsafe, shouldn't it be rejected? And if there wasn't a wholesale rejection, wouldn't a slowing of the program and a vigorous investigation of the problem be the bare minimum expected by an informed public?

Let's put this in another context. Let's say several members of the clergy of a church were accused of the sexual abuse of children. The church is then allowed to conduct its own investigation. It comes to the conclusion that none of their clergy are harming children and adds that children are positively influenced by their interactions with the clergy. The church then concludes that such allegations of sexual abuse are so devastating and dangerous to the spiritual health of its congregation that they will not investigate any future claims. Such an investigation would be so riddled with conflicts of interest that the public would never accept the report. But that is exactly what the CDC did in their MMR vaccine/autism investigation.

I think I'm like most people in that I look to science for answers, but for a few issues, I look to the human heart for the greater answers. I try to imagine Frank DeStefano in those moments before he took the stage at the Institute of Medicine conference of February 9, 2004. Who is he as a human being? I can easily discover the simple facts of his life, graduation from Cornell University in 1974 with a Bachelor of Science degree, a medical degree from the University of Pittsburgh in 1978, and a Master's in Public Health from Johns Hopkins University in 1984. I imagine a distinguished-looking man, a family who loves him, and many friends and colleagues. I imagine if we were to meet and have a conversation I'd probably have a favorable impression of him. I generally like people of intelligence and accomplishment. Under different circumstances I could easily imagine us being friendly toward each other.

And yet I can't help thinking of him in the moments before he took the stage at the Institute of Medicine conference of February 9, 2004, to conceal from the world the truth about the link between the MMR vaccine and autism. Did he dramatically close his eyes and take a deep breath as he prepared to tell his monstrous lies? Or was he talking and joking with his colleagues, the way athletes do before they are introduced to the crowd at a sporting event? How did he intellectually justify his actions? How does he justify them tonight, when he lays his head on the pillow and tries to fall asleep? DeStefano knew that Thompson found strong associations between earlier administration of the MMR vaccine and autism. These findings caused Thompson great psychological anguish. Did DeStefano tell himself that he would have greater emotional strength than Thompson? That he would not break? Was this the great noble lie he had to tell, because the ordinary citizens of the United States were incapable of making the decision? At what point does a believer cross the line into zealotry and become capable of terrible atrocities? Science has no provision for falsehood.

In the end I can give no answers. There is a darkness inherent in such a decision that I will never be able to comprehend. When I was a child, my Sicilian mother told me about the limits of her devotion to the safety of me and my older brother. I can still vividly recall her saying, "If somebody hurt you or your brother, and they didn't get punished, I'd get a gun and shoot them right between the eyes. And when the police came to arrest me I wouldn't deny it. I'd tell them, 'you're damn right I shot that son of a bitch!' And if they sent me away to jail for the rest of my life, I'd never regret it for a single moment."

As a child I felt comforted by the idea that the adults in my life put such a high value on my safety and well-being. And while I will never pick up a gun against my adversaries, I will pick up my pen and do exactly what my mother would have done to those who would harm children.

***

On August 26, 2014, the journalist Sharyl Attkisson interviewed Frank DeStefano regarding the Thompson allegations. Attkisson provided a transcript of her interview to Hooker and Wakefield, and they included some of that exchange in their complaint:

Attkisson: Were you aware of any of his concerns of, you know, have you been aware before today of any of his concerns about this?

DeStefano: Uh, uh, yeah, I mean I've continued to see, uh, uh, see him for over the past ten years and we've interacted fairly frequently, and uh, uh, no I wasn't aware of this.

Attkisson: So whoever he raised his concerns to, he didn't, he didn't raise it to you or anybody you knew of?

DeStefano: No, I mean the last time I saw him was probably about two months ago, and he didn't mention anything about this.

Attkisson: And at the time he didn't seem concerned when you said there was a consensus?

DeStefano: No, yeah, I mean at the time he did these analyses he did, you know, he did point out that in one group, you know, in that larger group the . . . the . . . measures of association [between MMR vaccine and autism] were higher than in the, uh, birth certificate group and, you know, we discussed that and for the reasons I mentioned, uh, we came to consensus that the, uh, birth certificate uh results were more valid. [Exhibit 23, emphasis added.]

Dr. DeStefano's account does not accord with either Dr. Thompson's current position [Exhibit 3] or that captured in the contemporaneous documentation [Exhibits 14 and 18]. The Group "came to a consensus" to conceal the valid "race" analysis, not because the "birth certificate results were more valid" but because they provided The Group with a convenient device for its research misconduct. Earlier in the same interview he sought to justify the use of the GBCC [Georgia Birth Certificate Cohort].

Dr. Frank DeStefano: I think what [Thompson's] saying there was a larger, um, uh, odds ratio or association among the-the larger group and that that there was not, uh, as strong an association among the birth certificate sample. And I mean, what I say to that, I think we discussed that, uh, as I recall, this was like you know, over ten years ago, and uh, I think at the time we had consensus among all co-authors that the birth certificate sample provided the more valid results because it could, uh, it had more complete information on, uh, on race for one. [Exhibit 23, emphasis added.]

For reasons described in detail above, Dr. DeStefano's response is incorrect. All "race" information was available in the school records. There appears to

> be no basis for Dr. DeStefano's contention, and no justification in any of Dr. Thompson's contemporaneous notes or data outputs, as to why The Group deviated from the Analysis Plan, and no explanation for the omissions in The Paper.[30]

It is difficult in reading the Attkisson interview to come away with any other conclusion than that DeStefano knew that some groups showed a higher association between earlier administration of the MMR shot and autism, and that they chose the group in a way to minimize that association. Additionally, the documentation that exists from that time does not indicate that the race information from the school records was inadequate.

The Hooker/Wakefield complaint claims that these actions constitute research misconduct that was not due to any "honest difference of opinion." The complaint asserts that "the decision not to report these significant results was made by management for 'political,' not scientific reasons, i.e., because of the cases pending in the Omnibus Autism Proceedings (OAP), the ongoing public controversy, and the accompanying fear that immunization rates might drop if causation were confirmed."[31] This was not science. It was politics. And it was deception, which impacted the lives of millions of children and their families.

In a different section of the complaint, Hooker and Wakefield lay out the extent of the deception:

> As a matter of fact, what we report here is not "honest difference of opinion," but consensus, agreement, and complicity between members of The Group to pervert the science.
>
> There was no "honest" difference of opinion; rather, there was a dishonest consensus to abandon the original Analysis Plan and omit from the public record, significant causation results on important autism subgroups. In Dr. Thompson's own words:
>
> "Oh my God" . . . I cannot believe what we did . . . but we did . . . It's all there . . . It's all there. I have the handwritten notes."
>
> In an email to Complainants, dated August 11, 2014, Dr. Thompson reaffirmed the dishonesty of The Group's actions, stating,
>
> "I was involved in deceiving millions of tax payers regarding the potential negative side effects of vaccines. I regret what I did."[32]

Corruption often seems to follow a similar pattern. I saw it regularly when I worked as an attorney. There is an attempt to do the right thing, to perform the job well, and then something unexpected happens. All of us have had the experience. There comes a moment of moral choice. Does one take the time to acknowledge the problem, even though uncomfortable, and fix what has gone wrong?

Sometimes people do the right thing, fix the problem, and are rewarded for their honesty and courage in speaking up. Other times they speak up, are ignored, and then are punished by people who don't want to fix the problem. But what if this reluctance to genuinely test the proposition that vaccines are related to autism is due to an even larger problem?

What if the very idea of vaccines is not compatible with human biology?

Without fear or favoritism, let's look at how your typical vaccine is created. A virus is isolated from a human host, then grown in tissue cultures of other animals, such as mice or monkeys (or in aborted human fetal tissue), then when the virus has gone through enough changes to provoke a mild immune response in humans, it is loaded back into a vaccine (and who knows which animal viruses may have also hitched a ride), accompanied by a myriad of chemicals such as aluminum, mercury, formaldehyde, and others, and injected into a broad array of human beings without any concern for genetic or immune status. This is "public" health, not personalized medicine. In order to sell that program you need to create a fiction that vaccines are extremely safe and any potential harm is outweighed by the massive societal benefits. And to maintain this fiction it is necessary to call anybody who demands proof of your claims "antiscience" and assert that they want children to die.

But what if it's not true? What if good, reasonable questions are being ignored and those in charge of public health are not concerned about real, hard-hitting, controversial science that has the potential to change people's lives, but politics instead?

I vividly remember asking this question of viral contamination to Dr. Frank Ruscetti, one of the founding fathers of human retrovirology, a thirty-five-year government scientist at the National Cancer Institute and head of a major lab (and an editor of my previous book, *PLAGUE: One Scientist's Intrepid Search for the Truth about Human Retroviruses, Chronic Fatigue Syndrome (ME/CFS), Autism, and Other Diseases*). I said, "Frank, I understand they grow these human viruses in animal tissue to weaken them, and maybe it works. But when they load up the virus they want, how do they make sure they don't get a whole bunch of other viruses from the animal that they don't want?"

He was silent for a moment, then said, "I asked a similar question myself when I was young. I was told that whatever came back from the animal, the human immune system was strong enough to handle it. We were the pinnacle of creation in their eyes, and our immune system was similarly superior. As an old man, I now understand the monumental arrogance of that statement."

What lies behind the frustration of so many autism parents is that we understand that honest science would determine whether our fears about vaccines are justified. Maybe all of these children with autism (currently estimated at one in fifty) were not diagnosed in previous generations. Maybe in your typical 1980s high school of about five hundred students there were ten students who didn't talk or had limited speech, couldn't look people in the eye, and often broke down into screaming tantrums. Maybe some of those ten were in diapers that needed to be changed by an unobtrusive aide that none of us can now recall, and occasionally these kids had seizures, but we were just too oblivious to notice. The inappropriate fart in English class we would recall decades later, but a grand mal seizure, who can be expected to remember that?

These students who couldn't talk and were often in diapers still managed to get through their classes, graduate, go to college, get married, and vanish into the great American public where we cannot find them to this day.

Hooker and Wakefield wrote in their concluding remarks of the complaint:

> We believe the facts presented here reveal a clear picture of research misconduct within the CDC with profound and far-reaching implications for public health, and in particular the wellbeing of children. This misconduct undermines the trust and reputation of CDC as a source for complete and reliable scientific information—so important to maintain the confidence of the public in the vaccine program. Honest risk communication may lead the public to demand (and industry to supply) safer vaccines, but lying to and misleading the public about safety risks threatens a permanent loss of this essential trust and confidence.
>
> The research misconduct involved scientists working in the National Immunization Program and the National Center on Birth Defects and Developmental Disabilities, right up to officials at the highest levels of the CDC, including the Director.
>
> The actions of those involved threaten not only the health of children but also the integrity of, and public confidence in, the US Public Health infrastructure.

> The alleged misconduct seriously undermines the ethical practice of medicine when pediatricians unwittingly obtain, and parents provide, informed consent to immunization based upon falsified data.
>
> The influence that this alleged misconduct has undoubtedly had on the IOM and, in turn, on the NVICP [National Vaccine Injury Compensation Program] cases, and the consequent injustices suffered by thousands of children who are victims of possible vaccine injury, constitutes, in our opinion, deliberate obstruction of justice. We urge that corrective action be taken at the earliest opportunity.[33]

Brian Hooker and Andy Wakefield filed their complaint on October 14, 2014. As of March 2016, no action has been taken, and to the best of their knowledge, the CDC, as well as the Department of Health and Human Services, has not interviewed a single person.

Why not? What could be more important than making sure that the children of America and their parents have the best information possible when making critical health decisions?

CHAPTER THREE

The Lipkin-Hornig Team

In my interview with Dr. Wakefield, he mentioned the work of Dr. Ian Lipkin, who along with his collaborator, Dr. Mady Hornig, is most associated in the scientific community with opposition to his work linking the MMR vaccine to autism and gastrointestinal disorders. I have covered the work of Drs. Lipkin and Hornig in my previous book, *PLAGUE*, cowritten with Dr. Judy Mikovits, a twenty-year government scientist and former director of the Lab of Anti-Viral Drug Mechanisms at the National Cancer Institute.

In that book, I had strong criticism for the work of Lipkin and Hornig in the XMRV (xenotropic murine leukemia virus-related virus) investigation, attempting to duplicate the findings of Dr. Mikovits. The Mikovits team, which included the prestigious Cleveland Clinic, the National Cancer Institute, and the University of Nevada/Reno, had found and reported in October 2009 in the journal *Science* that a significant number of patients with chronic fatigue syndrome/ME (68 percent) showed evidence of infection with this retrovirus. The easiest way to understand this issue is the way Dr. Mikovits views it, as akin to a nonfatal HIV virus, which does not kill the patient, but leaves them in a debilitated state where death would often be preferable. The book raised questions as to how this retrovirus jumped from mice to humans (some evidence and researchers have suggested it was from the use of mouse tissue to grow viruses that would later be used in vaccines). Second, retroviruses are known to hide out in the immune cells of the body, and thus any stimulation of the immune system (say, through a vaccination) might cause the virus to rampage out of control. Standard practice in medicine is that a child born to an HIV-infected mother will be put onto antiretrovirals and monitored as they are vaccinated to make sure their viral load does not go too high as a result of the immune stimulation and they develop AIDS. The fear of many about vaccines including the use of mercury and other metals, chemicals, and aborted human fetal tissue, as well as the combined MMR shot (the problem of the immune system responding to multiple viral challenges), has given headaches to public health officials and pharmaceutical companies. One might say our book added a third area of concern, the jumping of animal viruses from cultures used to grow the virus into the human population though vaccination.

Given my previous criticism of their work, it may then come as something of a surprise to many for me to say I consider the work of Drs. Lipkin and Hornig to provide some of the strongest scientific evidence for a link between vaccines and autism. In June 2004, Lipkin and Hornig published a paper titled "Neurotoxic Effects of Postnatal Thimerosal Are Mouse Strain Dependent" in the journal *Molecular Psychiatry*.[1]

The work looked at whether the genetic profile of various strains of mice, in particular a strain known to be susceptible to autoimmune problems, would have difficulty with the mercury preservative, thimerosal, at doses comparable to those found in the US immunization schedule. This is how the abstract describes the work (I have broken the abstract down into two parts, in order to more fully explain it):

> The developing brain is uniquely susceptible to the neurotoxic hazard posed by mercurials. Host differences in maturation, metabolism, nutrition, sex, and autoimmunity influence outcomes. How population-based variability affects the safety of the ethylmercury containing vaccine preservative, thimerosal, is unknown. Reported increases in the prevalence of autism, a highly heritable neuropsychiatric condition, are intensifying public focus on environmental exposures such as thimerosal. Immune profiles and family history are frequently consistent with autoimmunity.[2]

Lipkin and Hornig were pointing out issues that were familiar to those who work regularly in the biological field. The developing brain is vulnerable to mercury, and additional factors could involve age, metabolism, gender, nutritional status, and the propensity for autoimmune problems. The public was focusing on possible environmental triggers, and family histories often revealed a pattern of autoimmune problems among parents. Their abstract continued:

> We hypothesized that autoimmune propensity influences outcomes in mice following thimerosal challenges that mimic routine childhood immunizations. Autoimmune disease sensitive SJL/J mice showed growth delay; reduced locomotion; exaggerated response to novelty; and densely packed, hyperchromic hippocampal neurons with altered glutamate receptors and transporters. Strains resistant to autoimmunity, C57BL/6J and BALB/cj were not susceptible. **These findings implicate genetic influences and provide a model for investigating thimerosal-related neurotoxicity.**[3] [Bold added.]

Lipkin and Hornig used mice that were prone to have autoimmune problems and mimicked the thimerosal dosages of the American vaccination schedule. The result? They got mice with many of the traits of autism in humans.

On September 8, 2004, Dr. Mady Hornig, director of Translational Research, Jerome L. and Dawn Greene Infectious Diseases Laboratory, and associate professor of epidemiology, Mailman School of Public Health, Columbia University, gave testimony on this work to Congress, specifically the Committee on Government Reform, chaired by Congressman Dan Burton. This is her entire statement:

> Chairman Burton, Congressman Watson, and members of the Subcommittee, thank you for the opportunity to submit for the record this statement regarding our new animal model of the toxicity of thimerosal (ethylmercury preservative in vaccines) and its implications for public health. I regret that I am unable to personally present this testimony today due to a family medical emergency. Our work addresses whether genes are important in determining if mercury exposures akin to those in childhood immunizations can disrupt brain development and function. I also submit for the record an electronic copy of the first paper published on this animal model in the Nature Publishing group journal, *Molecular Psychiatry* (Hornig M, Chian D, Lipkin WI. Neurotoxic Effects of Postnatal Thimerosal are Mouse Strain Dependent. *Mol. Psychiatry* 2004; 9:833–845).
>
> The premise of our research is that if mercury in vaccines creates risk for neurodevelopmental disorders such as autism, genetic differences are likely to contribute to that risk. We built upon an extensive, existing literature on the toxicity of other forms of mercury in inbred mouse strains that affirmed the importance of specific genes controlling immune responses (major histocompatibility complex or MHC) in determining mercury-induced auto-immune outcomes in mice. Earlier studies, however, did not use the form of mercury

present in vaccines, known as thimerosal, and did not consider whether intramuscular, repetitive administration during early postnatal development, when the brain and immune systems are still maturing, might intensify toxicity. Based on reports of immune disturbances and family history of autoimmune disease in a subset of children with autism, we hypothesized that immune response genes linked to mercury immunotoxicity in mice would predict damage following low-dose, vaccine based mercury in our mouse model.

Our predictions were confirmed. Using thimerosal dosage and timing that approximated the childhood immunization schedule, our model of postnatal thimerosal neurotoxicity demonstrated that the genes in mice that predict mercury-related immunotoxicity also predicted neurodevelopmental damage. Features reminiscent of those observed in autism occurred in the mice of the genetically sensitive strain, including: general behavioral impoverishment and abnormal reaction to novel environments; enlargement of the hippocampus, a region of the brain involved in learning and memory; correlation of hippocampal enlargement with abnormalities in exploration and anxiety; increased packing density of neurons in the hippocampus; and disturbances in glutamate receptors and transporters. Only mice carrying the H-2s susceptibility gene showed these autismlike effects (SJL/J mice). Two mice strains with different H-2 genes (C57BL6/J mice, BALB/cJ mice, H-2d) did not demonstrate adverse consequences following thimerosal exposure.

It is important to emphasize that these animal studies do not provide conclusive evidence regarding a link between mercury exposure and human autism. Nonetheless, the finding that a specific genetic constraint profoundly alters the brains and behavior of thimerosal exposed mice confirms the biological plausibility of thimerosal neurotoxicity, provides critical guidance for the interpretation of existing epidemiological investigations into the potential association of thimerosal with neurodevelopmental disorders, and suggests important new avenues for future research. Our work implies that if genetic factors are operating in mediating a link between thimerosal and autism in humans, then studies that fail to consider genetic susceptibility factors will be compromised in their ability to detect a statistically significant effect even if one exists.

Recent findings, presented at scientific meetings but as yet unpublished, suggest that thimerosal neurotoxicity in susceptible mice involves the generation of autoantibodies targeting brain components. This autoimmune response persists long after the presence of mercury can no longer be detected. If confirmed, these findings will enable us to develop a human diagnostic test

> to determine whether some individuals with autism have similar autoantibodies present in their peripheral blood. Such work would not only bring us a step closer to identifying the genes associated with thimerosal neurotoxicity in humans, facilitating prevention programs, it would also validate the utility of this animal model for the development of safe and effective modes of intervention.
>
> It is highly likely that the neurotoxic effects of cumulative mercury burden, including exposure to other sources or forms of mercury (thimerosal in products other than vaccines; methylmercury in contaminated fish), follow similar patterns of genetic restriction; it is also likely that similar genetic factors influence the neurotoxicity observed following exposure to xenobiotics other than mercury (e.g., PCBs, the PBDEs used as flame retardants in computers, and infectious agents). Age and developmental status at the time of exposure, nutritional factors, and gender are known to influence outcomes. We have limited ability to explain the interplay of such factors in humans; consider the example of the disparate cognitive outcomes reported in children in the Faroe Islands and the Seychelles after similar prenatal methylmercury exposures. The reasons for this divergence remain unclear. The design of future epidemiologic studies must take into account the possibility of multiple xenobiotic exposures as well as the influence of factors that modulate risk. Our studies have important implications for understanding the role of gene-environment interactions in the pathogenesis of autism and related neurodevelopmental disorders.
>
> I refer Subcommittee Members to our recent publication in *Molecular Psychiatry* where experimental findings and their implication are discussed in more detail.
>
> Thank you for your attention.
>
> Mady Hornig, MD
>
> New York, NY[4]

In the opinion of this science teacher, that's the way science is supposed to be done. Mercury is suspected of being a problem in autism, but it doesn't seem to affect all children equally. In response, researchers come up with a hypothesis that explains the apparent difference. Specific genes related to the immune response may be related to how your body deals with the mercury to which you are exposed. Simply put, if your family has a history of autoimmune disease, you are likely to be more sensitive to the effects of mercury.

Lipkin and Hornig tested this proposition by using a strain of mice prone to develop autoimmune disorders. Mimicking the US immunization schedule, they found that mice with this genetic profile will suffer greater effects than mice with a different genetic profile. The changes experienced by the mice with a genetic profile that will render them more susceptible to autoimmune disorders go on to develop behaviors and brain architecture similar to that seen in autism. Hypothesis confirmed. Let's continue the investigation.

Isn't that the kind of research you expect would change the world and make the life of humanity better? Isn't that just the sort of information you'd expect to see splashed over the front page of newspapers all around the world? As for credibility, these scientists are from Columbia University and they testified in front of the United States Congress. I'll bet that for the vast majority of readers, even those who have been paying attention to the autism story, it is the first time they are learning about this research.

* * *

On September 19, 2013, I got to meet Dr. W. Ian Lipkin and Dr. Mady Hornig at a small cocktail party at the Soho Grand Hotel in New York City.

Even though I had strong criticisms about the quality of their work in the XMRV investigation with Dr. Judy Mikovits, and the autism community also considered their work regarding Dr. Andy Wakefield to be highly misleading, if not downright unethical, I wanted to make up my own mind. Lipkin had also shown some significant personal kindness to Dr. Mikovits, promising to find her some consulting work after the XMRV controversy died down. That counted for something with me.

There were also a number of samples that had been collected during that investigation, and if the money could be found to perform certain tests, it might move the research into chronic fatigue syndrome/ME forward. Lipkin and Hornig also had an amazing repository of autism clinical samples, known as the "Autism Birth Cohort" or the "Norway Project," which also needed money to perform certain tests. Since I believed that chronic fatigue syndrome/ME and autism were linked, any movement in one area was likely to bring me closer to answers for my daughter. He had the samples, the reputation, and the facilities, and I had a possible money connection. Despite what had transpired in the past, I wanted to pursue any opportunity that might lead to a brighter future for those with chronic fatigue syndrome/ME, as well as autism.

On December 7, 2012, Dr. Lipkin and I had a phone conversation about how a group I was involved with in Silicon Valley might come up with some funds to further his research. After we talked about that, our conversation moved to Dr. Mikovits.

Lipkin said:

> She's a good person. She's gotten a very bad shake. That said, I don't know how to include her. So what I've said we're going to do in the future is to make sure that anything that's done using samples we've obtained through this initial project, we're going to include her in authorship. We're going to give her an opportunity to look at the data, to comment on the data, and that's something I think we can clearly do which would be very helpful. Her problem right now is primarily financial and legal. It's a disaster. And I wish I knew how to fix that. But, and I did what I could to get her out of jail. But you know, it's a hard situation—well, I don't have to tell you. You know as well as I do, it's a miserable situation. She's bankrupt and it's terrible.[5]

I thanked him for his support of Dr. Mikovits and told him that it elevated him in my eyes to the very "top tier of medicine and as a human being."

Lipkin replied, "That's very kind." He gave a small laugh. "I wish I could do more for her. I just don't know what else there is to do. My concern is—you can imagine some people would say, we want to find some way to give her resources because we think she was important. There are other people who will say I'm not willing to invest in her to be a consultant unless she's willing to contribute something. And the same thing is true with Frank [Dr. Francis Ruscetti, her collaborator]. He's got a pension. He's fine. She's the one with the difficulty."[6]

We talked for a few more minutes and then ended the conversation. In the summer of 2013, he flew out to California to meet with some members of our Silicon Valley group, and he reciprocated by inviting me to a fund-raising event in New York City. I met him at the Soho Grand Hotel on September 19, 2013.

Because I was writing a book with Dr. Mikovits, including her experiences with Dr. Lipkin, I asked a number of people what they thought of both Ian Lipkin and Mady Hornig. I wanted to get a sense of them as people.

One former professor of medicine at Stanford University who had left to open his own practice told me Ian Lipkin and Mady Hornig had lived together for a few years, a fact that was common knowledge among their community. The personal relationship had apparently ended, but they continued their research collaboration. I also learned that Mady Hornig had a son with autism, which explained her great interest in the subject. This former Stanford University professor told me that most of his collaborators avoided Ian Lipkin and didn't trust him, but they liked Mady Hornig. He had even tried to recruit Hornig to his new practice when he left Stanford. According to him, Hornig had been tempted but chose to remain at Columbia, where it was widely expected she would take over the Center for Infection and Immunity when Lipkin retired.

My former supervising scientist at Lawrence Livermore Labs, who expressed such contempt for Dr. Andrew Wakefield, expressed a similar level of contempt for Ian Lipkin. According to him, any scientist who submitted samples to Lipkin's lab had better make damn sure that everything was labeled and identified correctly for fear that if anything interesting was discovered, Lipkin would attempt to take the credit.

A wealthy woman in Silicon Valley who facilitated a number of meetings for our philanthropic group and was friendly with both Andy Wakefield and Ian Lipkin told me a number of stories. She had introduced Lipkin to Larry Ellison, the billionaire CEO of Oracle Computers, and after the meeting had asked Ellison what he thought of Lipkin. Ellison reportedly replied, "He's the smartest man I've ever met."

The woman had for many years functioned almost as the press secretary for Andy Wakefield, and I asked how, given that relationship, she could also be on such good terms with Lipkin, given their very public feud. "I don't know," she replied. "I guess I just feel a little sorry for Ian. You know, he lives alone, and he's got elderly parents that he's taking care of. None of us has an easy life."

***

I must confess that I was nervous as I dressed in my New York hotel room for what was billed as a "Cocktail Reception to Celebrate the Work of the Center for Infection and Immunity Featuring W. Ian Lipkin, M.D., Columbia University, Mailman School of Public Health—Hosted by Emanuel Stern at the Soho Grand Hotel, 310 West Broadway, New York City." I couldn't help but feel like some barbarian tribal leader, summoned by Caesar to Rome for an audience. I dressed in black and when I looked at myself in the mirror thought I looked a little like Luke Skywalker in *Return of the Jedi*, surrendering myself to Darth Vader to be taken to the Death Star. I mean, isn't that what heroes are supposed to do? Venture into the very lair of the enemy? Luke to the Death Star? Frodo and Sam to Mount Doom?

I took a cab to the Soho Grand Hotel, entered the lobby, got directions to where the event was being held, and made my way to the location. I opened the doors, expecting to see some grand ballroom but instead saw a cramped little space with six to seven tables, a buffet and drink table set up at the back with folders on the work of the center, along with some cute little blue plastic circular hand sanitizers emblazoned with the crest of the Center for Infection and Immunity as well as a line drawing of a microscope, and a podium with a microphone at the front for the speakers. There were maybe thirty people in the room. I'd seen bigger crowds at a PTA meeting. "This is the grand citadel of science?" I thought to myself.

Looking around the room, I quickly identified Mady Hornig. She is an attractive woman, with a lean, angular face surrounded by ringlets of dark hair. I made my way over to her and introduced myself.

She smiled broadly, cocked her head a little to the side, and extended a hand. "Kent! It's so nice to finally meet you!"

"Really?" I asked, with a laugh. "It's nice to meet me?"

"Yes. Why not?"

"Well, I've written articles that have criticized some of your research."

"Were they mean? Or rude?"

"I don't think so. I didn't make it personal."

"Then we don't have a problem."

"Okay, Mady, I've got a question I've wanted to ask you for years."

I paused for a moment, and she gave me a look that seemed to say, "Proceed."

"I consider your article 'The Neurotoxic Effects of Thimerosal Are Mouse-Strain Dependent' to be one of the most important papers in autism. But you published that in 2004, and now it's 2013. Why no other papers?"

She rolled her eyes and said, "You don't know?"

"No."

She proceeded to tell me that after the publication of her paper, a blogger named Autism Diva had published many articles criticizing her work, referring to it as the "Rain Mouse" experiments,[7] and it had caused her a great deal of difficulty with the Columbia administration. "I felt like I was under probation for like five years after that paper came out."

"Did this *Autism Diva* person have any academic credentials that would make the administration sit up and take notice?" I asked.

"None."

"And they listened to her?"

"Yes."

We gave each other one of those looks that, in my mind, happens only when two autism parents get together, we tell our war stories, and then come to that inevitable point where further investigation should be done and it isn't and you both think, "That's autism. They really don't want to ask the important questions." We talked for a little while longer, and she told me she came from a socially prominent family who ran something called "Camp Hornig." I later found an article on "Camp Hornig" in the *New York Times* that described it in the following words:

> The invitation to Camp Hornig, as its patriarch—an investment banker and a triathelete—likes to call it, might include tennis at courts sunken into an apple orchard or kayaking out to the Atlantic from the dock across the road.
>
> It might be for a fund-raiser for Riverkeeper, the environmental group of which Mr. Hornig is a director, with Hampton blondes like Christie Brinkley meandering across the four acres punctuated by Japanese anemones, daphne and peonies shaded by stone walls and a pergola woven with purple wisteria.
>
> Or it might be for entrée into Ms. Hornig's salon on the veranda of the main house, with writers and artists holding court, and guests encouraged to converse on topics like politics and religion that typically zip lips in polite company.[8]
>
> It sounds like a nice place. I doubt I will ever receive an invitation.

I found an article on another member of the Hornig tribe in the pages of the *New York Times*. Donald Hornig had designed the detonators for the world's first atomic bomb, codenamed Trinity. After connecting the switches on the night of the final test and scampering down from the tower holding the bomb, the scientific director of the Manhattan project, J. Robert Oppenheimer, ordered him back, not wanting to leave the bomb alone during a lightning storm.

While waiting in the tower in the electrical storm with the world's first nuclear device, he passed the time by reading a collection of humorous essays. Babysitting an atomic bomb and reading a joke book. He sounded like just my kind of guy.

Around midnight the storm had moved through, and Hornig was allowed to come down from the tower to wait with the others in a small bunker located about five miles away. As recounted in the article, "The bomb was detonated at 5:29:45 a.m. on July 16 [1945] as Dr. Hornig and the others watched from the bunker. He later remembered the swirling orange fireball filling the sky as 'one of the most aesthetically pleasing things I have ever seen.'"[9] Hornig worked as a science advisor to President Lyndon Johnson from 1964 to 1969, advising him on space missions, atom smashers, and more mundane issues such as beds for Medicare patients and desalinization plans. He later became president of Brown University and was one of the youngest scientists ever elected to the National Academy of Sciences.[10]

As she talked about her family, it was easy for me to see how even though her thimerosal article had put her on "probation" for five years, her family standing would have in some way protected her. I got the impression that Mady Hornig had endured something of a Mexican standoff with Columbia over her research. Given her family connection, it would have been difficult for Columbia to get rid of her. But they could probably make her daily life unpleasant.

I found Dr. Lipkin a short time after that, as I had a present I wanted to give to him. Lipkin always reminded me of the actor Kevin Costner, with his intense, intelligent features, and although he looks thin, I was a little surprised to discover he wasn't more than five-foot-ten. I had expected him to tower over me, but we were pretty close to eye level. I gave him a hardcover copy of the book *7 Men and the Secret of their Greatness* by Eric Metaxas. I'd read an earlier biography by Metaxas of the English political figure William Wilberforce, who was on the way to becoming the prime minister of Great Britain when he made the choice to fight against slavery instead. The book I gave Lipkin profiled seven men, including George Washington, Jackie Robinson, and Dietrich Bonhoeffer, a Christian pastor who had stood up to Hitler and been executed in the final days of the war. Each of the men chose to surrender some personal advantage for what they believed to be a higher cause. Washington had given up the opportunity to become king of the United States after he beat the British. Robinson had decided not to respond to all of the racial insults thrown his way when he became the first high-profile African American to play professional baseball, even though he had a fiery nature. I wrote an inscription to Lipkin inside that read, "A man's past is not his future. I think you are a good man, but you could be a great one."

Prior to the evening's talk, I was able to spend a good deal of time speaking with Dean Linda Fried, as well as some of the other guests. Before the evening's presentations, I asked Dr. Lipkin if I could record the talk.

"It's probably going to be very boring," he replied.

"That's okay. I'd still like to record it if it's okay with you."

"That's fine," he said.

Dean Linda Fried opened up the evening by talking about the work of the Center for Infection and Immunity, mentioned that some of the people involved in the making of the film *Contagion*, a big-screen film with Matt Damon, were present. She also spoke with pride over how the Columbia team was the only American academic institution to be invited to the Kingdom of Saudi Arabia to figure out the source of the Middle East Respiratory Syndrome (MERS), a coronavirus. Lipkin served as master of ceremonies, bringing Mady Hornig out first to talk about autism. Hornig began by saying that "Twenty years ago, I would suspect that few, if any of you, knew of a family member, or of a family that was affected by autism. Today, I would be surprised if any of you didn't have some connection to a family that was affected by autism."[11]

Hornig talked for a little while longer about autism, the amazing repository of samples Columbia had, whether the microbes in the digestive system could be causing some chronic diseases, the possibility that vitamin D levels might be at issue, and then moved back and forth between talking about chronic fatigue syndrome/ME and autism, as if they were interchangeable problems. I was pleased that I had recorded the talk, because it was earth-shattering to me. I begin with Dr. Hornig in the middle of her presentation:

> **Mady Hornig:** . . . We have one of the largest collection of chronic fatigue syndrome samples in the country, if not the world. And are moving in this disorder, which like autism, has been thought of as a psychological disorder for such a long time. It's been stigmatized. People don't want this diagnosis because they feel they're going to be told it's all in your head and you should be trying harder. And it robs people of their vitality in the prime of life. There are a million people in the United States who are currently diagnosed with chronic fatigue syndrome. And we are looking for biomarkers and blood markers to tell us who might respond to certain types of therapy and who might have a response to a particular course of action that may really mitigate their diseases.
>
> We're really excited about that. It is really important work and addressing autism and chronic fatigue syndrome is really a vital function because there are so many individuals who are afflicted, who are often looked at as unsolved cases, things we can't do anything about. So we're really excited about that. And I'm happy to take any questions if you want.[12]

The room was silent. I wondered if I should stand to ask a question, but I was bursting with excitement inside and didn't want to blow it. The book I was writing at the time, *PLAGUE*, made the argument that autism was simply what chronic fatigue syndrome/ME looked like if the triggering event happened during infancy, and chronic fatigue syndrome was what autism looked like if the triggering event happened during adulthood.

I felt like I was in the room on probation, and I didn't want to get thrown out. None of the assembled guests seemed to have a question, so Lipkin stepped back to the podium:

Ian Lipkin: So many people are aware of the work we do with emerging infectious diseases. But we use the same tool kit to study chronic diseases. And autism, as Mady said, is a disease that's emerging, as is chronic fatigue syndrome. When we began working on these fifteen years ago they were described as psychological disorders. And they're clearly recognized now as biologically based. And the genetic tests and the protein studies that we're doing suggest we may have some clues to not only recognize them early, but to treat them as well. I'm going to turn this over to Mady for questions. Are there any questions that anybody would like to pose? [The room was silent for a moment.] You can also talk to her afterward.

Finally a woman raised her hand, and Lipkin called on her.

Ian Lipkin: Melanie.

Melanie: What about the increase in allergies? Is that related?

Mady Hornig: Very interesting point. Have you heard of the so-called "hygenie hypothesis"? There's an idea that perhaps we're too clean in our society. You've got some of these souvenirs, those hand sanitizers. There's an idea that perhaps we need to be exposed to certain bacteria and parasites, to some degree, in order to have a healthy immune system. That's one thought.

There are also stimulators that may also need to be considered. Heavy metals from the environment and other sources. We get it from our food and other sources, mercury as well as manganese—

"Stimulators?" Lipkin quickly chimed in from a few feet away. Was he worried about what I might be thinking when she started talking about mercury and "stimulators?"

Ian Lipkin: Coal-fired power plants!

Mady Hornig: Exactly. Coal-fired power plants are a very important source of mercury in the environment. And we know this can alter your immune system to actually respond to all sorts of exogenous agents, including various

> sorts of viruses, and so forth, in an autoimmune fashion. Also, probably interacts with genes.
>
> So, genetic susceptibility, the environmental exposure, and probably timing. The "three strikes" hypothesis. Genes, environment, and timing, which are probably very important in determining who might have these reactions to relatively common agents. Not everybody has autoimmune disease, not everybody has allergic disorders, but these are very important. We are just now doing an analysis in chronic fatigue syndrome and allergic signals and signatures in the peripheral blood as well as the spinal fluid, to see if we can predict a certain response, hopefully to different types of intervention.[13]

I sat at my table in the Soho Grand Hotel in New York City simply thunderstruck by what I was hearing. I was among the avatars of science, and they were saying the same thing that we, supposedly "antiscience" parents, were claiming. Children with autism, and also allergies, were probably being exposed to certain "stimulators" such as mercury, which then altered their immune response to "various sorts of viruses, and so forth." According to Lipkin, we should be concerned about the mercury floating in the air from coal-fired power plants but avoid like the plague the issue of the mercury that was directly injected into the bloodstreams through vaccination.

Genetics certainly played a role, but something new had come into the environment. The genes were the same as they'd been for thousands of years. When Donald Hornig scrambled back into the tower on the night of July 16, 1945, to wait for the storm to pass so that the world's first atomic bomb could be detonated, he wasn't wondering why 1–2 percent of his fellow countrymen couldn't fight the Japanese because they had autism.

A couple more questions were asked, and then I decided it was my turn to join the conversation. I raised my arm and Lipkin noticed me. Maybe it was just my perception, but I thought I saw pure, stark terror in Lipkin's eyes at what I might say. He didn't seem to know if he should call on me, but I was the only one in that small room with my hand raised. I gave him a little half smile and kept my arm in the air. Finally, after a few awkward moments, he called on me.

> **Ian Lipkin:** Kent.
>
> **Kent Heckenlively:** One doctor I've talked to suggests what may be happening in diseases like chronic fatigue syndrome and autism is that there's a massive powering down of the mitochondria, causing low mitochondrial energy. Is that a natural consequence of the things you're looking at?

> **Mady Hornig:** Very, very important. If you consider the blood cells that are important in your immune system, they all have mitochondria. Mitochondria are critically sensitive to oxidative stressors. Oxidative stress can happen from heavy metals, it can happen from viruses, it can happen from bacteria. And if you have a mitochondrial compromise, in other words, your mitochondria are the energy centers of all your cells.
>
> If this happens in your immune cells, you can shift that immune cell, that white blood cell, to where it is an autoimmune type of responsiveness. So that whereas before it may have been healthy and been able to respond to a virus by containing the infection and clearing it from your system. Instead, the mitochondria, if there's an oxidative stress reaction, you may actually shift it so that the virus induces an autoimmune response and that autoimmune response may lead to antibodies.
>
> That is, your body producing antibodies, which react against all sorts of body parts, including the brain. So in autism and chronic fatigue syndrome those may be particularly important types of autoantibodies that we want to try to detect, and try to understand the process that leads to them.[14]

It seemed that Lipkin decided Mady Hornig and I were having too much fun discussing the critical issues in chronic fatigue syndrome/ME and autism because he quickly stepped to the podium, thanked Mady for her presentation, and introduced the next speaker, Brent Williams, who would talk about the microbiome.

Williams began his presentation by asking the audience whether we should consider ourselves a man or a microbe. After people had shouted out various responses, he said we were both. He then launched into a very interesting discussion about the microbiome, which is the way scientists describe the different populations of bacteria that live inside of us. The current thinking is that these bacterial communities may play a vital role in our health and well-being:

> **Brent Williams:** So at the Center we are dedicated to trying to understand how these microbes and our collective microbiome is impacting human health. And to these ends we are looking at several factors that may be important, what microbes may be important. These include autism, colorectal cancer, as well as women's health issues. We were the first to show that in children with autism there is an alteration in the types of microbes inhabiting the intestine and the composition of those microbes.

And what we found is that there was actually this link to how children with autism were able to digest carbohydrates. And these changes in carbohydrate metabolism were actually altering the microbiome. And we have recently started looking at how microbes may be influencing cancer. And we have recently done a study with Steve Lipkin at Cornell, looking at very aggressive forms of colorectal cancer. And what we found is that there are very specific microbes that perform very particular functions that we really believe are driving the progression of these types of colorectal cancer.[15]

Dr. Brent Williams continued his discussion of the microbial world and its link to health, and I found it all to be very compelling. Laurie asked an interesting question, this time about whether the abnormal bacteria were a consequence of a bad diet that led to the abnormality and then disease, or whether the abnormal bacteria came first, leading to disease. (The short answer was: Good question—We don't know.)

Dr. Luc Montagnier, the French scientist who won the Nobel Prize for the discovery of the HIV retrovirus, had recently been looking at abnormal bacteria in autism, so I raised my hand again to ask a question. Lipkin seemed a little less wary when calling on me this time:

Ian Lipkin: Kent?

Kent Heckenlively: I know that Luc Montagnier, who has the Nobel Prize for HIV, is now treating autistic children with long-term antibiotics. Do you have any thoughts on that?

Brent Williams: So Luc Montagnier just came to our center recently, because he was interested in one of our findings in autism. That we had found a particular bacteria called setorella, that was found in more than 50 percent of the children with autism and absent in controls. But in regards to antibiotics, I think antibiotics can be very dangerous. Because in using an antibiotic you're not really targeting the bacteria that is the problem. And what you're ending up doing in some cases is wiping out all the beneficial gut flora, which might open you up to other infections. You know, definitely, as we start to understand which microbes cause which problem, then we can really begin to target those microbes rather than wiping out everything.[16]

I thought Dr. Williams made an excellent point. In addition to my concerns about vaccines, my daughter, Jacqueline, had several ear infections (which we think were connected to her cow-based milk formula) and was treated with several rounds of antibiotics. I have always wondered if that might have also contributed to her seizure disorder and autism.

The final speaker was Dr. Simon Anthony, and his presentation was about trying to create a world database of viruses, such as those that exist in seals or birds, bats, and rodents, even in remote places like the Amazon jungle. That way we might have some understanding of them before they come rampaging into the human population, like the Ebola and Marburg viruses. It was an ambitious and worthy project.

When the evening's presentations finished, an older man came up to me and extended a hand. "I don't know you, but you asked some really good questions. Are you a scientist?"

"No," I replied. "I'm just an autism parent."

There was a good deal of social mixing after the talks. I stood next to Ian Lipkin as he displayed for everybody what looked like an enormous plastic suitcase filled with scientific equipment. This was their rapid response kit, ready to be packed up and shipped anywhere in the world there was a suspicious viral outbreak. He was leaving the next morning for Saudi Arabia to investigate the Middle Eastern Respiratory Syndrome, eventually found to be linked to a coronavirus.

I had a wonderful and animated conversation with Dr. Jay Varma, a tall, good-looking Indian man who was deputy commissioner for Disease Control in New York City. I told him the story of Jacqueline's vaccine reactions, and he nodded, saying he'd heard a lot of similar stories and that "we need to find out what's going on with these kids." Even though I'd just met him, I had an overwhelming desire to hug the man.

As the evening wound down, I drifted back to Mady Hornig. She seemed to have enjoyed my questions during her talk, and I was still feeling good that I'd made at least one member of this distinguished gathering think I was a scientist. I told her about the book I was writing with Dr. Judy Mikovits and the XMRV investigation and asked if she'd let me interview her for it.

"Sure!"

"I might get you into a little bit of trouble," I warned her, knowing we would both be thinking of the difficulty she'd run into with Autism Diva and the "five years" she felt she'd been on "probation" at Columbia for the research she'd done on thimerosal. "You probably want to consider for a while before you answer."

"Okay, I'll think about it."

The night was coming to a natural close when Mady turned the conversation in an unexpected direction. "I've always thought I should write a book about my life in science."

I'm certain I could not conceal my surprise. "I'd love to help in any way I can," I said, shocked by this sudden revelation.

"I might need some."

"I'm happy to assist. I'll call you next week."

"Okay."

There was no doubt about it. I'd spent the night with a member of the scientific nobility, maybe even a queen. And even though some might consider me the barbarian leader of a bunch of nuts, I still had my manners. I took her hand, bowed slightly in deference like a nineteenth-century gentleman, and kissed it. "My lady, I take my leave of you now." I phoned several times the next week and left messages, but she never called back. I figured she must have finally come to her senses. I am a dangerous man. A collaboration with somebody like me would probably put her academic career at Columbia in great jeopardy. It was a new experience for me to be the "bad boy." Normally, I am the straight arrow, the Goody Two-shoes, the Dudley Do-Right. Maybe in times of injustice it is exactly those types of people who become the fiercest of rebels.

Regardless of what she thought of me, I will always have fond memories of that single night at the Soho Grand Hotel in New York, the greatest city in the world.

* * *

Now for my criticisms.

I want to focus on the Lipkin and Hornig investigations into three significant medical controversies: the question of the role of mercury in vaccines in autism, the potential involvement of retroviruses in autism and chronic fatigue syndrome/ME, and the possible contribution of the MMR vaccine to autism.

First, I think I've made it clear that I have nothing but praise for the work of Ian Lipkin and Mady Hornig in their joint publication "Neurotoxic Effects of Thimerosal Are Mouse-Strain Dependent." I consider it one of the best scientific pieces of evidence that the mercury in vaccines is linked to increasing rates of autism and other neurodevelopmental problems.

In addition to Mady Hornig's presentation of this seminal work to Congress on September 8, 2004, she also presented it several months earlier, at a special meeting of the National Academy of Sciences on February 9,

2004. Curiously, this is the same meeting at which CDC whistleblower Dr. William Thompson was so morally conflicted that he sent letters out to various colleagues, including CDC Director Julie Gerberding, that the findings would implicate the MMR vaccine in increased rates of autism.

That meeting is particularly important, as the documented attendees reads like a veritable list of the individuals I write about in this book. Congressman Dave Weldon, who wrote such a stinging letter to CDC Director Julie Gerberding, spoke from 8:15 a.m. to 8:30 a.m. Dr. Mady Hornig spoke from 8:30 a.m. to 9:00 a.m. Dr. Frank DeStefano, William Thompson's boss at the CDC, presented the allegedly false MMR data that so troubled Thompson from 10:00 a.m. to 10:30 a.m. Mark Geier, who would later work extensively with Dr. Brian Hooker, spoke from 12:15 p.m. to 12:45 p.m. Dr. Boyd Haley, the chairman of the Department of Chemistry at the University of Kentucky and one of the world's foremost experts on mercury, spoke from 3:30 p.m. to 4:00 p.m. And Dr. Jeff Bradstreet, who eventually became one of my daughter's doctors and the subject of significant government harassment, spoke from 4:00 p.m. to 4:30 p.m.

The only criticism I have of the Lipkin-Hornig team in this investigation is reflected in the question I asked Mady Hornig during that September 19, 2013, cocktail party at the Soho Grand Hotel in New York City. Why hadn't they done any follow-up research on mercury in vaccines and the effect on those with different genetic profiles?

And she gave me her answer.

They didn't do any further investigation because she got into so much trouble that she was, in her own words, placed on "probation" for approximately five years. This actually buttressed what I had heard from a former Stanford University medical professor, which was that Hornig's pursuit of the thimerosal question was actually one of the reasons for the breakup of her personal relationship with Ian Lipkin. In the words of the former Stanford University professor who had tried to get Hornig to join his practice, Lipkin believed Hornig had "soiled" herself by her pursuit of the mercury/autism question.

The failure was not one of science. The science was impeccable, well thought-out, and complete. The failure was a lack of courage.

* * *

The eighteen months I spent collaborating with Dr. Judy Mikovits on our book, *PLAGUE*, will rank as one of the greatest intellectual endeavors of my life. When she was later asked to describe what it was like to work with me, she declared, "Poor Kent! I'd just scream into the phone for hours about the terrible things they'd done to me and how corrupt and cowardly the scientific community was! And he just took it all down and turned it into a calm, sober book that still reflected all of the awful things I told him." She needn't have worried. I grew up around strong, opinionated women. My mother was a trailblazer, being one of the first three women to get a master's degree in education from the formerly all-male University of San Francisco, was the youngest county school supervisor in the state of California, helped found the local PBS station (for a time she was the "teacher on the television"), and also married my father and was an excellent mother to my older brother and me. She'd been getting her doctorate in education from Stanford University when I arrived on the scene, messing up her plans. She often referred to me as "her little doctorate."

Although you can read the longer account in our book, I want to give a condensed version of Dr. Mikovits's research into retroviruses, chronic fatigue syndrome/ME, and autism, as well as her interactions with Ian Lipkin and Mady Hornig. In May 2006, Dr. Mikovits found herself in Barcelona, Spain, at a conference on the HHV-6 virus, which had been discovered in the lab of Robert Gallo. Mikovits actually organized the conference. Near the end of the conference, she heard a presentation from Dr. Dan Peterson, a practicing physician from Incline Village, Nevada, who had been "on-scene" at what is considered the start of the modern-day chronic fatigue syndrome/ME outbreak, which began around Lake Tahoe in 1984–1985.

In his presentation, Dr. Peterson showed a slide of data from sixteen of his long-term chronic fatigue syndrome/ME patients who had developed rare cancers. The data showed some unusual arrangements in the T cells of their immune system, and Peterson said he was baffled by the findings. But it wasn't a mystery to Dr. Mikovits.

T cells normally eliminated viruses and cancer cells, but their arrangement meant they were prepped for battle against a single invader. That also meant they weren't ready to take on other pathogens or cancer cells. Maybe they were locked onto a phony target. While they thought they were repelling one invader, others were taking up residence.

Mikovits nearly sprinted to the stage to share her insights with Peterson. He was welcoming of her input and asked her to spend some time at his practice at Incline Village with his patients to see if she could shed some light on their condition.

She started with a systems biology approach, looking first for abnormal chemokines and cytokines, inflammatory markers in the immune system. The pattern she found suggested they were dealing with a retrovirus. Mikovits had performed pioneering work in HIV and also worked with other retroviruses. In her twenty-year career at the National Cancer Institute, she had risen to become director of the Lab of Antiviral Drug Mechanisms. She knew her way around viruses. By chance, she happened to meet Dr. Robert Silverman of the Cleveland Clinic, who had discovered a new human retrovirus in prostate cancer in 2006. The inflammatory markers of the two groups of patients, those with chronic fatigue syndrome/ME and prostate cancer, looked remarkably similar.

Maybe they had found their mystery retrovirus. She got the genetic sequence from Silverman and had her lab assistant, Max Pfost, perform PCR on several samples. There was no match. Just to be thorough, she had Pfost lessen the stringency of the PCR to look for related viruses. Bingo! In 2011 Silverman would admit that he had not isolated the genetic sequence from an actual retrovirus, the standard of practice in the field. He had stitched it together from three partial sequences and not disclosed this fact. Silverman had gotten in the neighborhood of the virus, but he had not bagged the beast.

In October 2009, the Mikovits team, along with their collaborators at the Cleveland Clinic and the National Cancer Institute, published their findings in the journal *Science*, the world's most prestigious journal of original research.[17] Their study showed that about 67 percent of those with chronic fatigue syndrome/ME showed evidence of infection with this retrovirus called XMRV (xenotropic murine leukemia virus-related virus), a number that would rise closer to 95 percent as Mikovits was able to refine her testing procedures.

But Mikovits wasn't satisfied with just explaining chronic fatigue syndrome/ME, especially not when so many of these women had children with autism. Was there a link to a retrovirus there, as well? She did what scientists are supposed to do. She tested her hypothesis. On October 14, 2009, she appeared on a television show called *Nevada Newsmakers* and said the following about the autism issue:

> It's not in the paper and it's not reported, but we've actually done some of these studies. And we found the virus present in a number, in a significant number of autistic samples that we've tested so far. [fourteen out of seventeen, or 84 percent] . . . There's always the hypothesis that my child was fine, then

> they got sick, and then they got autism. Interestingly, on that note, if I might speculate a little bit. This might explain why vaccines lead to autism in some children because these viruses live and divide and grow in the lymphocytes, the immune response cells, the B and T cells. So when you give a vaccine, you send your B and T cells in your immune system into overdrive. That's its job. Well, if you're harboring one virus, and you replicate it a whole bunch, you've now broken the balance between the immune response and the virus. So you could have had the underlying virus and then amplified it with that vaccine and then set off the disease, such that your immune system could no longer control other infections and created an immune deficiency.[18]

When I called up Dr. Mikovits to interview her for an article, she put it very simply: Standard practice for a baby born to an HIV-infected mother was to put that child on antiretrovirals prior to any vaccination, for fear that the vaccination would cause the HIV retrovirus to replicate out of control and cause AIDS. Instead of HIV-AIDS, we might be looking at something like XMRV-Autism, XMRV-chronic fatigue syndrome/ME, or XMRV-cancer.

I placed a call to the University of California, San Francisco Pediatric AIDS unit, and they confirmed the practice and even told me to consult their website on the issue:

> Activation of the cellular immune system is important in the pathogenesis of HIV disease, and that fact has given rise to concerns that the activation of the immune system through vaccinations might accelerate the progression of HIV disease . . . These observations suggest that activation of the immune system through vaccinations could accelerate the progression of HIV disease through enhanced replication . . . If feasible, it is preferable to have patients on anti-retroviral therapy (ART) prior to receipt of vaccination . . .[19]

Mikovits and others would find that roughly 4–8 percent of the healthy population showed evidence of infection with these types of retroviruses, suggesting anywhere from fourteen to twenty-eight million Americans could be at risk for a negative vaccine reaction due to a preexisting retroviral infection.

My historical investigation uncovered that the first reported outbreak of chronic fatigue syndrome/ME occurred among 198 doctors and nurses working at Los Angeles County Hospital in 1934–1935 who had been working during a polio epidemic. They had been given an experimental polio vaccine created by Maurice Brodie, who had grown the virus in mouse brain tissue, a new procedure at the time.[20] In addition, the doctors and nurses received an accompanying "immune boost" designed to stimulate their immune system.[21] This immune booster used a new mercury preservative known as "merthiolate," an early trade name for thimerosal. The experimental polio vaccine was also given to seven thousand children, just around the time the first children who would go on to develop autism were born.

My suspicion that early vaccine production had allowed for a mouse virus to jump into the human population (and perhaps the mercury was also a factor) was not mine alone. It was suggested by the historical evidence, and other scientists raised the possibility as well, such as these researchers who published in the journal *Frontiers in Microbiology* in 2011:

> One of the most widely distributed biological products that frequently involved mice or mouse tissue, at least up to recent years, are vaccines, especially vaccines against viruses
>
> . . . It is possible that XMRV particles were present in virus stocks cultured in mice or mouse cells for vaccine production, and that the virus was transferred to the human population by vaccination.[22]

To be clear, the research by Mikovits did not directly state that vaccines were an issue in chronic fatigue syndrome/ME or autism but can probably best be compared to a dangling string on a shirt that if pulled can unravel the entire item. Other researchers were making the connection. It was just a matter of time.

That is where, in my opinion, Ian Lipkin and Mady Hornig stepped into the fray to see if they could prevent the unraveling of the vaccine program.

* * *

One of the greatest things I learned from Judy Mikovits is how simply she views science. To her, it's just like taking your car to a mechanic. If your car rattles when you drive it over a bumpy road, you don't have your mechanic drive it on a smooth road. Taking a mechanic's view of science, let's review what Mikovits did to find her XMRV retrovirus, and what Lipkin and Hornig failed to do.

The first thing she did was look at their inflammatory markers, their cytokines and chemokines, to see if the results suggested any patterns. The pattern looked similar to that of a retrovirus. From her long experience with retroviruses, she knew they could hide in tissues, only showing up in blood in numbers high enough to be detected, during some kind of viral flare-up. So instead of taking just a single look at a patient's blood, she did several tests over regular intervals to see if she could find some evidence of viral infection. Cleveland Clinic researcher Bob Silverman suggested his XMRV retrovirus as the possible culprit, but his sequence didn't exactly fit. (It is not uncommon for viruses to show a wide genetic variation in patients, so this was not thought to be a significant problem.) What Mikovits and her team found was close to Silverman's sequence. Time to publish and get science ramped up to work on solving the remaining mysteries and helping patients, right?

Into the mix step Lipkin and Hornig, ready to confirm or refute with the imprint of official science. The first curious thing they do is dramatically change the patient population studied. Among the eight groups of people they excluded:

1. Patients with a medical or psychiatric illness that might be associated with fatigue;
2. Patients with abnormal blood serum characteristics; and
3. Patients with abnormal thyroid functions.[23]

This is absurd. The condition is called chronic fatigue syndrome/ME, and they are excluding patients with a medical or psychiatric illness that might be associated with fatigue? Even my sixth-grade science students would quickly pick out that mistake.

The second group of patients excluded is equally illogical. A patient is suffering from a disease so severe that they are often forced to spend the entire day in bed, they suffer from extreme light sensitivity, and they are often in great pain, and you think that their blood won't have some abnormal characteristics? Normally, one would look for some abnormal blood characteristics in a disease of this severity and call it a "clue."

The third group of patients excluded makes no sense but requires a bit more explanation to fully appreciate. Patients with chronic fatigue syndrome/ME are known to have thyroid problems as an accompaniment to their disease. One of the leading doctors in the field, who was present at the Lake Tahoe outbreak, Dr. Paul Cheney, told me that he estimates 85 percent of those with chronic fatigue syndrome/ME have thyroid problems.[24]

Mikovits did not exclude any of these groups in her study for the simple reason that it would have been insane. She wanted to find the problem and was therefore inclusive and thorough. By contrast, the Lipkin-Hornig study seemed designed NOT to find anything.

The other shocking thing Lipkin and Hornig did was to use Bob Silverman's stitched-together, Frankenstein, three-viral sample that was not present in actual human beings as the standard for diagnosing a retrovirus. Mikovits complained at the time but was overruled, and since she'd been fired from her job and briefly jailed (for allegedly stealing her own work), she wasn't really in a position to affect the outcome. To summarize, Lipkin and Hornig used the wrong patient population and then used the wrong viral sample for comparison. It's a bit like a botanist looking for a rare mountain flower in a desert and having a picture of a Redwood tree to help him with the identification.

Other than that, their work was flawless.

On September 16, 2012, the results of the Lipkin-Hornig study were revealed to the world. There was no XMRV infection of chronic fatigue syndrome/ME patients. Everybody could go home. Scientists still did not know what caused chronic fatigue syndrome/ME, or autism for that matter, but it wasn't the result of a retrovirus that might have been transmitted naturally, or through a vaccine. If there were some who didn't get the message, *Nature* magazine helped them know what to think with a profile on Lipkin titled "The Man Who Put the Nail in XMRV's Coffin."[25]

If that was the end of the Lipkin-Hornig spectacular train wreck of bad science, I might be more willing to forgive them. But Mikovits had found something, and I think Lipkin and Hornig knew it. They just had to figure out how to capture it and claim it for their own.

In a public conference call with the Centers for Disease Control on September 10, 2013, nearly a year after he had dismissed the XMRV issue, Lipkin sprang his trap:

> We found retroviruses in 85 percent of the sample pools. Again, it is very difficult to know whether or not this is clinically significant or not. And given the previous experience with retroviruses in chronic fatigue syndrome, I am going to be very clear in telling you, although I am reporting them in Professor Montoya's samples, neither he, nor we, have concluded that there is a relationship to disease.[26]

Judy Mikovits told me she had nothing but respect for the work of Professor Jose Montoya of Stanford University. The patient populations he used made sense. He didn't exclude from his chronic fatigue syndrome studies patients who had a "medical or psychological condition that might be associated with fatigue." That would be crazy, like excluding from a study on alcoholism any patients who reported buying more than a bottle of alcohol a week. Mikovits could only feel a bit of sorrow that a good researcher like Montoya was now mixed up with Lipkin.

The other startling revelation in the public conference call is Lipkin reporting he found evidence of retroviruses in 85 percent of the sample pools, well above Mikovits's first reporting of 67 percent. Did this mean that XMRV was now going to rise from the coffin that Lipkin had nailed shut and buried? Or was that body, like the shattered career of Dr. Mikovits, going to remain undisturbed?

And is Dr. Ian Lipkin, the John Snow Professor of Epidemiology at the Mailman School of Public Health of Columbia University, really saying that when he discovers evidence of a retrovirus in 85 percent of a sick patient population, he cannot make a determination of its possible significance? Even my struggling middle-school science students would rouse themselves for a moment and declare that required further investigation. And where is the public apology to Judy Mikovits?

Instead of focusing on the elephant in the room (or, more accurately, the mouse retrovirus, which seems to terrify the scientific community far more than any mouse ever terrorized any elephant), Lipkin chose to focus on the abnormal inflammatory markers, the cytokines and chemokines:

> So, I will confine my discussion to a number of cytokines and chemokines that are up or down in people who have disease. And again, this is important because I think we may find that there are drugs which can be used to modulate the levels of cytokines; and while these may not get at the cause of this disease, the primary cause which I still believe is likely to be an infectious agent, it may give us some insight in ways in which we can manage and decrease some of the disabilities associated with chronic fatigue syndrome.[27]

If one pays attention to what Dr. Lipkin, the John Snow Professor of Epidemiology at the Mailman School of Public Health of Columbia University, actually says in this statement, it is difficult to believe these are comments coming from a highly educated man.

Judy Mikovits went looking for a retrovirus because the pattern of abnormal cytokines and chemokines suggested a retrovirus lay at the heart of this disease. Lipkin confirms the abnormal pattern, as well as the retrovirus, but will not follow up searching for any pathogen, although he believes an infectious agent to be at the root of the disease. Instead, he'll focus on trying to modulate the inflammatory markers (which is not a bad idea), while completely ignoring any possible infectious agent (not only a bad idea, but an immoral one).

Drs. Ian Lipkin and Mady Hornig continue to work at Columbia University. Dr. Robert Silverman of the Cleveland Clinic, who concealed that his sequence for the XMRV retrovirus was made up of sequences from three different samples, also continues to work at his institution. Dr. Judy Mikovits had to declare bankruptcy, is unemployed, and is without good prospects for future employment. Lipkin's promise of consulting work for her, or future collaboration with her on papers in this field, have not panned out, either. It does not appear that the vaunted "self-correcting" mechanism of science has yet worked for Dr. Mikovits.

In a more just world, the intelligence and integrity of Judy Mikovits would be celebrated, and she would be offered high government positions of influence, such as head of the National Institutes of Health or the Food and Drug Administration or maybe even president of the Centers for Disease Control and Prevention. Under her administration, science would not be political, but dedicated to finding answers for the millions who suffer.

* * *

And finally we move to the main event, the investigation by Ian Lipkin and Mady Hornig into the claim by Dr. Andy Wakefield that the MMR vaccine is linked to the development of autism and gastrointestinal problems.

One of the most important concepts I teach my science students is the idea of replication. Toward that end, I have them write down their lab procedures in excruciating detail. I tell them that their instructions should be so clear they should be able to hand them to another student, and without even talking to that student, the other student should be able to perform the same experiment and get roughly similar results. One of my students engaged in preparation for a recent experiment looked at me and, as she rolled her eyes, asked, "Do these procedures have to be Heckenlively clear?" I laughed.

After Wakefield had published research showing an association between the MMR vaccine and Crohn's Disease in children, several parents approached him saying that their children developed not only gastrointestinal problems after the MMR vaccine, but autism. Wakefield found several of these cases, decided to give them a colonoscopy with the intention of taking biopsies, and looked for evidence of the measles virus persisting in the gut. Wakefield was crystal clear in his assertion that the measles virus in these children was found ONLY at specific areas of inflammation inside the terminal ileum, not any other area of the gastrointestinal tract. He maintains to this day that he had the proper approval, but for the purposes of the scientific issues, that is irrelevant. Lipkin and Hornig decided to do something different in their paper, published in 2008, "Lack of Association between Measles Virus Vaccine and Autism with Enteropathy: A Case Control Study."[28] I will detail why the Lipkin-Hornig work was NOT a replication study.

My criticisms of the Lipkin-Hornig work in chronic fatigue syndrome/ME to confirm the work of Dr. Mikovits breaks down into three main complaints:

1. Using the wrong patient population;
2. Not using the right tools or techniques to find the pathogen; and
3. After this spectacular train wreck of bad science, declaring the matter forever closed to further inquiry.

After looking at what they did in the Mikovits investigation, I can only conclude that they honed their technique in the Wakefield investigation. In criminal law, we would call this a "pattern of conduct." Considering that these efforts affected the lives of millions, it seems to me as nothing less than a crime against humanity.

Let's look first at the age and MMR status of the typical subject in the Lipkin-Hornig MMR study as they described it in their paper:

> Median age at receipt of first MMR was similar for cases and controls. The majority of study subjects were in the 3–5 year age stratum and below the age recommended for second MMR (4–6 years); expectedly, 80% of cases and 69% of controls received only one MMR prior to the study.[29]

The window of vulnerability described by the parents is before the age of two, and these were children who did not get their full round of MMR shots before the age of two. Considering this issue in light of the Thompson allegations that earlier MMR administration led to higher rates of autism, that alone is enough to throw suspicion on these results. But there's more. Wakefield's original case report contained information from twelve children with autism and gastrointestinal complaints, nine of who reported the appearance of symptoms after an MMR vaccination (77 percent). If you are attempting to do a replication study for an explosive finding such as

Wakefield's, one would expect to come very close to matching that percentage and use significantly larger numbers. That is careful, well-reasoned science. Here is how the Lipkin-Hornig team describes their group:

Only 5 of 25 subjects (20%) had received MMR before the onset of GI [gastrointestinal] complaints and had also had onset of GI episodes before the onset of AUT [autism].[30]

Wakefield had nine children who specifically fit the profile of MMR vaccination before the onset of autism and gastrointestinal complaints (77 percent) in his "Case Report" in *The Lancet,* and the Lipkin-Hornig team had only five in their confirmation study (20 percent), which would definitively answer for all time the questions about the MMR vaccine and autism.

And what about Wakefield's specific instructions that he only recovered the MMR vaccine virus from lesions inside the terminal ileum, and not in any other sections of the colon? It is difficult to say, because the authors are vague:

> Biopsy material was obtained from the terminal ileum and cecum under direct supervision of the team gastroenterologist. For analyses of MV [measles vaccine virus] RNA, four random samples were taken from the superficial mucosae of ileum and cecum. Additional specimens were acquired at sites indicative of inflammatory GI lesions, if present.[31]

According to Wakefield, Tim Buie, the scientist who obtained the biopsies for the Lipkin-Hornig team, was not good at getting into the terminal ileum, which is a challenging endeavor. Good science would have had the Lipkin-Hornig teams specifically searching for areas of inflammation in the terminal ileum. Does the above passage make you believe they were specifically looking for those areas of inflammation in the terminal ileum?

Even with all of those problems, Lipkin and Hornig reported a positive measles vaccine virus finding from a single child with autism, as well as one of the controls:

> Analyses in all three laboratories found two ileal biopsy samples with MV [measles vaccine virus] and H gene RNA; one from a boy in the AUT/GI group, the other from a boy in the control group . . . Sequence analysis confirmed that products of these samples were authentic.[32]

Despite these fundamental flaws in their research, Lipkin and Hornig managed to confirm at least part of Wakefield's findings. In some children with autism and gastrointestinal complaints, the measles vaccine virus was persisting in their gut. Was this child from the five out of twenty five (20 percent) from the Lipkin-Hornig group that matched Wakefield's population? Was that positive finding from an area of inflammation in the terminal ileum? I can't answer those questions because it's not in the Lipkin-Hornig article. And yet this shoddy research, which would be laughed out of every medical school classroom in America, is taken as gospel by the scientific community.

Can somebody please tell me what's going on here?

* * *

And what did CDC whistleblower Dr. William Thompson think of the MMR vaccine/autism study done by Ian Lipkin and Mady Hornig? Thompson had more than a passing interest in the study, as he was actually brought in to be the project officer when the investigation ran into financial trouble. In the fourth conversation Hooker taped, on July 28, 2014, Thompson laid out his frustrations with Ian Lipkin and Mady Hornig:

> **Dr. Thompson:** All right. There's one interesting study too. There's the NI . . . You and I have never talked about this study. There's a 2008 study that has Larry Pickering as an author.
>
> **Dr. Hooker:** [Affirmative response.]
>
> **Dr. Thompson:** Where they went and did the biopsies. The Larry Pickering, Ian Lipkin . . .
>
> **Dr. Hooker:** Oh yeah, I know all about this.
>
> **Dr. Thompson:** Okay. Did you and I talk about this study?
>
> **Dr. Hooker:** Ah, no. You and I have never . . . I've talked to Ian Lipkin about it. In fact he and I . . .
>
> **Dr. Thompson:** Okay.
>
> **Dr. Hooker:** . . . are not speaking right now, because basically he doesn't like me very much. Okay?
>
> **Dr. Thompson:** Right. I mean, Ian Lipkin . . .
>
> **Dr. Hooker:** [Unintelligible.]
>
> **Dr. Thompson:** Right. Ian Lipkin is one of those . . . Well, I'll give you an example. When I was trying to hold them accountable . . . It was funded by the CDC.
>
> **Dr. Hooker:** Right.
>
> **Dr. Thompson:** I don't know if you know that.

Dr. Hooker: Right.

Dr. Thompson: It was funded by the CDC; the money was sent to the NIH. **It was the worst mismanaged event of federal funds that I've ever seen,** um . . . [Bold added by author.]

Dr. Hooker: Wow.

Dr. Thompson: In terms of how the study was carried out. **If you looked at the original study design and the fact that they only ended up with twenty-five autism cases, it's just insane.** [bold added by author] So, I took over as project manager in the middle of that. And I kept trying to hold people accountable . . .

Dr. Hooker: [Affirmative response.]

Dr. Thompson: . . . for what they were doing with the money and, um, the project officer on their end eventually dropped off the study; she was so fed up and tired with it.

Dr. Hooker: Okay.

Dr. Thompson: In the middle, in the middle of the study, Ian Lipkin was asking for more money. And he actually, and I . . .

Dr. Hooker: [Affirmative response.]

Dr. Thompson: I don't think I kept the email. But it's the one email I wish I had kept where he said he was going to go talk to his congressman if we didn't uh . . .

Dr. Hooker: [Affirmative response.] That sounds like Ian.

Dr. Thompson: If we didn't give him more money.

Dr. Hooker: That sounds exactly like Ian Lipkin.

Dr. Thompson: No, I . . .

Dr. Hooker: Oh my goodness.

Dr. Thompson: No, he's an arrogant dick. And then the first author, Mady Hornig. I think she's first author on it.

Dr. Hooker: Yeah, yeah.

Dr. Thompson: I'm not sure. So anyway. So Mady Hornig, who was doing animal studies, is his significant other. So . . .

Dr. Hooker: Right, right, right. They're shacking up.

Dr. Thompson: So, you know, husband-and-wife team.

Dr. Hooker: They're shacking up. They're not married.

Dr. Thompson: Yeah. Yeah.

Dr. Hooker: But yeah. That's been historic.

Dr. Thompson: So anyway, that was criminal because they published that study with twenty-five autism cases and the power was like zero . . .

Dr. Hooker: [Affirmative response.]

Dr. Thompson: . . . and they tried to give the impression that they did a study of, you know [Unintelligible.]

Dr. Hooker: [Affirmative response.]

Dr. Thompson: I don't remember exactly.

Dr. Hooker: They ran PCR in the cases. They ran PCR in the controls. They found measles virus in several of the cases, and they found measles virus in several of the controls and they concluded there was no effect. But the actual conclusion of the study should be, "It was really a crappy study. We can't tell anything."

Dr. Thompson: It was the worst study ever.

Dr. Hooker: Thank you.

Dr. Thompson: It was the worst study ever.

Dr. Hooker: Thank you. When you talk to Ian Lipkin, he's like, "This is definitive. This shows there's no correlation."

Dr. Thompson: It was the worst study ever.[33]

Two scientists, Brian Hooker and William Thompson, both of whom had interacted with the Lipkin-Hornig team, had similar complaints about the quality of their work. Thompson certainly had the greater amount of personal experience with the Columbia University scientists, and his specific complaints about the MMR vaccine/autism study are worthy of note.

Lipkin had mismanaged the money, did not take responsibility for the mistakes, threatened to call his congressman to beg for more money, and then produced a study with no value. When one hears that research was conducted by scientists from Columbia University, the natural inclination is to give such a study great weight. But that respect lasts only as long as one does not look too closely at it. When one does examine the study, it quickly falls apart. Why has such scrutiny been so long in coming? Is it because so many in science do not understand the game that is being played? Or is it because those few who take the time to understand the glaring flaws realize that to raise such questions will bring them to the attention of powerful forces who will seek to drive them out of science?

***

Robert Kennedy, the former senator from New York and presumptive Democratic candidate for president before he was assassinated in 1968, said in a famous speech in South Africa: "Few men are willing to brave the disapproval of their colleagues, the wrath of their society. Moral courage is a rarer commodity than bravery in battle or great intelligence. Yet it is the one essential, vital quality of those who seek to change a world which yields most painfully to change."

I know there are many in the autism and chronic fatigue syndrome/ME community who have nothing but contempt for Ian Lipkin and Mady Hornig, because of the reasons I have outlined. Indeed, the catalog of hypocrisy and scientific double-talk is truly breathtaking, especially when one considers its impact on millions of people who suffer with chronic diseases around the world.

On the question of mercury and its contribution to autism, Lipkin and Hornig are on record believing this is a viable theory when you take into account genetic variation. But they don't talk about that research much anymore, preferring to point the finger at mercury coming from coal-fired power plants, rather than the mercury that has been in the past and continues to be injected into people's veins through vaccinations.

On the question of a retroviral contribution to chronic fatigue syndrome/ME suggested by Dr. Judy Mikovits, they are adamant that there is no connection. But wait, that opinion is based on a study that kicked out any patients who:

1. Had a medical or psychiatric illness that might be associated with fatigue;
2. Had abnormal blood serum characteristics; and
3. Had abnormal thyroid functions.

A different study with a patient population characterized by Dr. Jose Montoya of Stanford University found 85 percent of these individuals showed evidence of infection with a retrovirus. However, that was likely to be controversial, so they preferred to focus on the abnormal pattern of inflammatory markers, the chemokines and cytokines. This was exactly the finding made several years earlier by Dr. Mikovits, which led her to theorize that a retrovirus might be involved in the condition. But even Lipkin admitted that focusing on the abnormal immune markers was unlikely to get them to the root of the problem.

Was that a replication study of Dr. Wakefield's work? It wasn't really a replication study at all. And even with all of those problems, they found evidence to support his original findings. It was a classic "bait-and-switch" move, just like they did with the results of Judy Mikovits. Use a different population of children with autism, and then take samples from those areas of the gut that are unlikely to contain the measles virus. Then declare the matter settled. I cannot believe these are celebrated scientists with world-wide reputations.

I want to perform a thought experiment.

What if Ian Lipkin and Mady Hornig had stood up and declared to the world that mercury was likely to be connected to the autism epidemic and that was the hill upon which they were going to fight and die? What if Ian Lipkin and Mady Hornig had done a good job in the XMRV investigation and confirmed Dr. Mikovits's findings, bringing hope to millions around the world who suffer from chronic fatigue syndrome/ME, and possibly those with autism? What if Ian Lipkin and Mady Hornig had followed the basic rules for a replication study of Dr. Wakefield's work, using the same patient population and obtaining samples from areas of inflammation in the terminal ileum?

The simple truth is that if Ian Lipkin and Mady Hornig had been honest scientists, they would have faced the same persecution as Andy Wakefield and Judy Mikovits. They would have been removed from their positions at Columbia University and vilified throughout the scientific world. I do not believe Ian Lipkin and Mady Hornig have the "moral courage" of Andy Wakefield and Judy Mikovits. It really is that simple.

I do not mean this as a criticism. Few people have such courage. From my personal experience and research, Ian Lipkin and Mady Hornig are both highly intelligent, charming, and, at least in the case of Mady Hornig and what she tried to do with thimerosal and autism, possessed of some measure of courage.

Consider even what Ian Lipkin said in *Discover* magazine in an April 2012 article titled "The World's Most Celebrated Virus Hunter."

Could autism be another version of a PANDAS-like disease?

> It's possible, in some people. There is probably a group of people who have a genetic component to autism, and for them, there may not be much of a trigger or any trigger at all required. Another group is genetically predisposed, and if they encounter some factor or factors, individually or in combination, it could result in either the onset or the aggravation of neurodevelopmental disorder; by factors, I include everything from heavy metals to infection.[34]

PANDAS refers to pediatric autoimmune neuropsychiatric disorder, or in simpler language, an autoimmune problem provoking a neurological disease in children. Do you want to know what causes autism? If you listen to Ian Lipkin, it seems very simple.

Some people may have a genetic variation that does not react well to new components in the environment, specifically heavy metals, other chemicals, and infections, which would not normally have been encountered in previous generations. Where might we begin such an investigation? We all know the answer to that question.

But Lipkin and Hornig do not have the moral courage to face the enormity of this challenge. They are not willing to stand for the truth, regardless of the consequences. They want to find answers but are not willing to jeopardize their positions. They may speak the truth quietly in a magazine article or at a private cocktail party but will not launch a crusade to rescue those who suffer. They want to find treatment and cures but do not want to face the wrath of the pharmaceutical companies or the government scientists who push vaccines as an article of their medical faith. And they have made a decision to be the scientific "tip of the spear" against autism parents or chronic fatigue syndrome/ME patients who have opinions roughly similar to their own findings and public pronouncements.

I do not expect things to end well for Ian Lipkin and Mady Hornig. When the truth begins to unravel, as it inevitably must, the same forces that pushed them to be the tip of the spear will most likely break that spear and hand it to the patient communities as a peace offering. We will be told that if it were not for these "bad scientists" the truth would have been revealed years, if not decades, earlier. They will be scapegoated and proclaimed to be the locus of evil that has infected scientific debate. The outbreak of bad science will have been contained. Look at these two former scientists, we will be told, pulled down from their lofty academic positions, possibly put on trial, appearing daily in handcuffs, and if they are convicted, we will get reports of their incarceration. The greatest problem, though, is not with the tip of the spear, but with those who demanded it be wielded against the innocent.

I will not be able to raise a voice in defense of Lipkin and Hornig when that occurs. It will be justice of a sort, but it will not be the whole truth. In better times, I believe Ian Lipkin and Mady Hornig would have been better people. But we are not judged on our behavior during times of leisure and safety.

We are judged on how we acted in times of challenge and controversy.

CHAPTER FOUR

The Documents

On December 22, 2015, I contacted the office of Congressman William Posey (R, Florida) and asked for documents related to CDC whistleblower Dr. William Thompson. I made this request in response to an announcement by Congressman Posey's office on November 30, 2015, that they would make documents available to interested journalists. I talked at that time to congressional staffer George Cecala and identified myself as a founding contributing editor of the website *Age of Autism*, an attorney and science teacher, as well as coauthor of the book *PLAGUE: One Scientist's Intrepid Search for the Truth about Human Retroviruses, Chronic Fatigue Syndrome (ME/CFS), Autism, and Other Diseases*, with Dr. Judy Mikovits, a twenty-year government scientist.

Cecala took my information and said he would get back to me. The following day, December 23, 2015, Cecala called back with a fellow staffer, Anna Schartner. I had a pleasant conversation, asking them how things were progressing with their attempts to get Congress to subpoena Dr. William Thompson to testify about his allegations in a public forum. They indicated they were having difficulty getting other members of Congress interested in the story but were continuing to press the matter privately. They noted with some satisfaction that several members of Congress had requested the materials Dr. Thompson had turned over to Congress. George and Anna approved my request to view the Thompson materials, make copies of them, and make use of the documents in any way I thought appropriate.

George and Anna also suggested I speak with Beth Clay, a congressional investigator who had been reviewing the documents and had a long association with this issue. I found her expertise to be invaluable in the writing of this book. Whenever possible, I have sought to quote directly from the relevant documents so that the reader may determine whether my characterizations of the materials and conclusions are supported by the evidence turned over by Dr. Thompson.

On a personal note, it was often emotional for me to review the documents, especially those from 2001 and 2002, as it became clear that the CDC scientists were observing an unmistakable signal from the data that earlier administration of the MMR shot (measles, mumps, and rubella) was associated with an increased rate of autism, especially among African American males and children with what they termed "isolated autism" or, in simpler terms, every healthy and normally developing child on the planet. At that time, my wife and I were dealing with our three-year-old daughter, Jacqueline, who had a severe seizure disorder as well as autism.

On January 9, 2002, I took our eighteen-month-old son, Ben, to the pediatrician and because I thought the pediatrician missed something in my daughter's six-month checkup had him undergo a complete developmental screening. Ben had fifteen or twenty words, listened to people when they spoke, and had normal eye contact. Ben received his shots and we left. Three days later, my wife, Linda, a speech therapist, mentioned to me that Ben had stopped speaking. I had been aware of the claim that vaccines might be causing neurodevelopmental problems in children but dismissed such concerns, even though my daughter's problems had developed within a few weeks of her six-month pediatric checkup and her vaccinations.

I blame my failure to link my daughter's problems to her vaccinations due to my status as a new parent, her age, and an overwhelming belief in the American medical system. Surely our best scientific minds could not allow something so harmful to our children to exist in the United States, no matter how much money the pharmaceutical industries contributed to our politicians. I was not one of "those people."

In mid-January of 2002, over the Martin Luther King Jr. holiday, I happened to be in a Barnes and Noble bookstore when my hand fell on a book titled *Unraveling the Mystery of Autism and Pervasive Developmental Disabilities* by Karyn Serrousi. The story appealed to me, as the author was an excellent writer and her husband was a scientist. The book detailed her investigation into a wheat- and milk-free diet (often called the gluten/casein free diet), which seemed to work very well for some children with autism. Serrousi also stated her opinion that something about the vaccinations her son received had precipitated his autism, but the gluten/casein-free diet had brought him back. Both my son and daughter went on the diet the next day.

For my daughter, the diet had little effect (other than provoking constant tantrums), but for my son, the effect was profound. After twelve days, he started speaking again. As I watched him over the following weeks and months, it seemed that what he had previously learned had been wiped clean, but that he could learn again. My wife noted that he seemed to be about a year behind his peers in language at two years old, but by the age of three it seemed that he had caught up. For a period of about two years, it seemed that he had sensory issues, being unable to watch a movie in a theater, or having the ability to hear distant noises that we could only hear if we moved much closer to the source. Ben entered kindergarten as a normally developing child and, in the years that have passed, has developed into an excellent student and a fine athlete, and those who know him well comment that he has a wicked sense of humor like his father. My daughter, Jacqueline, still lacks the ability to speak.

I was presented with a choice when these events happened in my life. I could speak about them, risking the affection of friends and family members who still believed in the American medical system as fervently as I once did, or I could choose silence. My conscience would not allow me to remain silent. I became an activist. As a result, I have lost the close relationship I once had with many friends and family members who somehow thought I had become "antiscience." I have never regretted the choice I made, and it will always rank as one of the proudest decisions of my life. I believe, as Edmund Burke did, that "the only thing necessary for evil to triumph is for good men to do nothing." I might lose the fight, but I would not remain silent.

What has been most disturbing to me as I reviewed the documents released to me by Congressman Posey's office is the realization that at the same time I was struggling with these issues, top scientists at the CDC were struggling with them, as well. But the choices they made were starkly different. Even though tasked as the guardians of public health, they chose to hide the dangers of vaccines and actively thwart investigations into whether such medical interventions could be conducted safely.

Besides covering up evidence of harm from vaccines, these CDC scientists also introduced something uniquely venomous into the American body politic: the belief that those who raised concerns about vaccines and their effect on public health were themselves a threat to public health.

As a science teacher, I have often performed dissections with my students so that they more clearly understand the inner workings of living things. If we want science to regain its soul and its conscience, we must perform a similar dissection, searching for the truth that has been so assiduously hidden, and bring it out into the light. If we do not succeed in this task, we will have failed future generations for whom good health will be but a distant memory.

* * *

Before I became a science teacher, I was an attorney for several years. The practice of law often deals with uncovering the truth, and not just from the other side, but just as often from your own client. It is a mistaken belief that clients always tell the truth to their lawyers. "If a client tells you his mother loves him, get a second opinion!" is a common joke in the profession. I found that people are generally honest, up until the point that they realize something has gone terribly wrong, and it may be their fault or responsibility. So if you want to find some reliable evidence, go back before the time those involved believed there might be a problem.

The first draft analysis plan for "Autism and Childhood MMR Vaccine—April 3, 2001" is a good place to start. In April 2001, there was a new president in the White House, George W. Bush, who had won a contentious election from former Vice-President Al Gore. The terrorist attack of September 11, 2001, was still more than four months away. The first few paragraphs of the April 3, 2001, draft analysis plan clearly stated the issues that confronted the CDC scientists:

> Autism is a serious life-long developmental disorder characterized by marked impairments in social interactions, and communication skills; and repetitive, restrictive, or stereotyped behaviors. A recent review of studies conducted since 1985, shows an estimate of the prevalence to be 1–1.4 per 1,000 for classic autism, and possibly as high as 4–5 per 1,000 for all autism spectrum disorders (ASD) combined. While these rates are 3–4 times higher than rates found in studies conducted 15–20 years ago, there are several recent studies, including a study done by Baird et al. (2000) and an investigation in Brick Township, NJ, which suggested that the rate of autism may be higher still with rates of 3.1 per 1,000 and 4 per 1,000 respectively (CDC report, Baird study). These higher prevalence rates, coupled with reports of increasing numbers of children with autism being served by schools and service agencies (California report; DofED) have prompted concern that the rate of autism may be increasing.[1]

A few points stand out from this first paragraph. Autism is clearly noted to be a "lifelong developmental disorder," and, in 2001, the rate of autism was thought somewhere between 11.4 for every one thousand children for "classic autism" and "possibly as high as 4–5 for every 1,000 children for all autism spectrum disorders." These rates were "3–4 times higher" than studies conducted fifteen to twenty years earlier.

That means the rate for "classic autism" in 1980–1985 was 1–1.4 for every three to four thousand children. All autism spectrum disorders from 1980–1985 would be four to five for every three to four thousand children. But in 2001, the CDC was receiving reports of higher numbers of children with autism, including a study conducted in Brick Township, New Jersey, as well as reports from the California Department of Education. The CDC was well aware of concerns that the rate of autism was rising. And in the next paragraph, the CDC clearly laid out concerns of what might lie behind these startling increases:

> It has been suggested that vaccination, particularly with measles, mumps, and rubella (MMR) vaccine, may be related to the development of autism. The two main arguments that are used in support of a possible association are 1) the prevalence of autism is increasing at the same time that infant vaccination coverage has increased; and 2) in some cases of autism, there is an apparent temporal association in which autistic characteristics become apparent within a few weeks to a few months after receipt of the MMR vaccine. Although the prevalence of autism and similar disorders appears to have increased recently, it is not clear if this is an actual increase or due to increased recognition and changes in diagnostic criteria. The apparent onset of autistic symptoms in close proximity to vaccination may be a coincidental temporal association.[2]

For the typical autism parent who had noticed a change in their child after a vaccination (estimated to be about 40 percent of the autism parents), there was little to object to in the paragraph. The parents who had become activists were by and large highly educated people, professionals such as lawyers, doctors, teachers, policemen, and business people, not the kind who ever thought they would find themselves as members of any protest group. Prior to the birth of their children, we had religiously devoured books like *What to Expect When You're Expecting* or *The Girlfriend's Guide to Pregnancy*. We watched shows like *Friends*, *Seinfeld*, or *ER*, that thrilling medical drama created by Michael Crichton, produced by Steven Spielberg, and featuring that hot new actor at the time, George Clooney. There was no need to protest the system. We were proud members of the system. Most parents who had noticed a change in their children after a vaccination knew full well it might have been a coincidence. But they wanted the issue to be honestly investigated.

And it seemed like one medical doctor, Dr. Andrew Wakefield, a well-respected British pediatric gastroenterologist (a specialist in childhood digestive disorders), had already started on such an investigation:

> A study published in 1998 in the *Lancet* has led some to hypothesize that the MMR vaccine may play a role in recent trends in autism. This study was a case series of 12 children who were referred to a pediatric gastroenterology clinic because of chronic enterocolitis and the coexistence of autistic behavioral characteristics. Eight of the 12 children were reported by parental interview as first experiencing the onset of autisticlike symptoms following the MMR vaccine, and an additional child's onset occurred after measles infection, which led the investigators to hypothesize that the measles, mumps and rubella vaccine might be associated with the onset of autism. While suggestive, the clinical case study lacked evidence to evaluate a possible causal association between MMR and the occurrence of ASD [Autism Spectrum Disorder].[3]

The parents who became aware of Dr. Wakefield's work applauded his interest in trying to find the truth. If you are a proud member of a system, you believe that system will ferret out dangers, especially dangers to children. And Wakefield was pursuing a moderate course. Until this possible danger had been fully investigated, he suggested parents consider breaking up the shots, a single dose for measles, for mumps, and for rubella, an option available in England and the United States at that time.

By April 2001, at the time of the CDC's writing of their first draft of the "Autism and Childhood MMR Vaccine Analysis Plan," Wakefield had undertaken additional research and uncovered even more troubling information:

> Wakefield and his collaborators have since proposed that they have identified a new syndrome consisting of milder gastrointestinal conditions, predominantly ileonic lymphonodular hyperplasia and mild gastrointestinal inflammation, associated with behavioral regression. They have reported identifying laboratory evidence of measles virus genome in the peripheral white blood cells and bowel biopsy specimens of a few such patients.[4]

For those parents who believed in a system of public health that was interested in protecting their children from harm, regardless of the source, Wakefield seemed to embody science in its most noble incarnation.

The draft "Autism and Childhood MMR Vaccine Analysis Plan" of April 3, 2001, also laid out some strong evidence for increasing rates of autism in Wakefield's England. Reporting on the results from a study published in the *British Medical Journal*, "There was a significant increase in rates of autism between 1988 and 1999 from 0.3 per 10,000 person years in 1988 to 2.1 per 10,000 per person years in 1999." This is a sevenfold increase in rates of autism and suggests that the parental observation that "something" had caused their children to develop autism was correct. The increase was also noted in California: "It was noted that there was a marked increase in autism from 1980 to 1994, 44 per 100,000 in 1980 to 208 per 100,000 in 1994." This is almost a fivefold increase in rates of autism. And the CDC was saying they wanted to get to the bottom of what was causing the increase:

> In an effort to resolve the speculation about vaccinations and autism, the CDC in collaboration with the National Immunization program, has conducted a matched casecontrol study utilizing the Metropolitan Atlanta Developmental Disabilities Surveillance program to look at this potential relationship. The main objective of this study is to evaluate the association between the timing of the receipt of MMR vaccine and subsequent diagnosis of autism. The secondary objective will be to evaluate associations between other childhood vaccines and autism. This will include an examination of thimerosal exposure during the first year of life.[5]

The CDC was setting out an ambitious set of goals. First, they were going to investigate the time at which a child received their MMR vaccination and the effect that had, if any, on autism rates. Second, they were going to investigate associations between other childhood vaccines and autism. And third, they were going to examine the effects of the mercury derivative, thimerosal, on children, at least during their first year of life. It was an investigative agenda that would have met with wide approval from the parent community.

The agenda appealed to Dr. William Thompson, as well. But sadly, both Thompson and the parent community would be disappointed by the actions of the CDC and its superiors in the years that followed. The CDC had no interest in a complete and thorough investigation of vaccines and autism.

* * *

Probably one of the most disturbing documents I viewed among those released to me by Congressman Posey's office is one prepared by Dr. William Thompson specifically for congressional investigators titled "Events Surrounding the DeStefano et al. (2004) MMR-Autism Study" and dated September 9, 2014. The "Background" and "Conclusion" sections have been reproduced in full. There is also an extensive time-line section, which I will review later:

Background

My primary job duties while working in the Immunization Safety Branch from 2000 to 2006 were to lead or co-lead three major vaccine safety studies.

1. VSD Thimerosal Neurodevelopmental Study (Thompson et al., *NEJM*, 2007)
2. VSD Thimerosal Autism Study (Price, Thompson, et al., *Pediatrics*, 2010)
3. MADDSP MMR-Autism Case-Control Study (DeStefano et al., *Pediatrics*, 2004)

The MADDSP MMR-Autism Cases Control Study was being carried out in response to the Wakefield (1998) *Lancet* study that suggested an association between the MMR vaccine and an autism-like health outcome. There were several major concerns among scientists and consumer advocates outside the CDC in the fall of 2000 regarding the execution of the Verstraeten et al study (2003). The Verstraeten Study was the first study the CDC carried out to examine the association between thimerosal and neurodevelopmental outcomes including autism. Some of the major concerns included 1) many of the statistical analyses were carried out post-hoc after an initial set of analyses were run, 2) the study protocol evolved over time, and 3) the CDC did not share many of the internal study findings with individuals and constituents outside the CDC.

One of the important goals that was determined up front in the spring of 2001 before any of these studies started was to have all three study protocols vetted outside the CDC prior to the start of analyses so that consumer advocated could not claim we were presenting analyses that suited our own goals and biases.

My primary responsibilities for the MADDSP MMR-Autism Study were:

1. Lead the large majority of the study-related meetings with all coauthors.

> 2. Write all the SAS programs for all the statistical analyses associated with the paper.
> 3. Summarize and present the statistical results to the coauthors on a regular basis.
>
> In addition, all SAS programs and statistical analyses were reviewed by both Dr. Margarette Kolzcak and Dr. Andrew Autry. All data management work was led by Tanya Karapukar and she also reviewed the data management-related activities and decisions included in the SAS programs. All of my statistical analyses were run off of datasets cleaned and provided to me by Tanya Karapukar.
>
> On September 5, 2001, we finalized the vetted study analysis plan for MADDSP MMRAutism Study. (See Final Analysis Plan dated September 5, 2001). The study protocol included a timeline and the goal was to finish the analyses and submit the manuscript for publication to the *New England Journal of Medicine* by December 1, 2000. [Author's Note: I believe he meant to write 2001.] **The final analysis plan described analyses for the TOTAL sample and the BIRTH CERTIFICATE sample which included assessment of the RACE variable. (See pages 7 and 8 of the Final Analysis Plan.)** [Bold and capitalized words are as they appear in the original document prepared by William Thompson.] We hypothesized that if we found statistically significant effects at either the 18-month or 36-month threshold, we would conclude that vaccinating children early with the MMR vaccine could lead to autism-like characteristics or features. We never claimed or intended that if we found statistically significant effects in the TOTAL SAMPLE, we would ignore the results if they could not be confirmed in the BIRTH CERTIFICATE SAMPLE.[6]

The background section contains several important pieces of information. First, it clearly shows that William Thompson was a major part of the CDC's investigative unit regarding vaccines and neurodevelopmental problems. Second, it lays out many of the problems in the Verstraeten Thimerosal-Autism study including the running of "post-hoc" data analysis, the unexplained changes in the study protocol, and the fact that the CDC hid many of its internal study findings from the public. Third, the document makes clear that the CDC was supposed to present both the TOTAL sample numbers as well as those from the BIRTH CERTIFICATE SAMPLE. Last, Thompson was insistent that there was never intent to conceal information from the TOTAL SAMPLE if it could not be confirmed in the BIRTH CERTIFICATE SAMPLE. The conclusion section of the paper is much shorter than the background section, but its claims are truly breathtaking:

> **Conclusion**
> I believe we intentionally withheld controversial findings from the final draft of the DeStefano et al. (2004) *Pediatrics* paper. We failed to follow the final approved study protocol and we ran detailed in depth RACE analyses from October 2001 through August 2002 attempting to understand why we were finding large vaccine effects for blacks. The fact that we found a strong statistically significant finding among black males does not mean that there was a true association between the MMR vaccine and autism-like features in this subpopulation. This result would probably have led to designing additional better studies if we had been willing to report the findings in the study and manuscript at the time we found them. The significant effect of early vaccination with the MMR vaccine might have also been a proxy for the receipt of thimerosal vaccine early in life but we didn't have the appropriate data to be able to code the level of thimerosal exposure from the MADDSP school records.
>
> In addition to significant effects for black males, we also found significant effects for "isolated autism cases" and for the threshold of 24 months of age. If we had reported the 24 month effects, our justification for ignoring the 36 month significant effects would not have been supported. In the discussion section of the final published manuscript, we took the position that service seeking was the reason we found a statistically significant effect at 36 months. This was a post-hoc hypotheses regarding the findings after we confirmed one of our primary hypotheses. Because we knew that the threshold for 24 months was also statistically significant, reporting it would have undermined the hypothesis that service seeking was the reason we found an effect at 36 months. (See published paper.)[7]

The CDC's actions in this matter are so filled with mind-boggling deceptions that it can be difficult to unravel exactly how this public agency has lied to the American public. Sneaky lies from smart people who are presumed to be honest can be difficult to unravel, but I will detail the lies claimed by Thompson in the 2004 DeStefano paper. Once you separate the truth from the lies, you will be amazed at their audacity and lack of humanity.

The first lie is that they failed to report a significant effect for black males who received an MMR shot at twelve months versus thirty-six months.

The second lie is that they failed to report significant effects for "isolated autism cases," meaning the sudden development of autism in children who had no other health or behavioral problems.

The third lie is that they failed to report significant effects for those children who received the MMR vaccine at twenty-four months rather than twelve or thirty-six months.

The fourth lie is that even with the BIRTH CERTIFICATE SAMPLE, a statistically significant effect remained, but they claimed this difference was the result of "service seeking" and a mythical regulation that a condition of such services was the earlier administration of an MMR shot.

The fifth lie is that, by omitting the information about the significant effects observed among those children who received their MMR shots at twenty-four months, they strengthened the argument that the difference at thirty-six months was due to "service-seeking."

The CDC made a cold and sober decision to drive the American public exactly where it wanted, regardless of the truth. They wanted people to believe that vaccines and autism were not linked, and that those who made such claims constituted a menace to society. The wounds of our children made us outlaws. It was a crime against the children, a crime against the parents, a crime against humanity, and a crime against science. I don't think there has been anything comparable to it in all of human history. It almost makes me understand why so few friends and families believed us when we told them how our children had been injured.

***

The timeline of events prepared by Thompson listed thirty significant dates for this cover-up. I have narrowed the list down significantly, while keeping their numbers as they appeared in the original document:

2. On August 29, 2001, I outlined the method that would be used to code RACE for the TOTAL sample and the Birth Certificate Sample. (See scanned notes from 2001–2002.)
4. On October 15, 2001, I ran matched and unmatched analyses for whites and blacks. I would only do this if I had found statistically significant effects by RACE. (See 2001–2002 notes dated October 15, 2001.)
7. On November 2nd, I wrote in my notebook to run analyses for whites and blacks for the early-vaccinated and late-vaccinated

subjects. These analyses were run for the TOTAL sample. I would have only run those types of analyses if we had been attempting to explore why we had found significant RACE effects. (See 2001–2002 notes dated November 2, 2001.)

11. On May 22, 2002, all coauthors met and discussed analyses of the 24-month threshold for the Total sample. We did this because there were many statistically significant effects at the 24 month threshold. (See page 16 of Agendas attachment.)

13. In the Excel File named "describe_results_2002_0702.xis," Table 7 shows the RACE analyses that I had run using only the BIRTH CERTIFICATE sample—the unadjusted RACE effect was statistically significant. (OR=1.51, [95%Cl 1.02—2.24]). At bottom of Table 7, it also shows that for the NON-BIRTH certificate sample, the adjusted RACE effect statistically significance was HUGE. (OR=2.94 [95%Cl 1.48—5.81). That is the main reason why we decided to report the RACE effects for ONLY the BIRTH certificate Sample.

15. All the coauthors met and decided sometime between August 2002 and September 2002 not to report any RACE effects for the paper.

16. Sometime soon after the meeting where we decided to exclude reporting any RACE effects, also between August 2002 and September 2002, the coauthors scheduled a meeting to destroy documents related to the study. Dr. Coleen Boyle was not present at the meeting even though she was involved in scheduling that meeting. The remaining 4 coauthors all met and brought a big garbage can into the meeting room and reviewed and went through all our hard copy documents that we thought we should discard and put them in the large garbage can. However, because I assumed this was illegal and would violate both FOIA [Freedom of Information] laws and DOJ [Department of Justice] requests, I kept hard copies of all my documents in my office and I retained all the associated computer files. This included all the Word files (agendas and manuscript drafts), Excel files with analysis and results, and SAS files that I used to generate the statistical findings. I also kept all my written notes from meetings. All the associated MMR-Autism Study computer files have been retained on the Immunization Safety Office computer servers since the inception of the study and they continue to reside there today.

17. On or about September 3, 2002, I informed Dr. Melinda Wharton, the Division Chief for the Branch I worked in, that we had

concerning results from the MMR-Autism Study that we would like to discuss with her.

18. Dr. Melinda Wharton formally reprimanded Dr. Bob Chen, my Branch Chief, on September 18, 2002. As I stated in my emails to both Dr. Melinda Wharton and to Dr. Walt Orenstein, I believe this was an intimidating personnel action and threatened the credibility of the entire branch. It also put a big black cloud over our branch and demoralized many of the staff.
21. On October 20, 2002, I described to Dr. Orenstein the dilemma I was in regarding the concerning MMR-Autism Study results and the reprimand of Dr. Chen. I told him I felt intimidated by the move and I linked it to them knowing the results would be problematic if they were shared outside the CDC.
24. On January 8, 2004, I began to present draft PowerPoint presentations of the MMR-Autism Study for the Institute of Medicine meeting that I was scheduled to present on February 9, 2004 in Washington DC. I have copies of each of those PowerPoint presentations. During the next 30 days, I presented the results to the Division Director of ESD in the National Immunization Program, and the Director of the National Immunization Program. I would also present the results in the office of Dr. Julie Gerberding.
27. During the February 2 meeting with Dr. Cochi and Dr. Wharton, I also requested that Dr. Walter Orenstein be brought into the meeting because he had arrived in the building that morning. Dr. Colchi suggested that Dr. Orenstein was "heading off into the sunset" and that we shouldn't bother him with these issues.
29. On February 2, 2004, after meeting with Dr. Cochi and Dr. Wharton, I delivered my letter for Dr. Julie Gerberding regarding my concern regarding results from the MMR Study just before I had to present them to the Institute of Medicine on February 9, 2004. (See scanned letter to Dr. Gerberding dated February 2, 2004.)
30. On March 9th, I was put on administrative leave. In the Annex to the memorandum, they provided a list of my "inappropriate and unacceptable behavior in the work place" which included "you criticized the NIP/OD [National Immunization Program—Office of the Director] for doing a very poor job of representing vaccine safety issues, and claimed that NIP/OD had failed to be proactive in their handling of vaccine safety issues, and you requested

> that Dr. Gerberding reply to your letter from a congressional representative before you made your presentation to the IOM." (See scanned Memorandum dated January 9, 2004.) I stand by that statement and I do not think it was unacceptable to convey that to Dr. Gerberding.[8]

The allegations made by Thompson and the picture he paints of the CDC are deeply troubling. Thompson's allegations cover a three-year period, starting on August 29, 2001, when he first outlined the method that would be used to code racial information for Frank DeStefano's MMR-Autism study to his being placed on administrative leave on March 9, 2004.

During that two-and-a-half-year period, Thompson would observe his coauthors agree to hide significant effects of earlier administration of the MMR shot on autism among three groups: African American males, isolated autism, and children who received their MMR shot at twenty-four rather than thirty-six months. They would actively engage in actions to destroy these troubling findings by throwing their materials in a garbage can, and when Thompson sought to bring the troubling findings to the attention of his superiors, such as Melinda Wharton, Walter Orenstein, or CDC Director Julie Gerberding, he was punished for his actions.

What does a professional do when it seems that everybody around him has gone crazy? The situation may be analogous to the Great Recession of 2008 in which a deeply compromised system overestimated the strength of mortgage bonds and nearly collapsed the world economy. Many observers of the scene were confused as to why their fellow analysts didn't see the looming financial crisis. The weaknesses were clearly apparent to anybody who honestly investigated the issue. The financial crisis may have been a combination of several factors, greed certainly, but also a deeply held and often unacknowledged belief that the housing market was stronger than just about any potential shocks. To the vast majority of the public and scientists as well, vaccines are supposed to be the greatest scientific advance of the twentieth century.

When Galileo proposed his radical theory that the Earth and other planets revolved around the sun, the Catholic Church did not object because they had a theological devotion to celestial mechanics. They opposed his theories because of the philosophy that had grown up around their belief that the Earth, and by extension, humanity, was at the center of God's creation. Christianity did not collapse when the vast majority realized that Galileo's theory was correct, any more than when Darwin's theory of evolution gained wide acceptance. By the same token, the possibility that vaccines may be linked to the development of neurodevelopmental disorders in children does not herald the fall of science and rational thought. If there is one thing that science holds up as its most prized virtue, it is its insistence that all claims must be subjected to rigorous testing and analysis. If science is to have any integrity, it must seek the truth, find the truth, and speak the truth, even when it calls into question current medical practices. William Thompson seemed to possess this trait in greater abundance than his contemporaries. He was more of a scientist than his colleagues but was also willing to be silenced. Perhaps it is just one out of a hundred people, or one out of a thousand, who will refuse to be a part of any shading of the truth. Perhaps it is not fair to criticize the cowardice of others when it is questionable what we might have done when placed in a similar position.

Maybe Thompson was simply unable to believe that so many scientists did not want to look at vaccines and see harm, as legend has it that the first natives who saw the ships of Columbus on the horizon could not believe that they were looking at vessels with men. It is often said that we "see what we believe" in matters of great importance to us.

In his defense, it must be admitted that Thompson was a relatively junior member of the CDC hierarchy, and in some sense he was like the young boy in the Hans Christian Anderson story "The Emperor's New Clothes" who pointed out that the ruler was in fact wearing no clothes. He made the observations, was reprimanded for making them, and seemed to stew quietly for years over these lies.

And yet, maybe Thompson would have taken a different course if he knew that those who ruled over him, the very superiors to whom he was presenting these problems, had nearly a year earlier extensively discussed a similar issue and decided to forego their duty as scientists to tell the truth.

CHAPTER FIVE

The CDC Runs Away to Simpsonwood to Defend Mercury in Vaccines

June 7, 2000

Dr. Walter Orenstein, director of the National Immunization Program, looked out at the distinguished gathering of medical professionals he had summoned to the Simpsonwood Retreat Center in Norcross, Georgia, as he prepared to make his opening remarks. Orenstein's decision to have the meeting off-campus from the CDC's headquarters in Atlanta was based on his erroneous belief that such a meeting would not be subject to Freedom of Information Act requests.

This information would come to the attention of most of the public through an article written by Robert F. Kennedy Jr., nephew of former president John F. Kennedy and namesake of the former attorney general of the United States, US senator, and Democratic presidential candidate, Robert F. Kennedy, who was assassinated in 1968. The article "Deadly Immunity" appeared on June 16, 2005, in *Rolling Stone* and *Salon* and quickly became as controversial as Andrew Wakefield's findings about the MMR vaccine and autism in 1998.

How distinguished was this gathering of medical professionals, and is it accurate to consider them members of a "cabal," which is defined by the Oxford English Dictionary as "a secret political clique or faction"? Let's first try to answer how distinguished this gathering of medical professionals was by simply listing their names and titles as they gave them at the conclusion of Dr. Orenstein's opening remarks.

* * *

Dr. John Modlin, chair of the Advisory Committee on Immunization Practices and a member of the faculty at Dartmouth Medical School.

Dr. Paul Stehr-Green, associate professor of epidemiology at the University of Washington School of Public Health and a consulting epidemiologist for the Northwest Portland Area Indian Health Board.

Dr. Marty Stein, a general pediatrician, faculty member for pediatrics at the University of California, San Diego, and cochair of the American Academy of Pediatrics practice guideline on the diagnosis and evaluation of ADHD [Attention Deficit Hyperactivity Disorder].

Dr. Tom Saari, professor of pediatrics at the University of Madison, Wisconsin, and a member of the American Academy of Pediatrics Committee on Infectious Diseases.

Dr. Bonnie Word, professor at the State University of New York and a member of the Advisory Committee on Immunization Practices.

Dr. Peggy Rermels, a pediatric infectious disease specialist at the Center of Vaccine Development, University of Maryland, and a member of both the Advisory Committee on Immunization Practices and the American Academy of Pediatrics Committee on Infectious Diseases.

Dr. Isabelle Rapin, a neurologist for children at the Albert Einstein College of Medicine.

Dr. Kevin Sullivan, an epidemiologist at the Department of Epidemiology and Pediatrics at Emory University.

Dr. Tom Clarkson, from the University of Rochester, who claims he has had a longtime interest in mercury.

Dr. Loren Koller, pathologist and immunotoxicologist from Oregon State University.

Dr. Natalie Smith, director of the Immunization Program at the California State Health Department.

Dr. David Johnson, the State Public Health officer in Michigan and a member of the Advisory Committee on Immunization.

Dr. Richard Clover, chair of the Department of Family and Community Medicine at the University of Louisville, Kentucky, and a member of the Advisory Committee on Immunization.

Dr. Frank DeStefano, epidemiologist in the National Immunization Program and project director of the Vaccine Safety Datalink. [DeStefano would later become the lead author of the 2004 *Pediatrics* article asserting that earlier administration of the MMR vaccine was not associated with a higher incidence of autism in African American males and other groups. The decision of DeStefano and the other coauthors to conceal this information would haunt Thompson for years.]

Dr. Robert Chen, chief of Vaccine Safety and Development at the National Immunization Program at the CDC. [Chen was William Thompson's immediate supervisor. Thompson would fight for Chen against Walter Orenstein's letter of reprimand in 2002, and Chen would also supervise Thompson's administrative leave and "counseling sessions" in 2004.]

Dr. Robert Davis, an associate professor of pediatrics and epidemiology at the University of Washington.

Dr. Richard Johnston, an immunologist and pediatrician at the University of Colorado School of Medicine and the National Jewish Center for Immunology and Respiratory Medicine. [In his comments, Johnston said, "Adverse events related to vaccines have been of particular focus and interest for me mostly through serving on a series of committees dealing with the relationship between the vaccine and punitive adverse events."]

Dr. Roger Bernier, the associate director for science in the National Immunization Program.

Dr. Michael Gerber, medical officer at the National Institute of Allergy and Infectious Diseases at the National Institutes of Health and a member of the American Academy of Pediatrics Committee on Infectious Diseases.

Dr. Eric Mast, medical epidemiologist with the Hepatitis Branch at the CDC.

Dr. Barbara Howe, who was in charge of the clinical research group for vaccine development at Smith Kline Beecham in the United States.

Dr. William Phillips, a private family doctor from Seattle, Washington, who was representing the American Academy of Family Physicians and was their chair of the Commission on Clinical Policies and Research.

Dr. Vito Caserta, chief medical officer for the Vaccine Injury Compensation Program.

Dr. Xavier Kurtz, physician and epidemiologist from Brussels, Belgium, representing the European Agency for the Evaluation of Medicinal Products.

Dr. Robert Pless, medical epidemiologist with the Vaccine Safety and Development Branch at the National Immunization Program.

Dr. John Clements, from the Expanded Program on Immunization, World Health Organization, Geneva, Switzerland.

Dr. Ben Schwartz, from the Epidemiology and Surveillance Division at the National Immunization Program at the CDC.

Dr. Martin Myers, acting director of the National Vaccine Program office.

Dr. Harry Guess, head of the Department of Epidemiology at Merck Pharmaceuticals research labs.

Dr. Robert Brent, developmental biologist and pediatrician from Thomas Jefferson University and the Dupont Hospital for Children.

Dr. Michael Blum, from Safety and Surveillance at Wyeth Pharmaceuticals.

Dr. Jo White, who was in charge of Clinical Development and Research at North American Vaccine.

Dr. Bill Weil, who identified himself as "an old pediatrician who is representing the Committee on Environmental Health of the Academy [National Academy of Sciences] at the moment."

Paula Ray, the project manager for the Northern California Kaiser Vaccine Study Center.

Ned Lewis, the data manager at the Northern California Kaiser Vaccine Study Center.

Dr. Dennis Jones, a toxicologist and veterinarian who was the assistant director for Science, Division of Toxicology, at the Agency for Toxic Substances and Disease Registry.

Dr. William Egan, the acting director for the Office of Vaccine Research at the Food and Drug Administration.

Dr. Carolyn Deal, the acting deputy director of the Division of Bacterial Products at the Center for Biologics Evaluation and Research at the Food and Drug Administration.

Dr. Douglas Pratt, medical officer in the Office of Vaccines Research at the Food and Drug Administration.

Dr. Ted Staub, the global head of Biostatistics and Data Systems for Aventis Pasteur.

Dr. Tom Sinks, associate director for science at the National Center for Environmental Health at the Centers for Disease Control and Prevention, as well as the acting division director for the Division of Birth Defects, Developmental Disabilities and Disability Health.

Dr. Steve Hadler, medical epidemiologist at the National Immunization Program.

Dr. Alison Mawler, vaccine coordinator for the National Center for Infectious Diseases at the Centers for Disease Control and Prevention.

Dr. Lance Rodewald, a pediatrician and associate director for science in the Immunization Services Division at the Centers for Disease Control and Prevention.

Dr. Jose Cordero, deputy director of the National Immunization Program.

Dr. Susan Chu, deputy associate director for science at the National Immunization Program.

Dr. Philip Rhodes, a statistician at the National Immunization Program.

Dr. Thomas Verstraeten, from the Epidemic Intelligence Service at the National Immunization Program.

Dr. David Oakes, the chair of biostatistics at the University of Rochester.

Dr. Dixie Snyder, associate director for Science at the Centers for Disease Control and Prevention and the executive secretary for the Advisory Council on Immunization Practices.

Dr. Alex Walker, chair of the Department of Epidemiology at the Harvard School of Public Health.

And last, but not least, Wendy Heaps, a "health communication specialist" with the National Immunization Program![1]

* * *

There were a total of fifty-three individuals present at the meeting. Let's break down the composition of the group. There were fifteen employees of the CDC. The group had fourteen individuals with academic appointments often overlapping with other affiliations. Twelve individuals directly represented the National Immunization Program, including Walter Orenstein, Frank DeStefano, and Robert Chen. There were five representatives for the vaccine companies SmithKline Beecham, Merck, Wyeth, North American Vaccine, and Aventis Pasteur. There were four representatives of the American Academy of Pediatrics. Three individuals represented the Food and Drug Administration. Three individuals were not medical doctors: Paula Ray, the project manager for the Northern California Kaiser Vaccine Study Center; Ned Lewis, the data manager for the same Kaiser group; and Wendy Heaps, the "health communication specialist" with the National Immunization Program. Two individuals represented states: Dr. Natalie Smith, director of the Immunization Program at the California State Health Department; and Dr. David Johnson, the state public health officer for Michigan. There were two individuals representing foreign entities: Dr. Xavier Kurtz, on behalf of the European Agency for the Evaluation of Medicinal products;

and Dr. John Clements from the World Health Organization in Geneva, Switzerland.

Was there a single consumer protection group in this crowded meeting? No. Was there a single person representing the interests of parents who wanted answers as to why their children had been healthy and normally developing before a shot, but not afterward? No. Was any press release issued upon completion of the meeting, or was there any effort to inform the media of what had taken place? Again, the answer is no. They said they were doing this to protect the public.

If that is the case, why did they want to keep this information from the public?

***

Dr. Walter Orenstein opened the meeting, introducing himself as the director of the National Immunization Program at the Centers for Disease Control and Prevention, then saying:

> I want to thank all of you for coming here and taking time out of your very busy schedules to spend the next day and a half with us. Not only do we thank you for taking time out, but for taking the time on such short notice, and also putting up with what I gather those of us who are townies here didn't realize, but apparently the biggest meeting in Atlanta which has taken up all of the hotel space and all of the cars, so I think many of you had to take taxis here. We appreciate you putting up with this, but at least we did arrange the weather nicely and you can look out occasionally and see some beautiful trees. I think I am particularly impressed with the quality of expertise. We truly have been able to get at very short notice some of the most outstanding leaders in multiple fields. That will be important in interpreting the data. We who work with vaccines take vaccine safety very seriously. Vaccines are generally given to healthy children and I think the public has, deservedly so, very high expectations for vaccine safety as well as the effectiveness of vaccination programs. Those who don't know, initial concerns were raised last summer that mercury, as methylmercury in vaccines, might exceed safe levels. As a result of these concerns, CDC undertook, in collaboration with investigators in the Vaccine Safety Datalink, an effort to evaluate whether there were any health risks from mercury in any of these vaccines.[2]

Orenstein went on to discuss concerns about a "possible dose response effect of increasing levels of methylmercury in vaccines and certain neurologic diagnoses." He advised that this was not a policy-making meeting but would be something he described as "an individual simultaneous consultation."[3]

Dr. Roger Bernier, the associate director for science in the National Immunization Program at the CDC, presented some of the history of thimerosal:

> Basically there was a Congressional Action in 1997 requiring the FDA to review mercury in drugs and biologics. In December of 1998 the Food and Drug Administration had called for information from the manufacturers about mercury in their products. There is a European group of regulation authorities and manufacturers that met in April of 1999 on this, who at the time noted the situation, but did not recommend any change.[4]

Bernier continued:

> In the USA there was a growing recognition that the cumulative exposure may exceed some of the guidelines. There are three sets of guidelines that are much in discussion. One from the ATSDR [Agency for Toxic Substances and Disease Registry], the FDA [Food and Drug Administration], and one from the Environmental Protection Agency. These guidelines are not all exactly the same. There was a recognition that the cumulative exposure that children receive from vaccination may actually exceed at least one of the guidelines that is recommended, that of the EPA. That caused a concern which resulted in a joint statement of the Public Health Service and the American Academy of Pediatrics in July of last year, which basically stated that as a long term goal it was desirable to remove mercury from vaccines because it was a potentially preventable source of exposure. And if it was able to be removed, that it should be removed as soon as possible. That goal was agreed upon. In the meantime, there was postponement recommended for the Hepatitis B vaccine at birth. Also at that time, the FDA had sent a letter to manufacturers asking them to look at the situation with their products to see what could be accomplished as soon as possible.[5]

As a middle-school science teacher, I must simply interject for a moment as to the precautions I am required to follow if I break a mercury thermometer in my classroom, exposing my students to airborne mercury. I must evacuate the room. I must call a hazmat crew. Students may not return until the hazmat team has finished. Those at the Simpsonwood conference were talking about mercury being injected directly into the blood system of infants. The Public Health Service and the American Academy of Pediatrics stated the removal of mercury from vaccines was a "long term goal." How can these organizations take such a lax approach to prevent mercury from being injected into babies?

It gets worse.

Bernier was wrapping up:

> There was a public workshop on thimerosal in August of 1999. Dr. Myers will tell you a little bit about that this morning. In September of 1999, one of the Hepatitis B vaccines had removed thimerosal from the product, so the recommendation was made to resume use of the Hepatitis B vaccine at birth. Since that time, I believe in October of 1999, the ACIP [Advisory Committee on Immunization Practices] looked this situation over again and did not express a preference for any of the vaccines that were thimerosal free. They said the vaccines could be continued to be used, but reiterated the importance of the long term goal to try to remove thimerosal as soon as possible. Since then, I don't think there have been any major events. What has happened in the meantime is we have continued to look at this situation and that is what you are going to hear more about at this meeting. Are there any questions about this?[26]

Again, I have to put my middle-school science teacher hat on and imagine I presented my students with the following problem:

1. *Mercury is the second most toxic substance on the planet after plutonium, and*
2. *It is being injected into infants whose brains and nervous systems are in the process of development.*

If you had a choice between giving a mercury-free product or a mercury-laced product to an infant, which would be the better choice, and why?

I know all my students would answer that question correctly, even the ones who struggle. Based on what happened at the Simpsonwood conference, I think all of the CDC scientists would have failed my class.

The next to present was Dr. Dick Johnston, an immunologist and pediatrician at the University of Colorado, School of Medicine. He reviewed some of the history of the government's investigation of thimerosal and then gave some background:

> Thimerosal functions as an anti-microbial after it is cleaved into ethylmercury and thiosalicylate, which is inactive. It is the ethylmercury which is bactericidal at acidic PH and fungistatic at neutral and alkaline PH. It has no activity against spore forming organisms. There is very limited pharmokinetic data concerning ethyl-mercury. There is very limited data on its blood levels. There is no data on its excretion. It is recognized to both cross placenta and the blood-brain barrier. The data on its toxicity ethyl-mercury, is sparse. It is primarily recognized as a cause of hypersensitivity. Acutely it can cause neurologic and renal toxicity, including death from overdose.[7]

In plain language, Johnston said that thimerosal breaks down into ethylmercury, and even though they know it causes neurologic and renal toxicity, including death, and that it crosses both the placenta and the blood-brain barrier, they have little data on its blood levels in people, how quickly or slowly people may get rid of it through excretion, or even what it may do in the body.

Johnston continued his presentation:

> Because of the limited data for ethyl-mercury and its physical chemical similarities to methyl-mercury, it was the consensus of the meeting that in the absence of other data, that chronic exposure to methyl-mercury would need to be used to assess any potential neuro-developmental risk of ethyl-mercury, although it was recognized that we needed data specifically on ethyl-mercury. We learned a great deal about the toxicity of ethyl-mercury from animal studies, accidental environmental exposures, and studies of island populations who consume large amounts of predator fish that contain high amounts of ethylmercury. We learned that ethyl-mercury is ubiquitous and that assessments of exposure by infants would need to include environmental exposures, maternal foods, whether the baby was nursed or not, as well as their exposure to vaccines. Specialists in environmental health have extrapolated from those types of studies to establish safe exposure levels, and this is an important emphasis I would like to make on chronic, daily exposure to methyl-mercury that incorporate wide margins. That is three to ten-fold to account for data uncertainties.[8]

How does one make sense of the previous statement?

They don't know if ethyl-mercury has the same effects on biological organisms as methyl-mercury, but they'll simply assume it does. Does that sound like high-quality science to you?

The next part of Johnston's presentation demonstrated a remarkable degree of humility before the challenge of determining the effects of mercury and other metals. If he had made such a statement before parent advocacy groups, it might have laid the groundwork for a very productive relationship. Johnston said:

> As an aside, we found a cultural difference between vaccinologists and environmental health people in that many of us in the vaccine arena had never thought about uncertainty factors before. We tend to be relatively concrete in our thinking. Probably one of the big cultural events in that meeting, at least for me, was when Dr. Clarkson repetitively pointed out that we just didn't get it about uncertainty, and he was actually quite right. It took us a couple of days to understand the factor of uncertainty in assessing environmental exposure, particularly to metals.[9]

By their own admission, the researchers at the CDC have trouble with the idea of uncertainty. That's right, the top scientists in our public health system have trouble with the concept of uncertainty. They needed the concept pointed out to them "repetitively." Please insert your own joke about doctors believing they are God. These are the people who are making health recommendations for millions of people, and they have difficulty grasping the idea they may be wrong. Sometimes reality far surpasses any possible fiction.

Johnston finished up by talking about another confounding issue regarding mercury and vaccines, the often simultaneous presence of aluminum:

> Finally, I would like to mention one more issue. As you know, the National Vaccine Program Office has sponsored two conferences on metals and vaccines. I have just recounted a summary of the mercury, the thimerosal in vaccines. We just recently had another meeting that some of you were able to attend dealing with aluminum in vaccines. I would just like to say one or two words about that before I conclude. We learned at that meeting a number of important things about aluminum, and I think they are important in our consideration today. First, aluminum salts, and there are a number of different salts that are utilized, reduce the amount of antigen and the number of injections required for primary immunization. Secondly, they don't have much role in recall immunization, but it would represent a significant burden to try and develop different vaccines for primary and subsequent immunization. Aluminum salts are important in the formulating process of vaccines, both in antigen stabilization and absorption of endotoxin. Aluminum salts

> have a very wide margin of safety. Aluminum and mercury are often simultaneously administered to infants, both at the same site and at different sites. However, we also learned that there is absolutely no data, including animal data, about the potential for synergy, additivity or antagonism, all of which can occur in binary metal mixtures that relate to and allow us to draw any conclusions from the simultaneous exposure to these two salts in vaccines.[10]

The clear implication to be drawn from Dr. Johnston's comments is that although aluminum salts have been tested for safety, they've never been tested in combination with mercury, either with the simultaneous usage that might be expected from a series of vaccinations at a single pediatric visit, or with an infant who, for whatever reason, might have significant mercury retention because of parental lifestyle or environmental exposures.

A series of experiments conducted by Dr. Boyd Haley, former chairman of the Chemistry Department at the University of Kentucky, and published in the journal *Medical Veritas* in 2005 suggests reason for concern about the combination of mercury and aluminum:

> It is well documented in the literature that mercury toxicity is synergistic with other heavy metals such as cadmium and lead. It is also known that certain antibiotics greatly enhance the toxicity of thimerosal in ocular solutions and that antibiotics prevent test animals from excreting mercury. The major know difference between male and females is their hormones . . . Aluminum hydroxide alone at 500 nanomolars showed no significant death of cells at 6 hours and only slight toxicity over the 24-hour period. Thimerosal at 50 nanomolars effected only a slight increase in neuron death at 6 hours. However, in the presence of 50 nanomolars of thimerosal plus 500 nanomolars of aluminum hydroxide, the neuronal death increases to roughly 60%, an amazing increase and clearly demonstrates the synergistic effects of other metals on mercury toxicity and certainly thimerosal toxicity.[11]

While the public health experts gathered at the Simpsonwood Retreat Center in Georgia on June 6 and 7, 2000, may not have had any data on the potential catastrophe of the combination of mercury and aluminum in vaccines, they had some data by 2005, and the question remains why they have not more vigorously pursued this question in the ensuing twenty years since their ignorance of this subject was realized.

Another question to consider, even if every nanomolar of mercury was immediately taken out of vaccines, is the ubiquitous presence of mercury in the environment and in human bodies. As the researchers gathered at Simsponwood had already attested, we are exposed on a daily basis to mercury through various sources. How can any responsible scientist suggest that aluminum salts are acceptable in even mercury-free vaccines if the question of their effects in combination is unknown?

It is nothing less than Russian roulette with the human population.

After some discussion about what still remained unknown about chronic versus acute exposure, and specific health concerns related to metal exposure, Dr. Frank DeStefano (with whom Dr. Bill Thompson would work on the MMR vaccine-autism study) introduced the group to the Vaccine Safety Datalink, the same database to which Dr. Brian Hooker and the Geiers had been attempting to gain access for years. But among this group, the information was shared freely:

> The analyses you will be discussing for most of the morning come from the Vaccine Safety Datalink. I'm going to give you a quick overview of what the project is and some of the data. This is a project collaboration between the CDC's National Immunization Program and four large health maintenance organizations listed here, Group Health Cooperative in Seattle, Northwest Kaiser in Portland and Northern and Southern California Kaiser. They currently have an enrolled population of more than 6 million people.[12]

DeStefano spent some time recounting the history and design plan of the Datalink. Then he turned the meeting over to Dr. Thomas Verstraeten, who really was the main researcher whose findings had prompted the meeting.

Verstraeten got right into it:

> Good morning. It is sort of interesting that when I first came to the CDC as a NIS [National Immunization Service] a year ago only, I didn't really know what I wanted to do. But one of the things I knew I didn't want to do was studies that had to do with toxicology or environmental health. Because I thought it was much too confounding and it's hard to prove anything in those studies. Now it turns out that other people thought that this study was not the right thing to do. So what I will present to you is the study nobody thought we should do.[13]

***

Like a good scientist, Verstraeten began by detailing the strengths and weaknesses of his analysis, showing how the results would change if different assumptions were used, and stratifying the groups into seven categories, ranging from low exposure to thimerosal to children who received approximately seventy-five micrograms of mercury from a very aggressive vaccination schedule.

Verstraeten then said:

> For the overall category of neurologic developmental disorders, the point estimates of the categorized estimates suggest potential trends, and the test for trends is also statistically significant above one, with a P value below 0.01. The way to interpret this point estimates which seems very low is as follows: That's an increase of .7% for each additional microgram of ethyl-mercury. For example, if we would go from zero to 50 micrograms of ethylmercury, that would give us an additional increase of about 35%, which is pretty close to the point estimate for this category. Or for the overall, we would have to multiple 75 micrograms to .7 and that would give us about one and a half for the relative risk.[14]

The importance of this finding was difficult to ignore. Each microgram of mercury resulted in an increase of .7 percent in the odds that an infant would go onto have a neurological disorder. Given that it was estimated those who were most aggressively vaccinated would receive around seventy-five micrograms of mercury from the pediatric visits, the risk of a neurological disorder was increased by 50 percent.

There was a good deal of discussion about the assumptions used, the weaknesses, and uncertainties in the data, such as low birth weight, mercury exposure of the mother during pregnancy, and even genetic contributions, all of which are a proper part of scientific investigation. After much discussion and questioning, it fell to Dr. Verstraeten to summarize his opinion:

> The bottom line to me is you can look at this data and turn it around and look at this, and add this stratum, I can come up with risks very high. I can come up with risks very low, depending on how you turn everything around. You can make it go away for some and then it comes back for others. If you make it go away here, it will pop up again there. So the bottom line is, okay, our signal will simply not go away.[15]

Some talk ensued about testing premature babies, as they would be expected to have the highest risk from exposure to thimerosal. This could theoretically be done using the Vaccine Safety Datalink, or other sources. Again, it fell to Dr. Verstraeten to bring some structure to the discussion:

> Personally, I have three hypotheses. My first hypothesis is it is parental bias. The children that are more likely to be vaccinated are more likely to be picked up and diagnosed.
>
> Second hypothesis, I don't know. There is a bias that I have not yet recognized and nobody has yet told me about it. Third hypothesis, it's true, it's thimerosal. Those are my hypotheses.[16]

Verstraeten was subsequently questioned by Dr. Robert Brent, a pediatrician and developmental biologist from Thomas Jefferson University. He asked about the mechanisms and biological plausibility of mercury causing the observed neurodevelopmental disorders.

Verstraeten replied:

> When I saw this and I went back through the literature, I was actually stunned by what I saw because I thought it is plausible. First of all, there is the Faroe study, which I think people have dismissed too easily, and there is a new article in the same journal that was presented here, the *Journal of Pediatrics*, where they have looked at PCB. They have looked at other contaminants in seafood and they have adjusted for that, and still mercury comes out. That is one point. Another point is that in many of the studies with animals, it turned out there is quite a different result depending on the dose of mercury. Depending on the route of exposure and depending on the age at which the animals were exposed. Now I don't know how much you can extrapolate that from animals to humans, but that tells me that mercury at one month of age is not the same as mercury at three months, at twelve months, prenatal mercury, later mercury. There is a whole range of plausible outcomes from mercury.[17]

It was clear from Verstraeten's answer that he believed mercury could cause a host of neurodevelopmental problems.

Brent was inclined to agree with him and said:

> I think that is very helpful. I would add a couple of things in there and that is that there are three reasons why you might have the findings you reported. One is, and we don't have the data, that, with the multiple exposures you get an increasing level, and we don't know whether that is true or not. Some of

> our colleagues here don't think that is true, but until we demonstrate it one way of the other, we don't know that. The other thing is that each time you have an exposure there is a certain amount of irreversible damage, and with that exposure, the damage adds up. Not because of the dose, but because they are irreversible. And the third thing is that maybe the most sensitive period is later, like in the fifth or sixth month. In other words, the sensitivity period is not the same over the first six months.[18]

There was some more discussion of the issues, and the attendees were addressed by Dr. Roger Bernier, associate director for Science at the National Immunization Program at the CDC as the meeting came to a close for the day:

> Quickly if I may, I want to talk about the homework assignment for the consultants, and I would like to invite the other members of the meeting to feel free to fill it out. We would like to collect your opinions, although the people we are obviously looking to are the eleven consultants. I am sure you have seen the list of participants, you know who you are. I just want to read these questions in case there are any semantic issues, because we don't want to have any semantic problems when the questions are answered, and then, oh, that's not what I meant. So let's try to make sure that we understand clearly the questions. There are three questions altogether. As I mentioned this morning, I would like to suggest that you take your notes this evening and make notes on here as preliminary answers.
>
> Use tomorrow morning to make your comments because we will go around the room person by person. There are eleven consultants whose opinions will be solicited, and then after you hear those opinions, you may want to make some revisions on the final sheet you turn in.[19]

The attendees were then free to enjoy the good weather that Dr. Walter Orenstein had jokingly taken credit for in his opening remarks, look at the beautiful trees, and catch up with longtime friends or make new ones. They were the elite of the scientific community, the acknowledged experts from academia, government, and private enterprise, and they would decide not only what was best for the rest of society, but what facts or even questions the public would be allowed to consider.

***

June 8, 2000

The meeting on Sunday was shorter than the previous day's meeting, with many of the scientists, professors, and physicians planning to take an afternoon flight out of Atlanta so they could be back at work on Monday morning.

Dr. David Johnson, the state public health officer from Michigan, posed two questions to the first consultant, Dr. Paul Stehr-Green, an associate professor of epidemiology at the University of Washington School of Public Health and Community Medicine. Johnson asked:

> Do you think the observations made to date in the Vaccine Safety Datalink Project about a potential relationship between vaccines which contain thimerosal and some specific neurological developmental disorders, speech delay, attention deficit, ADHD and developmental delays constitute a definite signal? That is, are a sufficient concern to warrant further investigation? Paul?[20]

Dr. Stehr-Green replied:

> First I want to reiterate what others have said. I want to congratulate the folks who did the initial analyses for a tremendous amount of work, a lot of dedication, and very interesting results. In my judgment, these preliminary results are not compelling, but the implications are so profound that the lead should be examined further. My outstanding concerns and reasons for that statement really go to the validity and the accuracy of these results that revolve primarily around the issue of ascertainment bias or confounding, which I think is potentially a fatal flaw which was not dispelled by some of the clever analyses. Some other concerns I have deal with the uncertainties, as we talk about the low dose groups, and I think Dr. Rhodes [a statistician in the National Immunization Program] demonstrated these concerns very nicely. In effect that is closely related to the first issue of ascertainment bias.
>
> Another concern I have is the inconsistent, and in effect, mostly unknown case definitions. Again, even though Dr. Davis [an associate professor of pediatrics and epidemiology at the University of Washington and one of the investigators] did a very nice job of going back and showing that at least for some of the major outcomes, that the initial information on the electronic records were very closely supported by more detailed clinical follow up. I think there is a major case of ADD everything, at least as ascertained over this time period. Then finally, I think as Dr. Rhodes pointed out, the

> exclusion criteria may have introduced other biases that have altered our ability to draw inferences from this data.[21]

Even with all of these concerns, Dr. Stehr-Green voted "yes" to further research, drawing a rebuke from Dr. Robert Brent, a developmental biologist and pediatrician from Thomas Jefferson University, and setting off a lively discussion, which included Dr. Orenstein.

One of the concerns expressed was about the reported increase of developmental disorders. Dr. Isabelle Rapin, a neurologist for children at the Albert Einstein College of Medicine, said:

> Can I make one comment about the business of the increasing prevalence of developmental disorders? I think that this parallels increasing education and sophistication of people who examine children. I can tell from my own experience that 20 or 30 years ago I barely diagnosed autism unless it was so blatant that it stared me in the face, and now I see at least two new ones a week. And not so severe as the previous ones, so I think there is a tremendous change in the threshold of ascertainment. And yes, I have seen the California statistics which says it has increased 300 fold, but I would be interested to know whether it has increased 300 fold in areas where there are physicians who have been trained in this recognition, as opposed to areas where there are not.[22]

A couple of points stand out from Dr. Rapin's comments. Scientists are trained to be suspicious of their own recollections, especially over long periods of time. Her stated belief is that in the past she missed many cases of autism but is not missing them now. This strikes me as an unscientific method of determining the national or statewide presence of autism. A better method of determining the truth about past realities is to examine written records that have been reviewed, analyzed, and have gone through some process of peer review. She is referencing some findings by the State of California that have presumably gone through such a process, showing a "300 fold increase," but is choosing not to believe those findings.

Perhaps a better analysis is the one offered by Dr. David Johnson, the public health officer for the State of Michigan, right after her comments:

> In my opinion the evidence today is insufficient to determine whether or not thimerosal containing vaccines caused the neurological sequalae in question. The diagnosis, even in the hands of experts, and the number of diagnoses

> are too easily influenced by variations in parental and physician sensitivity and concern, and utilization of health care of similar merits. The underlying biologic, toxicologic and pharmacologic data are too weak to offer guidance one way or the other. That is the biologic plausibility component of this, in my opinion, is too badly defined. Now on the other hand, the data suggests that there is an association between mercury and the endpoints, ADHD, a well-known disability, and speech delay, as entered into the database. Then here comes an opinion, well it is all an opinion, but it expresses a flavor, so I think it related to what Dr. Bernier [associate director for science at the National Immunization Program] is trying to derive here. This association leads me to favor a recommendation that infants up to two years old not be immunized with thimerosal containing vaccines if suitable alternative prescriptions are available. I do not believe the diagnoses justifies compensation in the Vaccine Compensation Program at this point. I deal with causality, and it seems pretty clear to me that the data are not sufficient one way or the other. My gut feeling? It worries me enough. Forgive this personal comment, but I got called out at eight o'clock for an emergency call and my daughter-in-law delivered a son by C-section. Our first male in the line of the next generation, and I do not want that grandson to get a thimerosal containing vaccine until we know better what is going on. It will probably take a long time.[23]

Simply stated, Dr. Johnson does not believe the evidence is sufficient to convict or exonerate thimerosal as a cause of neurodevelopmental problems in children. But in the interest of safety, he suggests no child under two years of age should get a thimerosal-containing vaccine, and that regardless of what others do, he does not want his first grandson, "the first male in the line of the next generation," to get a thimerosal-containing vaccine until we "know better what is going on." He acknowledges that obtaining that information will "probably take a long time." One might say Dr. Johnson supports the guiding principle of the Hippocratic Oath to "First, do no harm." But this is only a superficial analysis. Johnson wants his own grandson to avoid a thimerosal containing vaccine until a thorough analysis has been performed, but for the rest of the public he thinks it is okay to give them thimerosal after the age of two. Can Dr. Johnson truly call himself a guardian of public health when he suggests one standard for "the first male in the line of the next generation" for his family but advises a different standard for the rest of society?

As the discussion continued back and forth, Dr. Bill Weil, who described himself as "an old pediatrician who was representing the Committee on

Environmental Health at the Academy" [Academy of Sciences], made a strong argument for accepting the implications of Verstraeten's findings:

> The number of dose-related relationships are linear and statistically significant. You can play with this all you want. They are statistically significant. The positive relationships are those that one might expect from the Faroe Island studies. They are also related to those data we do have on experimental animal data and similar to the neurodevelopmental tox data on other substances. I think that you can't accept that this is out of the ordinary. It isn't out of the ordinary.[24]

Weil, the old pediatrician, believed a compelling case had been made, from the records of the Vaccine Safety Datalink and the animal studies, that thimerosal was causing at least some problems in children.

Other attendees were concerned about the results, but not because of damage it might be inflicting on unsuspecting families, but what it might mean in the legal arena. Dr. Robert Brent, who had raised concerns earlier about the multiple uncertainties about mercury, and whether that meant they were over or underestimating the potential danger of mercury in vaccines, put it succinctly:

> The medical/legal findings in this study, causal or not, are horrendous and therefore it is important that the suggested epidemiological, pharmacokinetic and animal studies be performed. If an allegation was made that a child's neurobehavioral findings were caused by thimerosal containing vaccines, you could readily find a junk scientist who would support the claim with "a reasonable degree of certainty." But you would not find a scientist with any integrity who would say the reverse with the data that is available. And that is true. So we are in a bad position from the standpoint of defending any lawsuits if they were initiated, and I am concerned.[25]

Does Dr. Brent seem to be more terrified of the potential harm to children, or harm from lawyers trying to get justice for those potentially vaccine-injured children?

Dr. Martin Myers, the acting director of the National Vaccine Program Office, made a compelling argument that the group was setting its sights too narrowly by focusing only on thimerosal:

> Can I go back to the core issue about the research? My own concern, and a couple of you said it, there is an association between vaccines and outcome that

> worries both parents and pediatricians. We don't really know what that outcome is, but it is one that worries us and there is an association with vaccines. We keep jumping back to thimerosal, but a number of us are concerned that thimerosal may be less likely than some of the other potential associations that have been made. Some of the other potential associations are number of injections, number of antigens, and other additives. We mentioned aluminum and I mentioned yesterday mercury and aluminum. Antipyretics and analgesics are better utilized when vaccines are given. And then everybody has mentioned all of the ones we can't think about in this quick time period that are part of this association, and yet all of the questions I hear we are asking have to do with thimerosal. My concern is we need to ask the questions about the other potential associations because we are going to thimerosal-free vaccines. If many of us don't think that is a plausible association because of the levels and so on, then we are missing looking for the association that may be the important one.[26]

It is difficult to find fault with the reasoning of Dr. Myers, or his call that the investigation should move forward on several different fronts.

As the meeting was drawing to a close, an extraordinary speech was delivered to the assembled group. Dr. John Clements—seventy-seven years old at the time and working for the Expanded Program on Immunization for the World Health Organization and the 1994 winner of the Lasker Award for Clinical Medical Research (often referred to as the American Nobel Prize for medical research) and later the 2008 Pollin Prize for Pediatric Research (the only international pediatric award for research)—was called upon to give his comments on "the implications of dealing with the composition of vaccines for the international community."[27]

Clements pushed himself up from his chair to address the gathering. "Thank you, Mr. Chairman, I will stand so you can see me." Everybody else at the gathering could see him, as well. Few had stood to deliver their comments, and those who did were generally presenting slides in front of the group. He paused for a moment to make sure all eyes were upon him and his gaze swept the room. He said:

> First of all I want to thank the organizers for allowing me to sit quietly at the back. It has been a great privilege to listen to the debate and to hear everybody work through with enormous detail, and I want to congratulate, as others have done, the work that has been done by the team. Then comes the BUT. I am really concerned that we have taken off like a boat going down one arm of the mangrove swamp at high speed, when in fact there was not

> enough discussion early on about which way the boat should go at all. And I really want to risk offending everyone in the room by saying that perhaps this study should not have been done at all. Because the outcome of it could have, to some extent, been predicted. And we have all reached this point now where we are leg hanging [over the side of the boat?], even though I hear the majority of the consultants say to the board that they are not convinced there is a causality direct link between thimerosal and various neurological outcomes. I know how we handle it from here is extremely problematic. The ACIP [Advisory Committee on Immunization Practices] is going to depend on comments from this group in order to move forward into policy, and I have been advised that whatever I say should not move into the policy area because that is not the point of this meeting.[28]

However, Dr. Clements was not going to let anybody tell him the areas into which he could venture.

As he continued, his comments began to veer into territory that might be expected in a dystopian science-fiction movie:

> But nonetheless, we know from many experiences in history that the pure scientist has done research because of pure science. But that pure science has resulted in splitting the atom or some other process, which is completely beyond the power of the scientists who did the research to control it. And what we have here is people who have, for every best reason in the world, pursued a direction of research. But there is now the point at which the research results have to be handled, and even if this committee decides that there is no association and that information gets out, the work has been done, and through freedom of information that will be taken by others and will be used in other ways beyond the control of this group. And I am very concerned about that as I suspect it is already too late to do anything regardless of any professional body and what they may say.[29]

It was clear from his comments that Clements considered the research into vaccine side effects to be comparable to the threat from nuclear weapons, and that those who might use that information to protect their children were akin to rogue nations or terrorists.

Clements began to move to his final points:

> My mandate as I sit here in this group is to make sure at the end of the day that one hundred million are immunized with DTP, Hepatitis B and if possible

> HiB, this year, next year and for many years to come, and that will have to be with thimerosal-containing vaccines unless a miracle occurs and an alternative is found quickly and is tried and found to be safe. So I leave you with the challenge that I am very concerned that this has gotten so far, and having got this far, how you present in a concerted voice the information to the ACIP [Advisory Committee on Immunization Practices] in a way they will be able to handle it and not get exposed to other traps which are out there in public relations. My message would be that any other study, and I like the study that has just been described very much, I think it makes a lot of sense, but it has to be thought through. What are the potential outcomes and how will you handle it? How will it be presented to a public and a media that is hungry for selecting the information they want to use for whatever means they have in store for them? I thank you for that moment to speak, Mr. Chairman, and I am sorry if I have offended you. I have the deepest respect for the work that has been done and the deepest respect for the analysis that has been done. But I wonder how on earth you are going to handle it from here.[30]

And with that, Dr. Clements, working for the Expanded Program on Immunization, World Health Organization, Geneva, Switzerland, recipient of the Lasker Award, and future winner of the Pollin Prize for Pediatric Research, took his seat.

One might call the comments of Dr. Clements the vaccine two-step. On the one hand, he compliments the researchers on their work, but then reverses course by questioning whether the research should have been done because "the outcome of it could have, to some extent, been predicted." Has there ever been a comparable moment in the history of science, with the health of so many people at stake?

Can you imagine a gathering of Roman Catholic bishops in the seventeenth century saying to Galileo, "Really first-class observations and mathematics, but you went wrong when you said the Earth revolves around the Sun. I understand I can't disprove your theory with the evidence you've submitted, but it's wrong anyway." The bishops then decided to enforce their edict by forbidding people from observing the night sky. Even the entitled princes of the Catholic Church in the Middle Ages did not go as far as the scientists gathered at the Simpsonwood Retreat Center for those two days in June of 2000.

Dr. Walter Orenstein brought the meeting to a close by thanking his National Immunization Program staff, singling out Robert Chen, Frank DeStefano, Phil Rhodes, and Thomas Verstraeten, who came in for especially high praise: "I have seen him in audience after audience deal with exceedingly skeptical individuals and deal with them in a very calm way in answering their questions and doing the analyses and I think you are mature beyond your years."[31] And what conclusions can be drawn by this review of the Simpsonwood Conference at Norcross, Georgia, which reviewed Vaccine Safety Datalink information on June 7 and 8 of 2000?

What seems to be unmistakable is that Thomas Verstraeten believed the data showed a strong signal indicating that thimerosal was associated with neurodevelopmental problems in children. As a group, there did not seem to be a strong belief that the case had been proven. But the information was so concerning that it generated several possible courses of action, some of which were at odds with each other. Many seemed to believe that the question should be pursued, while others worried about its falling into the hands of lawyers or members of the public. The opinion of Dr. Clements that things had come to such a point that they could no longer be contained proved to be illusory. The story could, to a great extent, be kept off the public radar.

In June 2005, Robert Kennedy Jr. published his article on the Simpsonwood Conference in *Rolling Stone* and *Salon*. In addition, Kennedy's allegations about the Simpsonwood conference had attracted the attention of Jake Tapper at CNN, and his investigative team spent several weeks confirming Kennedy's account. As Kennedy wrote in an email to an autism list group on September 22, 2015, about the incident:

> In 2005, when I published my *Rolling Stone* article about Simpsonwood, Jake Tapper spent two weeks with me preparing an exclusive story for his network endorsing all of the points in my article. He spent a lot of time on independent research due diligencing all my *Rolling Stone* assertions. Tapper and his entire production crew were entirely enthusiastic about the piece and the segment and told me that their story paralleled everything [in] my article. Three hours before the piece was supposed to air on the evening news, a shocked Jake Tapper called me and said the piece had been pulled. "Corporate killed it." Audibly shaken, he said that in all of his years on television nothing like that had ever happened to him. His network was deluged with hundreds of emails from angry members of the autism community furious that the truth they had longed for was once again torn from them by the power of Big Pharma. Tapper was also furious and vowed to get to the bottom of [the] scandal. But somebody gave him a career memo [and] two days later he was back on the network reservation. He stopped talking to me or taking my phone calls and the last conversation I had with him, he was backtracking on his original statements . . .[32]

In the opinion of many in the autism community, the same forces that caused CNN reporter Jake Tapper to pull his well-researched story on Simpsonwood three hours before broadcast would also lead *Rolling Stone* and *Salon* to later pull the Kennedy article.

Do these facts and allegations justify calling the group that met at Simpsonwood and their allies a "cabal"? The Simpsonwood meeting was devoid of any press or representatives of autism families. Was the rest of the country being "managed" by these scientists from the CDC, industry, and academia? Is that the way a democracy is supposed to function?

***

While Drs. Orenstein, Chen, DeStefano, and Verstraeten were preparing for the Simpsonwood Conference, I was making preparations of a different nature.

The two-bedroom townhouse my wife, Linda, and I lived in on Central Avenue in Alameda, an island in the San Francisco Bay, had become too small for us as we awaited the birth of our second child, Ben. Alameda has often been described as having a small-town feel, but with a definite urban flair. The island is accessible by two bridges and a tunnel. The streets were broad and tall trees grew along them, and it was about a two-block walk to good restaurants, shops, and, of course, my favorite place in the world, a bookstore.

When we got married, I told my wife that she never had to worry about finding me in a bar with another woman. If there was ever a question about where I was, all she had to do was search the nearest bookstore or movie theater. Chances were good that's where I could be found.

But our wonderful place on Central Avenue was too small. I found a three-bedroom apartment on Walker Avenue, in Oakland, just behind the fabulous old-time art deco movie palace, the Grand Lake Theater. We didn't have San Francisco Bay anymore, but peaceful Lake Merritt was just three blocks away. This was the time between being a lawyer and becoming a science teacher, when I was trying to become a movie director. I'd always been interested in writing, penning my first novel when I was in college, then in my law practice I started representing some local screenwriters, and after I saw how they put a script together, I wrote some of my own. I got an agent in Hollywood, and even though I wrote more than twenty screenplays, and while I got some interest with a few producers telling me they "were a fan" of my work, none of the scripts sold.

Those were the days of the "independent filmmaker," as new technology brought down the cost of making a film dramatically, and people like Ed Burns, with ten thousand dollars, were making their own independent films, such as *The Brothers McMullen*, which would be purchased at the Sundance Film Festival for a couple million dollars and lead him to eventually work with figures such as Steven Spielberg.

I went to Sundance twice, watched the films, talked to some of the players, and went to many different classes trying to figure out the best way to break into the business. Like Burns, I figured the way to break in was to shoot a film at a single location, the beautiful property my grandparents owned in the Napa Valley. My film was called *Fruit of the Vine*, and I imagined it as something of a cross between the TV show *Friends* and the movie *The Big Chill*. The main theme of the movie was how a bunch of friends put away the foolishness of youth, grew up, and started to concentrate on the next generation. I've always thought that true maturity is a progression from focusing on the self to focusing on others and the greater world.

Then came Jacqueline, with her seizure disorder, her autism, and her incessant tantrums. I'd been prepared to be a good father, but I'd never imagined a test like that. Suddenly there were visits with neurologists, EEGs, different medications, some that caused her to swell up like a little beach ball and others which did nothing. Finally, we found a medication in Canada, Vigabatrin, that stopped the seizures but still did little to stop her incessant tantrums. It wasn't her fault. She seemed to be in pain, a pain that would not go away. But we wanted more children. Linda and I had planned a large family, maybe four children. And so we got pregnant again and had to find a new place to live.

The Simpsonwood Conference ended on June 8, 2000.

Eighteen days later, on June 26, 2000, our son Ben was born, and like all those other good parents, we gave him a Hep B shot filled with thimerosal on that first day. We had no worries about vaccines at that time. I try to imagine how our lives would have been different if, in the days after the Simpsonwood Conference, the CDC scientists had acted the way in which guardians of public health are expected to act.

I imagine Linda and me sitting down to watch the evening news in our new three-bedroom apartment in Oakland, in the weeks before Ben is born. Dr. David Johnson, the state public health officer for Michigan, appears on the screen. He is an older man, a grandfather after all, and he appears on camera and speaks to an audience of millions. I imagine he says something like, "Although we must investigate further, there is grave reason for concern about children receiving a thimerosal-containing vaccine because of a possible link to neurodevelopmental problems. I will not let my grandson get one of these shots and I do not think any child in America should get one until we understand the process to ensure safety. I must tell you that this information is incomplete, and we have concerns about aluminum, the number of shots, how young a child is when they get these shots, and many other concerns for which we do not yet have adequate answers. As a precaution, we are telling people to proceed slowly with immunization, and by all means, avoid those vaccines that contain thimerosal."

Linda and I would have immediately snapped to attention. She has a master's in speech pathology, and I have a law degree. We would have quickly figured it out: Jacqueline's problems might have been caused by her vaccinations. We would not have given her any more until we knew better what was going on. And we would certainly not have given any to our newborn son.

"It may be a long time until we figure this out," Dr. Johnson says as he finishes up. There was a moment in June 2000 when the health authorities of this country and the pharmaceutical companies could have done the right thing. But that moment never came.

Instead, our son was born later that month, and we let him get a thimerosal-containing Hepatitis B shot. And we continued vaccinating our neurologically impaired daughter. And we remained in the dark.

In my reminiscing, I fast-forward to January 9, 2002, when I took Ben to his eighteen-month well-baby visit. Even though we had moved to Oakland from Alameda, we kept our kids at the same pediatrician's office. Our pediatrician was a pleasant-faced woman who was kind and full of reassurances. I recall Ben on that day, nervous around new situations, retreating to the safety of my arms as the doctor came close for her examination and the full developmental check-up, which he passed with flying colors. As we finished up, the last thing to do was for him to get his shots. He would have retreated again to my arms, I would have told him everything was fine, and I would have handed over my son for his shots. And I was wrong. Wrong in the worst possible way.

Three days later, Ben was mute and pounding his head on the floor. On Martin Luther King Jr. Day of 2002 I bought the book *Unraveling the Mystery of Autism and Pervasive Developmental Disorder* by Karyn Serrousi, and it became clear what happened to our daughter. And what was happening to our son. The gluten/casein-free diet Serrousi had recommended would rescue Ben from the abyss of autism, but Jacqueline would remain locked in her silent prison. I think about those days a lot, eighteen years later, and what they have wrought in me. I think of the family and friends who have turned away, not because they think I am wrong, in the way you might disagree with somebody's politics or religion, but because I am fundamentally dangerous, at war with the modern world in the manner of some religious zealot. Perhaps I will become a terrorist. I must confess to being shocked at how quickly those who have known me all my life succumbed to this view. It is a subject they will simply AVOID.

But what would have been the response of those family and friends if they had read in their morning paper the email sent by lead author Dr. Thomas Verstraeten a little more than a month after the Simpsonwood Conference, when my son, Ben, was still less than a month old?

> Dear Dr. Grandjean,
>
> Thank you for your very rapid response. I apologize for dragging you into this nitty gritty discussion, which in Flemish we would call "muggeziften." I know much of this is very hypothetical and personally I would rather not drag the Faroe and Seychelles studies in this entire thimerosal debate, as I think they are as comparable to our issue as apples and pears at best. Unfortunately, I have witnessed how many experts, looking at this thimerosal issue, do not seem bothered to compare apples to pears and insist that if nothing is happening in these studies than nothing should be feared of thimerosal. I do not wish to be the advocate of the anti-vaccine lobby and sound like being convinced that thimerosal is or was harmful, but at least I feel we should use sound scientific argumentation and not let our standards be dictated by our desire to disprove an unpleasant theory.
>
> Sincerely,
>
> Thomas Verstraeten.[33]

I went to a Catholic high school, De La Salle High School in Concord, CA, run by the Christian Brothers. In religion class, we were often challenged with moral questions. The saints we admired had refused to renounce their belief regardless of torture or death. Would we be so brave? Would we stand up for what we believed? What kind of moral courage would we possess in our lives? I have not endured torture, but I have stood firm in telling the truth about what I saw, regardless of what people may think of me. I don't get invited to many baby showers, I can tell you that. An expectant mother doesn't want to hear that the system she is depending on to take care of her newborn child is fundamentally corrupt and dangerous. And I am left puzzling over even more challenging questions. Of whom were the scientists and industry representatives at the Simpsonwood conference truly afraid? They talked about lawyers, parent groups, even the media, but I think there was one group that terrified them more than any of these, even including pesky congressmen.

I think they were afraid of other scientists. The Cabal liked to believe it represents science. But it betrayed science. They did not even share the truth with fellow members of their profession. I think of our soft-spoken pediatrician. She would be horrified to learn of the lies that were told to her and the damage those lies did to her patients. I know that in the very marrow of my bones.

Beth Clay, a former senior professional staff member of the House Government Reform and Oversight Committee whom I interviewed, made the point that the members of the Cabal didn't even trust the other scientists they assembled at Simpsonwood:

> What happened at Simpsonwood is that they were given fraudulent data. They were not given the first review, the Generation Zero data. Verstraeten and DeStefano basically lied to those folks from the very beginning. They were not told the truth about the initial findings. That information showed an eleven-fold increase in the risk of autism. The CDC had that information, never wrote it up, and didn't present it at Simpsonwood, even though they had it. They manipulated the study, basically committed fraud by changing the study, intentionally hid and washed out the data at least four times before publication.[34]

When this former staff member was asked what would have happened if the Generation Zero data had been presented to the group assembled at Simpsonwood, she had little doubt: "We wouldn't be looking at thimerosal existing right now. We would have had an immediate recall. I don't know how anybody who sees that elevenfold increase risk of autism couldn't conclude that you didn't need to get it out immediately."[35]

Does this mean that those who attended the Simpsonwood Conference for two days in mid-2000 should not be considered members of the Cabal? Were they, like many other medical and scientific researchers, deceived by this group of CDC scientists? I will let the readers draw their own conclusions. The information with which the attendees were presented clearly showed there were significant questions about the use of thimerosal in vaccines, as well as the use of aluminum, and that the increasing vaccination schedule was harming a significant number of children.

I think of what our sweet, soft-spoken pediatrician would have done if she had been at that conference. I think of all the wonderful medical professionals I have met through my wife's work as a speech therapist at our local hospital. Even if they had been presented with the "washed" data, I am convinced that they would have called a halt to what was currently being done and immediately rushed to the nearest news station to broadcast the news. For all the people I have ever known in medicine are healers. They went into the profession to alleviate human suffering, armed with the latest scientific knowledge. They assumed they could trust what government scientists told them. They never imagined they would need to fight their own government in order to protect their patients.

In addition to the public, the Cabal has betrayed all of its fellow colleagues in medicine and science.

CHAPTER SIX

The View from Congress

Before Beth Clay was hired by Congress to work for the Committee on Oversight and Government Reform, she spent seven years working at the National Institutes of Health.[1] At the NIH, Clay worked for the Fogarty International Center, in the Rare Diseases Section, and finally in the Office of Alternative Medicine. She spent several years, from 1975 to 1984, working directly with Dr. Richard Krause, the legendary former director of the National Institute for Allergy and Infectious Diseases (NIAID) and a man best remembered for predicting the AIDS outbreak. Clay felt that during her seven years at the NIH she received the equivalent of a Master's in Public Health from some of the brightest scientific minds on the planet. If she had a scientific or medical question, she'd simply ask one of the research staff, and they'd sit her down and have a half-hour conversation about the details of a clinical study, or tutor her on various facets of the scientific process. Of her time working at the NIH, two memories stand out for Clay.

In the Rare Disease section of the NIH, Clay would often take calls from citizens who were suffering from a rare disease and wanted guidance. One day, a woman called up because her son had recently been diagnosed with autism. At that time, Clay knew exactly one person with autism, a young man who went to her church. Clay went to ask her director how to assist the woman. "Autism used to be a rare disease," the director told her. "But it's not anymore." He gave Clay the contact information for some physicians, and she passed it along to the caller.

Another strong memory, which would have a crucial bearing on later events in her life, was when she was in the office with Dr. Krause, and he was talking about the Swine Flu vaccine, released around 1976, which had been such a disaster and was suspected of causing more than eight hundred cases of Guillan-Barre syndrome and was later withdrawn. "That vaccine was so rough it still had chicken feathers in it!" he said.

Shortly after Congress created the Office of Alternative and Complementary Medicine, Clay expressed an interest in working for it, and her superiors happily obliged. The creation of the Office of Alternative and Complementary Medicine was opposed by the National Institutes of Health, but championed by many in Congress, most notably Senators Charles Grassley and Tom Harkin of Iowa. Clay's official title was project assistance and committee management officer for the Office of Alternative and Complementary Medicine, which meant she reported directly to the director of the newly created office. It was in this position she first became acquainted with Congressman Dan Burton, who would later go on to head the House Oversight and Reform Committee and with whom she would work for five years.

Burton had a long history of outsider efforts to make the medical system pay attention to new therapies that came from outside the mainstream. In 1977 as a state representative in Indiana, Burton led the fight to approve laetrile, a purported cancer treatment made from peach pits. In her time with Burton, Clay estimates she heard the story more than twenty times of the fight Burton had led on that issue.

While in the state legislature, Burton carried a bill that would have made laetrile available to the citizens of Indiana. At the time, the Food and Drug Administration was cracking down on the treatment through the use of the commerce clause that gave it jurisdiction over products that crossed state lines. Burton's bill provided that laetrile would be produced and marketed in Indiana but could not cross state lines, thus cutting the legs out from under the government's argument that it had authority over the issue. The governor let it be known he was going to pocket-veto the bill, meaning he would neither sign nor reject the measure, meaning the bill would not become law. Burton threatened to take the bill back to the legislature to get a veto-proof majority, as well as to contact the cancer patient community and to plan to protest at the governor's office. The governor relented and let it be known that he would sign the bill. In a later conversation with Burton, the governor told Burton he signed the bill because he didn't want to be part of a spectacle in which his office was besieged by cancer patients demanding a drug they believed would save their lives. For Burton, it was a story of how the creative application of political pressure and skill in understanding the system could bring about significant change.

At the time Clay started with the Office of Alternative and Complementary Medicine, it had a budget of approximately two million dollars a year but would increase to twenty million a year by the time she left in 1999. Clay's years at the newly created office would give her an unparalleled look at how the relationship between Congress and its agencies is supposed to work and how often that system breaks down. In theory, there is supposed to be a smoothly functioning system among the president, Congress, and federal agencies, which results in the greatest possible benefit to the public.

After the State of the Union Address, the president sends Congress a request for appropriations, which details how much he wants for each federal agency and lists the tasks those agencies are expected to accomplish. According to the Constitution, the House of Representatives is then supposed to create an appropriations budget for each of the federal agencies, hold hearings with the directors of the various agencies, and ask specific questions of those directors to understand the needs and challenges of their agencies. After such an investigative process, a Congressional committee will develop a budget for that agency, vote, and pass that budget in the committee, and then submit it for a full vote in the House. In addition to the budget, the committee attaches a report that may contain some very specific language, detailing what the agency can or cannot spend money on. A similar process takes place in the Senate, and if there are differences between the two budgets, they are reconciled by a House-Senate committee and sent to the president's desk for his signature.

Clay started to notice a change in how the scientists in charge of the various agencies began to treat Congress, starting in the 1990s. Congressional members and their staff were well aware of how commonly admirals and generals would show up to testify with a chestful of medals and stars as they pleaded for increased budgets or the newest weapon system. These military members would insinuate that Congressional members just weren't capable of fully appreciating the threat and that the money requested should simply be given without too many questions being asked.

Clay first became aware of this shift among scientists with Harold Varmus, Nobel Laureate, who served as head of the National Institutes of Health from 1993 to 1997.[2] In her opinion, Varmus was the scientist who started to tell Congress that in effect they shouldn't "get involved in science" and that they "should just give them the money and trust them to do the research."[3] Varmus was assisted in this effort by his close personal friendship with Bill and Hillary Clinton. Clay observed that these scientists could be pretty persuasive. Added to the fact that the directors of the institute were often getting media training in the powers of persuasion, it was easy to see how Congress could be as dazzled by a prize-winning scientist as they were by a three-star Army general.

In early 1999, Congressman Dan Burton was investigating how the Food and Drug Administration (FDA) was unfairly targeting doctors who were engaged in cutting-edge research and treatment. He came to this issue because of an immunological cancer therapy pioneered by Dr. George Springer, who was using it to treat women with breast cancer. Burton had a personal interest in this matter, as his own wife suffered from breast cancer. In the ensuing years, immunological therapy (boosting the body's own immune system to fight cancer) has become part of the accepted regimen to treat breast cancer. But in the 1990s, the therapy was still controversial.

Burton had a young staffer working on this issue for the House Government Oversight and Reform Committee that Clay had met in a conference. The woman was planning to leave the committee and told Clay she should apply for the job. Clay competed with two other applicants, one of whom later became a Harvard professor, and she was thrilled when Congressman Burton picked her for the job. Clay took over the former staffer's investigation into FDA bullying of those who were advocating for immune therapies, and from her years at the NIH's Office of Alternative Medicine and Complementary Medicine, she quickly understood the issues and interests involved.

Even though she had been working for the government for several years, her new position at the House Government Oversight and Reform Committee was thrilling, and she looked forward to her first hearing in February 1999.[4] The subject was alternative medicine, and it should have been a relatively subdued affair. Well-known cardiac doctor and author Dr. Dean Ornish was testifying about the effect of diet and exercise on heart health, along with the well-known actress Jane Seymour, who had recently finished a several-year run on the popular television series *Dr. Quinn, Medicine Woman.* But right out of the gate, Congressman Burton was under attack by another congressman on the committee, Henry Waxman of California. Clay was well aware that the two congressmen had been at odds in the past year, Burton playing a leading role in the 1998 impeachment of President Bill Clinton for lying under oath in a sexual harassment case about his relationship with White House intern Monica Lewinsky, while Congressman

Waxman had been one of Clinton's strongest defenders. But surely the battle over impeachment shouldn't flow over into something as nonpolitical as health. In Clay's opinion, Waxman seemed to want to take the opposite side of anything Burton was interested in, purely out of spite. In her later investigations into vaccines and autism, Waxman continued to play a similar role. Waxman's defense of the vaccine industry shouldn't have been a surprise, since he was one of the guiding forces behind the 1986 National Childhood Vaccine Injury Act, and Clay later found a document on a slideshow by the FDA/CDC that referred to Waxman as the "Godfather" of the so-called "Vaccine Court."

The House Oversight and Government Reform Committee was investigating the anthrax vaccine used in the military when Clay first became aware of parent complaints that vaccines might also be contributing to autism. Clay considers herself a person who believes in coincidences and synchronicity, but what happened to her in 1999 left her profoundly shaken, as if she were being given a direct command from a higher power to pursue this question.

Clay was meeting in her office with Barbara Loe Fischer, a longtime vaccine safety advocate and founder of the National Vaccine Information Center. Normally, she liked to meet with people in one of the Committee conference rooms, but those rooms were in use. And besides, Clay had a lot of work to do. She thought she'd take a quick meeting with Fisher and then dive back into a pile of unfinished work. Clay found herself taken by the soft-spoken Fischer, whose oldest son had suffered a convulsion and collapsed in 1980 after his fourth DPT shot when he was two-and-a-half years old. He was left with multiple learning disabilities and attention deficit disorder. They were finishing up the meeting and Fisher was saying, "Beth, you really need to be looking at this autism issue," when the phone rang.

It was Congressman Burton. He'd been back home in his district in Indiana and talking to his daughter, Daniella. Daniella had taken her son, Christian, to a pediatric appointment where he had received seven shots. He had become ill over the course of the next few days, spiking a fever, arching his back, and beating his head against a wall. Congressman Burton said to her, "Beth, my grandson has autism as a result of his vaccine injuries. I need you to look at this." Clay recalls that she literally stopped and said to herself, "Okay, God, I get it. This is really important."[5]

When Clay started looking at autism, she did a PubMed search and found the publication of Dr. Andrew Wakefield in *The Lancet* and called him up. She found him to be delightful, kind, and gracious. He talked about the firestorm his research in the United Kingdom had provoked as well as the technology he was using, including PCR replication to identify the measles virus found in the gut of the children who had developed autism as being from the MMR jab. One could look at the evidence in two different ways. The first assumption is that the measles virus was directly at fault for the development of autism. The second is that something in the system of autistic children was preventing them from clearing the vaccine strain of the measles virus from their digestive tract. Either way, it was an important clue, and one that Dr. Wakefield felt should be vigorously investigated. Clay was familiar with the advanced technology Wakefield was employing and could not understand the anger his research generated among certain segments of the scientific community. Didn't the scientific community want to help these suffering children?

Clay was also surprised by the actions of the man who had become Wakefield's most vocal opponent, Dr. Brent Taylor. Taylor was actually a fellow colleague of Wakefield's at the Royal Free Hospital, but he refused to share his data and samples. In one of the hearings, Congressman Burton directly asked Taylor to share his raw data, but Taylor refused. It is common practice in science that, after a paper is published, the scientist releases his raw data, so that others can do an analysis to confirm or dispute the results.

Congress and the rest of the country had recently been rocked by the Clinton impeachment, and Clay couldn't help but see that some of the same tactics used against Monica Lewinsky and the Republican questioners were being used against Dr. Wakefield. The first thing they did was to point the finger at Wakefield and claim he was taking money from lawyers interested in a certain outcome. This was not true. The second line of attack was that his lab was contaminated, another false charge.

The campaign of intimidation continued against Dr. Wakefield, led in a bizarre fashion by a freelance journalist, Brian Deer, who was allowed to publish his accusations in the *British Medical Journal*.[6] Articles appearing in the *British Medical Journal* are supposed to be solely from scientists and medical professionals, and Deer was neither. Clay had the opportunity to observe what she later described as Deer's "stalker-like behavior" when Wakefield addressed a meeting of parents at the Autism One Conference in Chicago in 2004. Deer had been chasing Wakefield around the world, and many questioned where this freelance reporter was getting the money to pursue his quarry like a modern-day Captain Ahab. Security at the hotel removed Deer from the venue, and Deer seemed to be having something of a mental breakdown as he sat outside. Wakefield left the hotel, spoke to

Deer, seeming to talk the man out of a near psychotic episode, and then escorted him into the hotel and gave him a seat of honor for his talk. Clay recalls Wakefield going through an amazing presentation of the scientific process and how the original twelve children had come to him and what had led him to his conclusions.

The later attacks on Wakefield as a substandard scientist with an ax to grind against the pharmaceutical companies have continued to gall Clay, as she knew that before his MMR/autism findings, he'd actually been paid by Merck to conduct research. The situation was complicated because, at least in the United States, Merck had been able to retain a monopoly on the use of the MMR vaccine by claiming an effectiveness rate above 95 percent. This claim is now the subject of a separate civil whistleblower action by two former Merck scientists who claim that Merck falsified their claims that the vaccine's effectiveness was above 95 percent.[7]

Added to Merck's balance sheet is the fact that the former head of the CDC, Dr. Julie Gerberding, left to work for Merck as head of the vaccine division, and it's easy to understand why Clay is suspicious of anything that comes out of the CDC, or that makes it into the mainstream media regarding this issue. In political parlance, this is known as "agency capture," and no group was as successful as the pharmaceutical industry. In addition to the corporate public relations departments of pharmaceutical companies, the National Institutes of Health (NIH) has a website where they prepare news stories for TV stations or newspapers that are essentially ready for broadcast or print. Clay is supportive of a bill that recently passed the Appropriations Committee requiring that when a media outlet runs one of these stories, they have to disclose it as "government prepared." However, Clay is not optimistic of the bill passing Congress and making its way to the president's desk.

Clay's concerns about how the damage from vaccines is being kept from the public are reinforced by an insular media culture that readily accepts what the government, the scientific community, and the pharmaceutical industry tells them:

> In Washington, DC there's a very active social scene. The White House brings in the media people and under this current administration [Obama] it's a very liberal social world. It's not uncommon to have Anderson Cooper [CNN anchor] and Harold Varmus [Nobel Laureate and former head of the National Institutes of Health] at the same party. Or Frances Collins [current head of the National Institutes of Health], who is a very good friend of the

> White House. So you have the DC and New York social scene where all of these folks come together. They are friends and then they have meetings. And they feed the media people these storylines. So if you have Anderson Cooper, or any of these national news people, they're going to have a sit down with people who will prep them. It's even branched out to top universities. Columbia University actually has a press team which promotes their doctors for the local and national news.[8]

In addition to assistance from Dr. Andrew Wakefield, Clay relied heavily on the group Safe Minds for guidance and analysis of critical information. (After Clay left Congress, she would do consulting work for Safe Minds in order to assist their ongoing efforts to get the truth out about vaccines and neurological harm.) As Clay began her investigation, she had a team of staffers around her, all very bright, well educated, and often parents of young children. They were the kind of parents who read food labels, dressed their children in clothing as free from chemicals as possible, and read just about everything they could get their hands on to keep their kids healthy. She recalled asking them what kind of conversation they had had with their pediatricians before immunizing their children, or whether they'd actually read the package inserts of the vaccines they were allowing their doctors to inject into their children. One hundred percent of them responded with blanks looks, as if to say, "What conversation?" When they did talk about their experiences, they'd mutter something like, "Well, that's what my doctor told me to do." Their belief in their doctors and the medical community was so complete and overwhelming, it would have been the envy of any religious faith.

As the investigation continued and they started peeling back the layers, it seemed as if several different parts of Clay's life were coming together. She reconnected with a friend from high school who had a vaccine-injured son. "How many people is this happening to?" Clay found herself continually asking. They started subpoenaing experts and documents from the National Academy of Sciences, and the move seemed to send shock waves through the organization. It was the first time in their history that they had ever been subpoenaed. Clay noted at first the Academy didn't really seem to understand what they were looking for, and their documents would arrive as they were supposed to be presented to Congress, in binders, with tabs, and annotated so the investigators could efficiently review them.

But as the Academy became aware that they were looking at the vaccine issue, they would send Congress boxes of unorganized files, not in binders, not tabbed and labeled, making the job of the committee much more difficult. One item in particular that Clay wanted to review was the recording of a meeting at the National Academy of Sciences in which the scientists had discussed Wakefield and his findings. There was supposedly no written transcript of the discussion, but a recording had been made. Clay was then told that the tape had "accidentally" been erased in the process of trying to copy it. Of Dr. Wakefield, Clay says, "I believe Dr. Wakefield is credible, and having looked at his research I believe he was onto something. I know other people have done the same research and found the same thing. He was following science to its truth, and that's all any of us wanted. We didn't get involved in this investigation to find vaccines guilty. We got into this investigation to find the truth."[9]

Clay goes on to describe in great detail her growing disillusionment with the public health authorities and their practices as a result of her investigations:

> I've become a little bit cynical about any time the CDC publishes something. I just don't trust it any more. And that's the sad state of affairs we're in as a nation, that there's a huge portion of the population that no longer trusts that the public health community is honest. And they violated the public trust in this process. And I kept saying to them from the very beginning, like the issue with thimerosal, you're handling this wrong. I'd say this to folks at HHS [Health and Human Services]. You're handling this wrong and you need to say there's a problem here and you need to take it out. Because what you've done is you've violated the public trust and you won't ever regain that trust if you don't fix this issue. And not only did they not fix the thimerosal issue, now we know with the MMR issue that they knowingly hid data where a portion of the population has been left at an increased risk of autism. How is that okay?
>
> There are people who refuse to let the truth come out and they're willing to destroy somebody else's reputation as a way of protecting their agenda. Protecting the current policies on vaccines was more important to government and industry officials than the reputation of one scientist.[10]

Clay's low opinion of government and industry science on the safety of vaccines is not much different from many of the parents of children with autism, although their views are not heard in the mainstream media, which excludes them on the grounds that they are not responsible voices.

Although Clay's views and those of the parent community might not get much media attention, the twenty-eight-page "Mercury in Medicine

Report" issued on May 20, 2003, by the Subcommittee on Human Rights and Wellness, Committee on Government Reform, held little back. The executive summary began with an explanation of the problem and the investigation of the committee:

> Vaccines are the only medicines that American citizens are mandated to receive as a condition for school and day care attendance, and in some instances, employment. Additionally, families who receive federal assistance are also required to show proof that their children have been fully immunized. While the mandate for which vaccines must be administered is a state mandate, it is the Federal Government, through the Centers for Disease Control and prevention (CDC) and its Advisory Committee for Immunization Practices that make the Universal Immunization recommendations to which the majority of states defer when determining mandates. Since the early to mid-1990s, Congress has been concerned about the danger posed by mercury in medical applications, and in 1997, directed the Food and Drug Administration (FDA) to evaluate the human exposure to mercury through foods and drugs.
>
> In 1999, following up on the FDA evaluation and pursuant to its authority, the House Committee on Government reform initiated an investigation into the dangers of exposure to mercury through vaccination. The investigation later expanded to examine the potential danger posed through exposure to mercury in dental amalgams. This full committee investigation complemented and built upon the investigations initiated by two of its subcommittees. In January 2003, the investigation continued into the newly formed Subcommittee on Human Rights and Wellness.
>
> A primary concern that arose early in the investigation of vaccine safety was the exposure of infants and young children to mercury, a known toxin, through mandatory childhood immunizations. This concern has been raised as a possible underlying factor in the dramatic rise in rates of late-onset or "acquired" autism. The symptoms of autism are markedly similar to those of mercury poisoning.[11]

The opening three paragraphs of the committee report did a good job of laying out the nuts and bolts of vaccine policy in the United States. The CDC develops vaccine guidelines through its Advisory Committee on Immunization Practices (ACIP), and the states generally follow those recommendations. In the early-to-mid-1990s, Congress became concerned about mercury in food and medical products, including vaccines and dental amalgams, and began an investigation. As a result of that investigation, they also became aware of similarities between mercury poisoning and autism.

The report gave a sharp rebuke to the FDA's position that it was better to be more concerned about the known risks of infectious diseases than the theoretical harm caused by mercury in childhood vaccines:

> This argument—that the known risks of infectious diseases outweigh a potential risk of neurological damage from exposure to thimerosal in vaccines, is one that has been continuously been presented to the Committee by government officials. FDA officials have stressed that any possible risk from thimerosal was theoretical; that no proof of harm existed. Upon a thorough review of the scientific literature and internal documents from government and industry, the Committee did in fact find evidence that thimerosal posed a risk. The possible risk for harm from either low dose chronic or one time high level (bolus dose) exposure to thimerosal is not "theoretical," but very real and documented in the medical literature.
>
> Congress has long been concerned about the human exposure to mercury through medical applications. As a result of these concerns, in 1997, Congress instructed the FDA to evaluate the human exposure to mercury through drugs and foods. Through this Congressionally mandated evaluation, the FDA realized that the amount of methylmercury infants were exposed to in the first six months of life through their mandatory vaccinations exceeded the Environmental Protection Agency's (EPA) limit for a closely associated compound, methylmercury. The FDA and other Federal agencies determined that in the absence of a specific standard for ethylmercury, the limits for ingested methylmercury should be used for injected ethylmercury. The Institute of Medicine, in 2000, evaluated the EPA's methylmercury standard and determined that based upon scientific data that it, rather than the FDA's, was the scientifically validated safe exposure level.[12]

A number of important points are covered in these two paragraphs of the Congressional report. The first is that government scientists were telling Congress there was no proof of harm from mercury exposure, a claim the investigators believed was plainly false from the documented medical literature. The second is that Congress has been concerned since at least 1997 about mercury in food and medical products. The third point was that all these federal agencies agreed that NO safety information existed for ethylmercury, the chemical that was being injected directly into the bloodstream of American children, often on the first day of life. Instead, they would use the standards for methylmercury, a related compound. And perhaps most astounding of all, these federal agencies claimed that the safety of any substance was the same whether you took it orally or it was injected directly into your bloodstream. Maybe next time you're feeling sick you should try injecting that chicken soup into your bloodstream and seeing which results you get.

In all, the Congressional report listed seventeen findings and seven recommendations, many of which would shock a public that has been led to believe that no reasonable concerns existed about vaccines and neurological disorders:

> A. Findings
>
> Through this investigation of pediatric vaccine safety, the following findings are made:
>
> 1. Mercury is hazardous to humans. Its use in medicinal products is undesirable, unnecessary and should be minimized or eliminated entirely.
> 2. For decades, ethylmercury was used extensively in medical products ranging from vaccines to topical ointments as preservative and an anti-bacteriological agent.
> 3. Manufacturers of vaccines and thimerosal (an ethylmercury compound used in vaccines), have never conducted adequate testing on the safety of thimerosal. The FDA has never required manufacturers to conduct adequate safety testing on thimerosal and ethylmercury compounds.
> 4. Studies and papers documenting the hyperallergenicity and toxicity of thimerosal have existed for decades.
> 5. Autism in the United States has grown at epidemic proportions during the last decade. By some estimates the number of autistic children in the United States is growing between 10 and 17 percent per year.
> 6. At the same time that the incidence of autism was growing, the number of childhood vaccines containing thimerosal was growing, increasing the amount of ethylmercury to which infants were exposed threefold.
> 7. A growing number of scientists and researchers believe that a relationship between the increase in neurodevelopmental disorders of autism, attention deficit hyperactivity disorder, and speech or language delay, and the increased use of thimerosal in vaccines is plausible and deserves more scrutiny. In 2001, the Institute of Medicine determined that such a relationship is biologically plausible, but that not enough evidence exists to support or reject this hypothesis.
> 8. The FDA acted too slowly to remove ethylmercury from over-the-counter products like topical ointments and skin creams. Although an advisory

committee determined that ethylmercury was unsafe in these products in 1980, a rule requiring its removal was not finalized until 1998. [Author's Note: That's eighteen years and four presidential administrations!]

9. The FDA and the CDC failed in their duty to be vigilant as new vaccines containing thimerosal were approved and added to the immunization schedule. When the Hepatitis B and Haemophilus influenza Type B vaccines were added to the recommended schedule of childhood immunizations, the cumulative amount of ethylmercury to which children were exposed nearly tripled.
10. The amount of ethylmercury to which children were exposed through vaccines prior to the 1999 announcement exceeded safety thresholds established by the Federal government for a closely related substance—methylmercury. While the Federal Government has established no safety threshold for ethylmercury, experts agree that the methylmercury guidelines are a good substitute. Federal health officials have conceded that the amount of thimerosal in vaccines exceeded the EPA threshold of 0.1 micrograms per kilogram of bodyweight. [Author's Note: A twenty-pound infant weighs a little more than nine kilograms, meaning that such an infant should not get more than 0.9 micrograms of thimerosal—a twenty-sevenfold overexposure.] In fact, the amount of mercury in one dose of DTaP or Hepatitis B vaccines (25 micrograms each) exceeded this threshold many times over. Federal health officials have conceded that this amount of thimerosal in vaccines exceeded the FDA's more relaxed threshold of 0.4 micrograms per kilogram of body weight. [Author's Note: Under this standard, a twenty-pound infant should not receive more than 3.6 micrograms of thimerosal—a nearly sevenfold overexposure.] In most cases, however, it clearly did.
11. The actions taken by HHS to remove thimerosal from vaccines in 1999 were not sufficiently aggressive. As a result, thimerosal remained in some vaccines for an additional two years.
12. The CDC's failure to state a preference for thimerosal-free vaccines in 2000 and again in 2001 was an abdication of their responsibility. As a result, many children received vaccines containing thimerosal when thimerosal-free alternatives were available.
13. The influenza vaccine appears to be the sole remaining vaccine given to children in the United States on a regular basis that contains thimerosal. Two formulations recommended for children six months of age or older continue to contain trace amounts of thimerosal. Thimerosal should be

removed from these vaccines. No amount of mercury is appropriate in any childhood vaccine.

14. The CDC and the National Immunization Program in particular are conflicted in their duties to monitor the safety of vaccines, while also charged with the responsibility of purchasing vaccines for resale as well as promoting increased immunization rates.
15. There is inadequate research regarding ethylmercury neurotoxicity and nephrotoxicity. [Author's Note: Nerve toxicity.]
16. There is inadequate research regarding the relationship between autism and the use of mercury-containing vaccines.
17. To date, studies conducted or funded by the CDC that purportedly dispute any correlation between autism and vaccine injury have been of poor design, underpowered, and fatally flawed. The CDC's rush to support and promote such research if reflective of a philosophical conflict in looking fairly at emerging theories and clinical data related to adverse reactions from vaccinations.[13]

This Congressional report may be one of the most important in American history for the picture it paints of a government agency that is supposed to protect and promote the health of the population but seems fundamentally incapable of performing the task.

If the CDC and FDA were corporations and behaved in such a manner, it's clear they would have been subject to both civil and criminal prosecution. Consider the charges made in the Congressional report: No safety testing was performed on thimerosal, even though evidence of such toxicity had been in medical journals for decades. The CDC and FDA remained quiet as the autism epidemic exploded, were slow to remove thimerosal from pediatric vaccines, and the studies they did perform were "of poor design, under-powered, and fatally flawed." In addition, the report pointed out "a philosophical conflict in looking fairly at emerging theories and clinical data related to adverse reactions from vaccinations." These are supposed to be the guardians of the public's health.

Imagine if you were the head of a corporation and these were your employees. Suppose they had a design flaw in one of your products and people were injured as a result. Trust and integrity are important concepts. Civilization does not function without them. Society breaks down. Division replaces community. Distrust takes the place of friendly relations. If any leader failed to remove employees who acted in such a cavalier fashion, he would invite destruction of his organization. That is exactly what our politicians have invited by not dealing with this situation. They have failed us by harming our children under the guise of protecting them. It is difficult to imagine a more heinous dereliction of duty.

The Congressional report laid out a series of seven recommendations in its 2003 report. From the vantage point of 2020, it is difficult not to imagine how the lives of so many hundreds of thousands of children and their families might be different today if these recommendations had been implemented:

B. Recommendations

1. Access by independent researchers to the Vaccine Safety Datalink database is needed for independent replication and validation of CDC studies regarding exposure of infants to mercury-containing vaccines and autism. The current process to allow access remains inadequate.
2. A more integrated approach to mercury research is needed. There are different routes that mercury takes into the body, and there are different rates of absorption. Mercury bioaccumulates; the Agency for Toxic Substances and Disease Registry (ATSDR) clearly states: "This substance may harm you." Studies should be conducted that pool the results of independent research that has been done thus far, and a comprehensive approach should be developed to rid humans, animals, and the environment of this dangerous toxin.
3. Greater collaboration and cooperation between federal agencies responsible for safeguarding public health in regard to heavy metals is needed.
4. The President should announce a White House conference on autism to assemble the best scientific minds from across the country and mobilize a national effort to uncover the causes of the autism epidemic.
5. Congress needs to pass legislation to include in the National Vaccine Injury Compensation Program (NVICP) provisions to allow families who believe that their children's autism is vaccine-induced the opportunity to be included in this program. Two provisions are key: First, extending the statute of limitations as recommended by the Advisory Commission on Childhood Vaccines from 3 to 6 years. Second, establishing a one to two year window for families, whose children were injured after 1988 but who do not fit within the statute of limitations, to have the opportunity to file under the NVICP.
6. Congress should enact legislation that prohibits federal funds from being used to provide products or pharmaceuticals that contain mercury, methylmercury, or ethylmercury unless no reasonable alternative is available.

7. Congress should direct the National Institutes of Health to give priority to research projects studying causal relationships between exposure to mercury, methylmercury, and ethylmercury to autism spectrum disorders, attention deficit disorders, Gulf War Syndrome, and Alzheimer 's disease.[14]

But President George W. Bush never announced a "White House conference on autism to assemble the best scientific minds from across the country and mobilize a national effort to uncover the causes of the autism epidemic." Other recommendations were ignored, as well. Independent researchers are NOT given access to the Vaccine Safety Datalink, although it was created by the use of our tax dollars. And federal research into the effect of heavy metals, or any other environmental cause of autism, is virtually nonexistent.

In the opinion of Beth Clay, this is a problem in both of our political parties. When interviewed in January 2016, she remarked, "The challenge is that who's in the White House matters. When President Obama was 'candidate Obama' he said he was going to have an 'Autism Czar.' And he also said he thought vaccines were involved in the autism epidemic. Then as soon as he's elected, and he's listening to the people at Health and Human Services, his story completely changes. And he never delivers the 'Czar' and he never follows through on any of it. And while 'candidate Trump' talked about vaccines and autism, he's not talking about it very much now. I don't know if he would follow through on it if he's elected and do anything, either. It's challenging because if you don't have integrity among the people who are supposed to be managing safety now, how do we know we're going to have integrity among the people who might manage it in the future? It gets down to integrity."[15]

At the time this report came out, it was of great importance to people like Dr. William Thompson, who, several months later, would send his email to Dr. Julie Gerberding, asking her to meet with him to discuss his troubling results in the MMR/Autism study. And these issues remained important to Congressman Dave Weldon, who, on May 29, 2004, gave an address titled "Something is Rotten, But Not Just in Denmark" to a group of parents gathered for the Autism One Conference in Chicago, Illinois.

After thanking the parents for their passion, noting that he generally supported the use of vaccines but also believed in an open and transparent process for looking at problems, he gave an address that dealt with much of what had transpired since the release of the Congressional Report. Many were interested in the recent publication of the Institute of Medicine conference findings on vaccines and autism, the one at which Dr. Thompson had originally been scheduled to present his MMR/Autism findings. As with the publication of the "Mercury in Medicine" Congressional Committee Report, if the American media had focused on Congressman Weldon's speech at Autism One, the world might look very different today. From the main body of Congressman Weldon's speech:

> Is it any wonder that the CDC has spent the past two years dedicating significant funding to epidemiology while starving funds for clinical and biological research? The IOM notes in their report that the epidemiology studies they examined were not designed to pick up a genetically susceptible population. Yet, they attempt to use these five flawed and conflicted statistical studies to quash further research into the possible association between vaccines and autism. The report is extreme in its findings and recommendations. The IOM process became little more than an attempt to validate the CDC's claims that vaccines have caused no harm while quashing research to better understand whether or not and how the MMR or thimerosal might contribute to the epidemic of neurodevelopmental disorders, including autism.
>
> I would like to turn now to the specifics of these five studies.
>
> **Verstaeten Study—*Pediatrics*, November 2003**
>
> The Verstraeten study has been the subject of considerable criticism. This study, published in November 2003 in *Pediatrics* the journal of the American Academy of pediatrics was released with much media fanfare and public relations "spin." Much has been written exposing the study's methodological problems, findings, and conclusions. Most importantly however, is that this study did not compare children who got thimerosal to those who did not. Instead, its CDC-employed authors focused primarily on a dose response gradient.
>
> In addition to the study itself, it is important to note the public relations "spin" surrounding this study. On the day the Verstraeten study was released, a top CDC researcher and a coauthor of the study was quick to declare to the news media that, "The final results of the study show no statistical association between thimerosal vaccines and harmful health outcomes in children, in particular autism and attention-deficit disorder." Let me repeat that, "The final results of the study show no statistical association between thimerosal vaccines and harmful health outcomes in children, in particular autism and attention-deficit disorder."
>
> The newspaper headlines of the day read:
>
> - "Study Clears Vaccines Containing Mercury" *Associated Press* and *USA Today*
> - "CDC Says Vaccines are Safe . . ." *The Seattle Times*

While that was the spin of the day, allow me to quote from the study. ". . . we found no consistent significant associations between TCVs [thimerosal containing vaccines] and neurodevelopmental outcomes. In the first phase of our study, we found an association between exposure to Hg [mercury] and some of the neurodevelopmental outcomes screened. In the second phase, these associations were not replicated for the most common disorders in an independent population." They did find associations, but as they changed the study, most of the associations of the study, but not all, disappeared. Furthermore, in a January 2004 article this lead co-author was forced to admit that many children in the study were too young to have received an autism diagnosis. He went on to admit that the study also likely mislabeled young autistic children as having other disabilities thus masking the number of children with autism. [Bold and underline added for emphasis.]

The message from the CDC to media was that there is nothing to be concerned about, but the study said something somewhat different. The news media to a large degree took the CDC's spin hook, line, and sinker, and chose not to read the study itself. Five months after the article was published, and largely after the IOM report had been written, the lead author of the study, Dr. Thomas Verstraeten broke his silence in a letter to *Pediatrics* stating: "The bottom line is and has always been the same: as association between thimerosal and neurological outcomes could neither be confirmed nor refuted, and therefore, more study is required."

Dr. Verstraeten, the lead author of the study, says that an association between TCVs and NDDs cannot be refuted based on his study, yet the IOM in their assessment of the same study state it is a basis for concluding that "there is no association between thimerosalcontaining vaccines and autism."

The IOM acknowledges that Verstraeten would not have picked up an association in a genetically susceptible population. The IOM also noted that this study was limited in "its ability to answer whether thimerosal in vaccines causes autism because the study tests a dose-response gradient, not exposure versus non-exposure." It is also critical to note that the Verstraeten study cannot be validated. The earlier datasets have been destroyed and the only datasets the CDC will make available to outside researchers are the ones they have already manipulated. The raw, unaltered data is not available. Additionally, outside researchers are held to a much more restrictive access to information than are CDC researchers. Only one independent researcher has been granted access to the CDC's VSD database and the CDC has kicked those researchers out based on ridiculous reasons. They claimed their research methods might infringe on privacy. Yet the database contains no names. The researchers so

> not even know what HMO the patient is enrolled in. Nor do they know the state the subjects live in. There is no way for an individual to be identified through their research.[16]

Congressman Weldon can seem almost like a prosecuting attorney in his review of the Verstraeten study, but his dual profession, that of a medical doctor and a lawmaker, give him a unique perspective on this question of whether government scientists are presenting an honest picture to the public. The first criticism that Weldon levels against the Verstraeten study is the "spin" that was placed on the results. It is difficult to close the door any more definitively than by saying "the final results of the study show no statistical association between thimerosal vaccines and harmful health outcomes in children, in particular autism and attention deficit disorder."

And yet when one actually read the study, it was difficult to reconcile that definitive statement with the acknowledgment that "in the first phase of our study, we found an association between exposure to Hg [mercury] and some of the neurodevelopmental outcomes screened." The assertion that the public had nothing to fear from mercury-containing vaccines was further undercut by a January 2004 article, in which the lead coauthor admitted that many children were too young to have received an autism diagnosis, and that it was "likely" that many autistic children were mislabeled as having other disabilities. The final retreat from their reassuring message was delivered by Verstraeten's letter to *Pediatrics* stating: "The bottom line is and has always been the same: an association between thimerosal and neurological outcomes could neither be confirmed, nor refuted, and therefore, more study is required." One wonders why anybody even bothered.

It gets even worse when Weldon notes that even the Institute of Medicine admitted that the Verstraeten study "would not have picked up an association in a genetically susceptible population" and that the study was limited because it did not test a mercury exposure group against one that was not exposed to mercury. It is difficult to see how such shoddy research could have won even an honorable mention in a middle-school science fair, much less set health policy for the nation. And if even all those glaring inadequacies weren't enough to convince you that something was wrong, maybe your suspicions would be raised by the destruction of the original datasets. Or the blocking of access to independent researchers like the Geiers, with whom Dr. Brian Hooker worked for several years. Or kicking out the one group of researchers who did get access by claiming their efforts might infringe on privacy in a database that contained no names.

In Congressman Weldon's speech at Autism One, he continued his review of the studies that had supposedly cleared thimerosal-containing vaccines of any link to neurological problems:

Hviid Study

The IOM cited the 2003 study by Hviid of the Danish population as one of the key studies upon which it bases its conclusions.

Let's consider first the conflict of interest of the principal author. Hviid works for the Danish Epidemiology Science Center which is housed at the Statum Serum Institute (SSI) the government owned Danish vaccine manufacturer. Also, all of his coauthors either work with him at the Center or are employed by SSI. Statum Serum Institute (SSI) makes a considerable profit off the sale of vaccines and vaccine componments and the U.S. is a major market for SSI. SSI has $120 million in annual revenues and vaccines are the fastest growing business segment accounting for 80% of its profits. Both the U.S. and the U.K. are important export markets for SSI's vaccines and vaccine components.

Furthermore, if Hviid were to find an association between thimerosal and autism, SSI with which he and his Center are affiliated would face significant lawsuits. These facts are important and are critical when evaluating this study. Furthermore, this study only looked at autism and not neurodevelopmental disorders as a whole.

Mercury exposures in the Danish population varied considerably from thos in the U.S. Danish children received 75 micrograms of mercury by 9 weeks and another 50 micrograms at 10 months. By comparison, children in the U.S. received 187.5 micrograms of mercury by 6 months—nearly 2 ½ times as much mercury as Danish children in just the first 6 months of life.

Dr. Boyd Haley has said that comparing the exposures in the U.S. to those in other countries is like comparing apples and cows. I think there is a lot of truth in that.

Hviid states that the rate of autism went up after they began removing thiermosal from vaccines in 1992. The numbers in the Hviid study are skewed in that they added autism diagnosis to the number after 1992. The IOM notes other limitations of the study including the differences in the dosing schedule and the relative genetic homogeneity of the Danish population.

Yet even with these serious limitations, the committee concludes that this study has a "strong internal validity," finding an increased in autism after the removal of thimerosal.

> Like the Verstraeten study, Hviid would not be able to pick up a group of children who were genetically susceptible to mercury toxicity.
>
> Danish autism rate is about 6 in 10,000 vs. 30 in 10,000 in the U.S.—once again we are comparing apples and cows. Indeed, I believe it can be legitimately argued that the lower rate of autism in Denmark is attributable to the lower exposure to mercury in their population.[17]

For those with the slightest acquaintance with law or science, the bias and the flaws in this study immediately become apparent. The principal author and all of his coauthors are physically housed in the buildings of the government-owned Danish vaccine manufacturer. Weldon points out that "if Hviid were to find an association between thimerosal and autism, SSI [the government-owned Danish vaccine manufacturer], with which he and his Center are affiliated, would face significant lawsuits." It would be a little like living in your parent's house as a college student, getting some money from them for school and maybe eating whatever's in the fridge, then deciding to report your parents to the IRS for tax fraud.

Other criticisms go directly to the quality of the scientific conclusions drawn. First, they are looking at dramatically different mercury exposures, 75 micrograms in Denmark vs. 187.5 micrograms in the United States during the first six months of life, a more than twofold difference. Second, the rate of autism at the time in Denmark was six in ten thousand vs. thirty in ten thousand in the United States, a fivefold difference. Third, Denmark is a genetically homogenous population, meaning it may not pick up a difference in mercury toxicity among different ethnic groups. The last point deserves a little more explanation. Prior to 1992, autism was diagnosed in a hospital setting. After 1992, an autism diagnosis could be made on an outpatient basis. This may have dramatically increased the number of autism diagnoses after 1992, just at the same time thimerosal was being removed from childhood vaccines.

Congressman Weldon next turned his attention to the Madsen study, which, like Hviid, had been performed in Denmark:

> **Madsen Study**
>
> Next the IOM relies on the study by Madsen et al., once again examining virtually the same population that Hviid examined. Again, the relevance of the Danish experience to the U.S. is limited in that the Danish population is genetically homogenous and had significantly lower thimerosal exposures than children in the U.S.

> Let's consider the conflicts of interest with this study. First of all, two of Madsen's coauthors are employed by the Staten Serum Institute. Additionally, like Hviid, two of Madsen's coauthors work directly for the Staten Serum Institute (SSI)—the Danish vaccine manufacturer which exports vaccines and vaccine components to the U.S. and which faces liability if an association is found. Madsen works for the Danish Epidemiology Science Center—which is affiliated with SSI.
>
> This study, like Hviid, added outpatient cases into the number of cases of autism after 1995. The authors acknowledged that this addition might have exaggerated the incidence of autism after the removal of thimerosal. The IOM acknowledged that this limits the study's contribution to causality.[18]

Weldon is pointing out the rotten Denmark data again, the blatant conflict of interest among the researchers, and the addition of outpatient cases after 1995, a weakness so glaring that even the Institute of Medicine acknowledged how it limited the evidence for causality. The next study involved the Danish population again but also mixed in a dramatically different ethnic population, the Swedes (at one point, Denmark and Sweden are separated by a mere 3.3 miles):

> **Stehr-Green Study**
>
> The IOM relied on the Stehr-Green study which examined the Danish population (do you see a pattern, yet?) and Swedish populations and attempted to compare that to the U.S. population. Furthermore, a key coauthor in this study is employed by the Danish vaccine manufacturer—Staten Serum Institute.
>
> I will not repeat the problems with the Danish data again, but with regard to Sweden it is important to note the children there received even less thimerosal than children in Denmark—receiving only 75 micrograms by age 2. Furthermore, the authors included only in-patient autism diagnoses in the Swedish population. The IOM notes that the ecological nature of this data "limits the study's contribution to causality." But they cite it anyway.[19]

Again there is the issue of researchers with enormous conflicts of interest, a homogenous ethnic population, dramatically different thimerosal exposures, and the authors counting only autism diagnoses made in hospitals. Even the Institute of Medicine did not seem to be impressed by this study.

And finally, Weldon considered the Miller study, which, although it was not yet published, the CDC was brandishing as yet one more piece of evidence that thimerosal-containing vaccines had nothing to do with autism:

> **Miller et al.**
> The Miller study examines the population of children in the United Kingdom. This study is still unpublished which limits a critical and public evaluation of its findings.
>
> Dr. Miller has actively campaigned against those who have raised questions about vaccine safety. She and her department receive funding from vaccine manufacturers, and she reportedly serves as an expert witness on behalf of vaccine manufacturers who are being sued.
>
> This study, like the Verstraeten study is a dose response study which is limited in that it does not compare children who received thimerosal to those who did not.
>
> Children in the U.K. were exposed to up to 75 micrograms of mercury by 4 months of age. This represents about one-half of what children in the U.S. would have been exposed to by this age, plus children in the U.S. got another 50 micrograms two months later at age 6 months for a total exposure in the first six months of life of nearly 2 ½ times what children received in the U.K.
>
> The author concludes that the study found no association between increasing exposures to thimerosal and autism.[20]

The CDC was playing a crooked game. Not only were they touting studies that were demonstrably weak, and with authors who had significant conflicts of interest, they were citing research that hadn't even been published. Was it any wonder that members of Congress, the parent community, and groups like Safe Minds were so strident in their opposition to what was taking place? Weldon continued his speech by sharing his analysis of the epidemiological studies:

> **Conclusion on Epidemiological Studies**
> You can clearly see why the IOM is on very shaky ground in drawing the conclusions they did. They based their decision on five epidemiology studies:
>
> - Three of them examining the genetically homogenous population of Denmark.
> - At least one employee of the Staten Serum Institute serves as a coauthor of at least 3 of the studies.

- Only one study examining the U.S. population—and that study did not compare those with no mercury exposure to those with exposures.
- Four of them with populations receiving less than half of the mercury exposure that children in the U.S. received.
- None of them with any ascertainment of prenatal and postnatal background mercury exposures.
- None of them considering prenatal exposure of children.
- None of them able to detect a susceptible subgroup that may have had a genetic susceptibility to mercury toxicity.
- Three of them failing to address how the addition of outpatient cases of autism in Denmark might have perilously skewed the results.
- Four of them examined populations with autism rates considerably below that in the U.S.
- One of the studies has not been published and not subjected to public review.[21]

Perhaps a good analogy to the games of the CDC regarding vaccines and autism is the comparison to a driver under the influence. Maybe a single drink isn't enough to put somebody over the edge. Just like a single puff of pot may not be enough to make somebody dangerous. But add to that a line of coke, maybe a hit of heroin, and some crystal meth, and you have a menace on the road.

Then it is practically assured that our hypothetical driver will cause a serious accident. One might say that the scientists at the CDC, especially after the revelation of Simpsonwood, Congressman Burton's investigation, and the continued persistence of Congressman Dave Weldon, knew that they were traveling a dangerous road. It may explain why they seemed so interested in pursuing epidemiological evidence rather than biological evidence in actual living organisms. Congressman Weldon's speech at Autism One also took the government scientists to task for this irrational decision:

> **Bio/Clinical Research—Thimerosal**
>
> Since the release of the IOM's [Institute of Medicine] report in 2001, public health officials in the U.S. virtually ignored the biological and research recommendations. While the CDC had no trouble funding epidemiology studies—all with their flaws and inadequacies—several critical biological and clinical research recommendations were starved of funding.

The IOM recommended that the following studies be done, but the CDC and the NIH failed to dedicate the resources to fund these studies:

- Identify primary sources and levels of prenatal and postnatal background exposures to thimerosal, including Rho (D) Immune Globulin in pregnant women and other forms of mercury (fish) in infants, children and pregnant women—NOT DONE.
- Compare the incidence and prevalence of NDDs before and after removal of thimerosal from vaccines. NOT DONE and the CDC tells me they will not begin such studies until 2006.
- Research how children, including those with NDDs, metabolize and excrete metals—particularly mercury—NOT DONE.
- Conduct research on theoretical modeling of ethylmercury exposures, including the incremental burden of thimerosal with background mercury exposures from other sources—NOT DONE.
- Conduct careful, rigorous and scientific investigations of chelation when used in children with NDDs, especially autism. NOT DONE though in their latest report they urge that this be highly restricted.
- Conduct comparative animal studies of the toxicity of ethylmercury and methylmercury to better understand the NDD effects of thimerosal—ONLY PARTIALLY DONE—but with very little federal support.

In 2001 the IOM stated that it is "unclear whether ethylmercury [from vaccines] passes readily through the blood-brain barrier . . ." The IOM recommended several biological and clinical studies to answer this question and whether this mercury could cause developmental problems. These studies were in large part never done. Yet IOM chose to ignore the need for this research and instead has focused its analysis on the data available today, most of which is statistical data.

There is much more research that needs to be done before it can definitively be said that thimerosal does not contribute to NDDs. Even today, the IOM cannot tell you with any degree of certainty what happens to ehylmercury once injected into an infant. Does it go to the brain? Does it cause developmental problems? Who knows?[22]

In any discussion of a relationship between two parties, the fundamental issue of trust must be addressed. Parties who trust each other can have an amicable and friendly relationship, overlooking mistakes, as long as both groups trust they are committed toward agreed-upon ends. But the slightest disagreement between parties who do not trust each other can quickly escalate into conflict. And the question must be asked if the other party is worthy of trust.

Weldon's speech points out that the CDC is well aware of the large parent community who claim that vaccines have harmed their children but is not performing the biological and clinical research that would answer such questions. And it wasn't as if other respected medical organizations hadn't brought up these concerns. The Institute of Medicine had said it was uncertain whether mercury from vaccines passed through the blood-brain barrier of infants and remained in their brains, causing neurodevelopmental problems. Didn't getting the answer to that question qualify as a public health emergency?

The next topic Weldon covered was the recent Institute of Medicine conference that had a truncated discussion of the MMR vaccine/autism issue. It is ironic that Weldon pointed out this issue, not even knowing that the results of the CDC's own study had troubled Dr. William Thompson, who broke the chain of command by directly contacting CDC Director Dr. Julie Gerberding to discuss this very issue. Instead of welcoming the conversation, Dr. Thompson was placed on administrative leave. Weldon continued:

> **MMR-Autism Association**
>
> Allow me to touch briefly on the IOM's analysis of the MMR-Autism issue. They devoted only one hour of discussion to this topic at the February meeting and failed to invite those who were most intimately involved in this research to present to the IOM.
>
> As with thimerosal, the IOM relied almost exclusively on epidemiology. They made their decision about whether or not measles may be related to autism in children, by reviewing 13 statistical studies in which many of the authors have conflicts of interest. Some of these authors have been openly hostile in their assessments, which calls into question their objectivity. Also, remember it is epidemiology that reigns supreme in this review—even if the studies are flawed in their design.
>
> The IOM still cannot answer the question as to why measles is in the intestine of some autistic children. Why is it there? What is it doing? How did it get there? Is it contributing to autism? The IOM attempts to explain this issue away by saying it's likely that the presence of measles could be just a co-morbidity to autism. This cavalier attitude of the IOM, the CDC, and others in the public health community is unacceptable. We have a moral obligation to fully support research to understand why vaccine strain measles is in the

> intestines and CSF [cerebral spinal fluid] of these children. The government mandated vaccination. The least we should do is fund research to understand why measles is persisting in these children, what harm it might be causing, and how we might best treat these children.
>
> The NIH is only now attempting to duplicate the work of Dr. Andrew Wakefield. Despite being vilified for the last 6 years Dr. Wakefield's work has been demonstrated to be correct. Practitioners across the U.S. and in many other parts of the world are finding the same inflammatory bowel disease he first described in *Lancet* in 1998. Drawing "conclusions" at this time time is counterproductive. Statistical studies are of little benefit. Only a clinical pathological study will lay this issue to rest.[23]

It really is remarkable to review this speech from 2004, talking about a controversy that had existed since 1998, and realize that in 2016 this issue was still just as hot. The situation described by Weldon describes a scientific community that has its priorities backward. If you suspected you had cancer, your doctor wouldn't do a survey of your neighborhood. They'd biopsy and test the suspect tissue.

Weldon took some time to discuss the negative atmosphere created by the Institute of Medicine, including some significant failures of public health agencies, such as the 1989 study by the National Institute of Child and Human Development that missed the link between folic acid deficiencies and neural tube defects. The Institute of Medicine had also recently reversed a long-standing finding that chronic lympocytic leukemia was not due to Agent Orange (a defoliant used in the Vietnam War) exposure. Weldon also discussed some legislative proposals he had recently made, which would eventually go nowhere. He also took direct aim at the CDC and their conflicted efforts to monitor vaccine safety and at the same time promote immunization to the public:

> The CDC has the greatest responsibility in this area. Unfortunately, they also have the greatest conflict of interest. The CDC's vaccine safety program amounts to about $30 MILLION a year, and half of this goes to pay HMOs for access to the Vaccine Safety Database.
>
> The biggest conflict within the CDC is that they are also responsible for running a $1 BILLION dollar a year vaccine promotion program. The CDC largely measures its success by how high vaccination rates are. Here lies the largest conflict. Any study raising concerns that there might be adverse reactions is likely to result in safety concerns leading to lower vaccination rates.

> Lower vaccination rates are in direct conflict with the CDC's top measurement of success. Clearly, due to its overwhelming size and the manner in which the agency measures its success, the vaccine promotion program overshadows and influences the CDC's vaccine safety program.
>
> In fact, rightly or wrongly, the vaccine safety office within the CDC is largely viewed by outside observers as nothing more than another arm of the vaccine promotion program, giving support to vaccine promotion policies, and doing very little to investigate and better understand acute and chronic adverse reactions.
>
> Further complicating the CDC's role and undermining their research is the fact that the vaccine safety studies produced by the CDC are impossible to reproduce. External researchers are not granted the same level of access to the raw datasets that the CDC's internal researchers are granted. The bottom line is that the CDC's studies related to vaccine safety cannot be validated by external researchers—a critical component in demonstrating the validity of scientific findings.[24]

None of these problems cited by Weldon would be unexpected by your typical lawyer, or someone with even a passing familiarity of the behavior of large organizations. Our current justice system believes that no individual or organization is above scrutiny. What follows from that belief is the expectation that every system must have a rigorous arrangement of checks and balances if it is to remain fair. Any person or group of persons given too much power will inevitably abuse it, regardless of how many academic honors they possess, or how many scientific discoveries they have made. Let's look at the conflicts built into the vaccine system, according to Congressman Weldon.

As Weldon moved toward the end of his speech, he brought up another troubling issue. The CDC clearly understood the trouble Congress was having with this closed loop of science, so they took measures to create what looked like an independent review board:

> **Brighton Collaboration**
>
> Finally, I want to turn my attention to something known as the Brighton Collaboration.
>
> I am very concerned about the development of the Brighton Collaboration which began in 2000. This is an international group comprised of public health officials from the CDC, Europe, and world health agencies like WHO [World Health Organization], and vaccine manufacturers.

> The first task of the Brighton Collaborations, created several years ago, is to define what constitutes an adverse reaction to a vaccine. They have established committees to work on various adverse reactions to vaccines. Particularly troubling is the fact that serving on the panels defining what constitutes an adverse react to a vaccine, are vaccine manufacturers. What is even worse is the fact that some of these committees are chaired by vaccine manufacturers. It is totally inappropriate for a manufacturer of vaccines to be put in the position of determining what is and is not an adverse reaction to their product.
>
> Do we allow GM, Ford, and Chrysler to define the safety of their automobiles? Do we let airlines set the safety standards for their airlines and determine the cause of an airline accident?
>
> Do we allow food processors to determine whether or not their food is contaminated or caused harm?
>
> Then, why I ask, are we allowing vaccine manufacturers to define what constitutes an adverse reaction to a vaccine?
>
> This collaboration is fraught with pitfalls and merges regulators and the regulated into an indistinguishable group.
>
> It is critical that the American public look at what is going on here and how this entity may further erode their ability to fully understand the true relationship between various vaccines and adverse reactions.[35]

There can be few better definitions of institutional corruption than an industry that is allowed to determine what constitutes an injury from the use of one of its products. A car company is not allowed to certify whether its cars are safe, an airline isn't allowed to perform the sole investigation into a plane crash, and food producers aren't allowed to determine whether their food is contaminated. All of these investigative functions are carried out by an independent group so that the public will be assured it is getting the truth. How are vaccines any different, simply because they are administered by doctors?

As Weldon finished his remarks, it was perhaps not surprising that many in the audience felt they were listening to a modern-day Jimmy Stewart and the decent character he played in the classic film *Mr. Smith Goes to Washington*. They could genuinely believe that a true public servant was speaking to them from the podium. He continued:

> **Concluding Remarks**
>
> Finally, autism is a difficult challenge for our nation. We have made considerable progress through groups like Autism One and other organizations

represented here. The work you are doing is work that must continue. I commend each of you.

I commend the researchers who are engaged to develop a deeper understanding of what is going on with these children and how we might improve their treatments. I am hopeful that the folks down at the NIH, the CDC, and the IOM will be more supportive of your work. I will do all that I can to see that critical research in all areas of autism research continue to receive increased funding.

I commend the parents who have failed to give up on their children. I commend you for your dedication to want the best for your children and the sacrifices you have made for them.

I urge each of you to take your story to your Member of Congress and your Senator. Share your struggles with them. If I, along with the few others who have made defeating autism a top priority are to be successful, it is critical that every Member of Congress know what autism is and that they have constituents who are watching them and asking for their help.

I urge you to tell your local television reporters and newspaper reporters your story and your struggles. Tell everyone who is willing to listen. It is through your testimony that others will know of this devastating epidemic plaguing our children.

I also urge you to share with others what is working in the treatment of your children. You are blessed with resources that are available to you at this conference. Listen and learn from the providers here who have a lot to offer.

Finally, let me know what I can do to help. I stand in partnership with each of you. Thank you for inviting me to join you today. It has been a great honor.[26]

Congressman Weldon received thunderous applause from the crowd, and they stood, cheering his speech, many pumping their fists in the air or with tears streaming down their faces. It was unlike any reaction he had ever received in all his years in Congress. It seemed for a brief moment that the parents might actually win this war and begin to get answers for their children.

***

Things became even more rotten in Denmark. In 2011, it was reported that one of the leading scientists involved in the production of the Danish studies, Dr. Poul Thorsen, had embezzled nearly two million dollars of the research money he'd been paid by the CDC to study the relationship between thimerosal and autism.[27] Thorsen's 2003 study had shown a twentyfold increase in autism after the removal of mercury in vaccines, prompting many to believe the suspected link had been soundly disproved.

On April 13, 2011, the United States attorney for the Northern District of Georgia released a statement about the Thorsen case and their prosecution of the scientist:

> POUL THORSEN, 49, of Denmark, has been indicted by a federal grand jury on charges of wire fraud and money laundering based on a scheme to steal grant money the CDC had awarded to governmental agencies in Denmark for autism research.
>
> United States Attorney Sally Quillian Yates said of the case, "Grant money for disease research is a precious commodity. When grant funds are stolen, we lose not only the money, but also the opportunity to better understand and cure debilitating diseases. The defendant is alleged to have orchestrated a scheme to steal over $1 million in CDC grant money earmarked for autism research. We will now seek the defendant's extradition for him to face federal charges in the United States."[28]

The investigation had been a collaborative effort between the United States Attorney's office and the IRS criminal investigation office to follow the money trail in this case of apparent fraud. The press release from the US Attorney's Office went on to detail some of the specifics of the case:

> In the 1990s, THORSEN worked as a visiting scientist at the U.S. Centers for Disease Control and Prevention (CDC), Division of Birth Defects and Developmental Disabilities, when the CDC was soliciting grant applications for research related to infant disabilities. THORSEN successfully promoted the idea of awarding the grant to Denmark and provided input and guidance for the research to be conducted. From 2002 to 2009, the CDC awarded over $11 million to two governmental agencies in Denmark to study the relationship between autism and exposure to vaccines, between cerebral palsy and infection during pregnancy, and between childhood development and fetal alcohol exposure. In 2002, THORSEN moved to Denmark and became the principal investigator for the grant, responsible for administering the research money awarded by the CDC.
>
> Once in Denmark, THORSEN allegedly began stealing the grant money by submitting fraudulent documents to have expenses supposedly related to

> the Danish studies to be paid with the grant money. He provided the documents to the Danish government, and to Aarhus University and Odense University Hospital, where scientists performed research under the grant. From February 2004 through June 2008, THORSEN allegedly submitted over a dozen fraudulent invoices, purportedly signed by a laboratory section chief at the CDC, for reimbursement of expenses that THORSEN claimed were incurred in connection with the CDC grant. The invoices falsely claimed that a CDC laboratory had performed work and was owed grant money. Based on these invoices, Aarhus University, where THORSEN also held a faculty position, transferred hundreds of thousands of dollars to bank accounts held at the CDC Federal Credit Union in Atlanta, accounts which Aarhus University believed belonged to the CDC. In truth, the CDC Federal Credit Union accounts were personal accounts held by THORSEN. After the money was transferred, THORSEN allegedly withdrew it for his own personal use, buying a home in Atlanta, a Harley Davidson motorcycle, and Audi and Honda vehicles, and obtaining numberous cashier's checks, from the fraud proceeds. THORSEN allegedly absconded with over $1 million from the schemes.[29]

In a more just world, one would imagine that Thorsen would be quickly extradited, forced to stand trial, and, if he was found guilty, sent away for a long prison term. If convicted on all twenty-two counts, Thorsen was looking at the possibility of 260 years in prison and $22.5 million in fines. But as the years passed by, and Thorsen continued to work in his new position at Sygehus Lillebaelt Hospital in Kolding, Denmark, many wondered if the US Attorney's Office had forgotten about him. On April 14, 2014, the group Safe Minds put out a statement noting the three years that had passed since Thorsen's indictment and calling for his extradition and prosecution:

> Safe Minds calls upon the US Department of Justice and the US Department of State to bring Dr. Thorsen back to the United States for justice. We also call upon Chairman Darrel Issa and the House Committee on Oversight and Government Reform to take up this issue as an oversight activity this year. We believe this warrants a Congressional hearing to understand this failure to fully address the allegations, to determine if others at the Centers for Disease Control and Prevention (CDC) were complicit, and to address the failure of the CDC and the scientific community to investigate all of the studies from this project while holding current findings in deferral until fully investigated.[30]

The failure of government authorities to extradite a suspect who was alleged to have stolen more than a million dollars in research money intended to benefit disabled children is truly appalling and calls out for an explanation. Thorsen had not fled his known haunts in Denmark and continued working in plain sight. Why was there such a lack of political will to bring this man to justice?

Five months later, in August 2014, the failure to extradite this alleged criminal attracted the attention of former CBS reporter and five-time Emmy-winning newswoman Sharyl Attkisson, who wrote about the case:

> A former Centers for Disease Control (CDC) researcher, best known for his frequentlycited studies dispelling a link between vaccines and autism, is still considered on the lam after allegedly using CDC grants of tax dollars to buy a house and cars for himself... Poul Thorsen, listed as a most-wanted fugitive by the Department of Health and Human Services Office of Inspector general, was discredited in April 2011 when he was indicted on 13 counts of wire fraud and nine counts of money laundering... Thorsen co-authored studies in the *New England Journal of Medicine* and *Pediatrics* concluding there is no link between autism and thimerosal used in vaccines nor between autism and the MMR vaccine.[31]

As of the update of this book in late 2020, Dr. Thorsen still remains "on the lam" living openly and working in Denmark.

* * *

The normal expectation is that when Congress acts on a matter, federal agencies will quickly respond, if for no other reason than wanting to make sure their budget doesn't get cut, or to avoid unflattering media stories. But the issue of vaccines and autism seems to play out in a starkly different manner. While one might expect that representatives from different political parties, like Dan Burton and Henry Waxman, might carry powerful grudges based on previous political battles, the expectation is that such things would be set aside in dealing with matters of national importance such as public health. But in the view of House Oversight and Government Reform committee senior staffer, Beth Clay, congressman Waxman showed little interest in anything outside of conventional medicine, and his own efforts in drafting and passing the 1986 National Childhood Vaccine Injury Act made him openly hostile to any safety concerns regarding vaccines.

One might conclude that the investigation by the House Oversight and Government Reform Committee, chaired by Congressman Burton, did succeed in striking fear into the hearts of CDC scientists regarding this issue. They ran to Denmark, obtaining poor-quality science as detailed by Congressman Dave Weldon, an MD, and probably not realizing at the time that the principal investigator on the grants, Dr. Poul Thorsen, was embezzling nearly two million dollars from them.

But did the members of the Cabal really care that Thorsen had embezzled nearly two million dollars from them? Did they care that the money that was supposed to go to research on the potential causes of autism was stolen? What does that say about the quality of the research? Shouldn't Thorsen's research at least be considered suspect until we have a better picture of what has taken place? These questions may sound hypothetical, and they are, but in the absence of an attempt to extradite Thorsen and place him on trial, we do not have any other choice. Despite the efforts of brave congressmen like Dan Burton, Dave Weldon, and now, William Posey, nothing has happened. There was an effort to get Congressman Jason Chaffetz, the head of the House Government Reform and Oversight Committee, to subpoena CDC whistleblower, Dr. William Thompson, but no hearings have been called as of this writing in 2020.

One might ask the old philosophical question of "if a tree falls in a forest and nobody hears it, is there a sound?" in a different context. If a member of Congress says there is a crisis and sounds a call to arms, but the media refuses to cover it, is there any emergency?

CHAPTER SEVEN

The Legal View

Senior House Reform Committee staffer Beth Clay suggested that if I wanted to better understand the "Vaccine Court," I should talk to a former longtime special master (the equivalent of a judge in the traditional legal system) who might be willing to speak with me for this book. I called him up, explained what I was doing, and he agreed, but with the provision I not use his name. I agreed to this condition and will refer to him as Special Master X. I spoke with him in two separate interviews, totaling a little under two hours. He was warm and friendly and helpfully pointed me to several sections of federal law and legislative history, as well as Stanford Law Professor Nora Freeman Engstrom, who had written a long article about the Vaccine Injury Compensation Program.[1]

I spoke with her at length and include that interview later in this chapter, as well as sections from her law review article. Special Master X was able to give me a great deal of background on the Vaccine Court, as he had worked in it for more than two decades. While Special Master X was willing to give me a great deal of information on what he saw as the strengths and weaknesses of the system, there were several times when I asked the ultimate question as to whether the Vaccine Court was adequately protecting the public from vaccine injuries, as well as whether the awards took into full account the number of injuries suffered by the public.

He declined to answer those questions, suggesting those were assessments that should be made by others. Special Master X struck me as an honorable man but reminded me of the generals who fought the Vietnam War and believed their duty began and ended with giving the president their advice. It was against their personal code of honor to share the private opinions they had provided to the commander-in-chief. I will leave it to the readers of this book to determine whether that fits the definition of a "public servant." And for those who believe I am doing the wrong thing in agreeing to conceal his identity, I can only reply that he gave me long and candid answers to my many questions and pointed me to other sources that I believe reveal his own opinions about the Vaccine Court.

***

I began by asking Special Master X how a nice guy like him ended up in Vaccine Court. The answer was simple from his perspective. He was good with money and had an interest in administration.[2] Those two traits were in short supply. He'd been hired by Chief Judge Alex Kazinsky to help out with tax and money cases. Kazinsky is now the chief judge on the Ninth Circuit. The National Childhood Vaccine Injury Act was passed in 1986, and the new court got funding on September 30, 1988. Special Master X started setting up the office, hiring staff, renting furniture, and finding office space. In January 1989, they began to hear their first cases.

As the Vaccine Court started to function, Special Master X was exceptionally busy, having meetings with Congressional staff, lobbyists for the petitioners, industry, the American Academy of Pediatrics, and the Vaccine Advisory Commission. One of the surprising things he learned in those early days was how the National Childhood Vaccine Injury Act was first proposed, and who were its most enthusiastic backers:

> They developed the program for the DPT vaccine. And the story I heard is it all began with a pediatrician who gave a DPT to a friend's son and the son suffered a reaction. And the friend did not want to sue the doctor because they were friends. They basically settled with whatever insurance coverage the doctor had. And the doctor thought that was grossly unfair. And that was the genesis of the idea. A lot of people are under the impression that the manufacturers were the big pushers of the legislation. But the biggest backer of the program was the American Academy of Pediatrics. Through the years the manufacturers have been very careful politically, in interacting with me and the court.[3]

If Special Master X's account is to be believed, it was the American Academy of Pediatrics that was pushing the creation of the Vaccine Court. If this is accurate, it is certainly an unexpected foray by the medical community into lawmaking and the justice system.

One of the claims I have heard made by many individuals is that the National Childhood Vaccine Injury Act of 1986 came about because of frivolous claims brought by parents. So I asked Special Master X about this question. He was adamant that there was no concern about frivolous parent complaints, and that in actuality the program was designed to speed compensation to the families of those who suffered vaccine injury: "The original act created a table of events. If you basically met the table of events, you won. There was a presumption that the vaccine caused the injury. The vaccine was not even a consideration in determining whether you won or lost. If you received the vaccine and you suffered a particular injury, as defined in the table within a certain time frame, you won."[4] This "table of events" remained in place from 1989 until about 1995, when it came under criticism from the medical community as being too loose and was tightened up. Special Master X believes a lot of the problems that have arisen in the Vaccine Court have been a result of this "tightening up," which he believes fundamentally changed the Act.

While the Act had been designed to speed recovery to the families of vaccine-injured children, it also had benefits for the manufacturer:

> One would say the Act supports the manufacturers as well because they wanted to cut off litigation, which the Hill wanted as well. The Act was designed to cut down financial pressure on the manufacturers, to create an environment for the research and development of new vaccines. To do that, they created this table that was over-inclusive. And one would argue that the manufacturers were very supportive of that, because it has done exactly what it was designed to do, and that is to cut off litigation against manufacturers.[5]

Special Master X continued:

> It makes it extremely hard to pursue one of these cases against a manufacturer. The Act itself came under outside pressure from the medical community because in their mind it did not reflect science. Which always struck me as odd because it wasn't designed to reflect science. It roughly reflected science, but it was done in a way that gave the benefit to the petitioner. And then furthered the policy. Let's bring them in here, so they don't sue the manufacturer. Pay them off so at the back end they have that election as to whether to accept the Vaccine Act judgment, or pursue a judgment. If they

> get a positive judgment from the Vaccine Act, they're not going to pursue litigation against the manufacturer.
>
> It's a win-win for everybody.[6]

Perhaps the most startling revelation from Special Master X was the disposition of autism cases from the very earliest days of the program. I told him I had heard that many children with autism had been compensated in the Vaccine Court for their injuries, but under a theory of encephalopathy (brain swelling), which preceded the development of their autism. It explained why children with autism often engaged in self-injurious behavior such as head banging, because their brains were swollen inside their heads, causing significant pain that they tried to alleviate by hitting themselves.

Special Master X understood my question but put a little different spin on the answer:

> Before the autism cases were filed, we had compensated cases in which a child suffered an encephalopathy, and then went on to exhibit symptoms of autism, or had autism. That was a table case. If you got the covered vaccine, got the encephalopathy within three days of the vaccine, you were compensated. Now where the autism came in was not on the causation side of the equation. Autism came in on the damages side, in determining what the lifetime needs of the child were. So long before the autism cases were filed, autism cases were compensated, but they were compensated as a byproduct of the encephalopathy.[7]

He said that this understanding was even written up in a document known as General Order #1, of which Chief Special Master Gary Golkewicz kept a copy in his pocket so he could refer to it if any questions arose.

As you can imagine, I was somewhat stunned by this revelation and made sure I understood exactly what Special Master X was saying.

He was adamant in saying that it was a mistake to say that the Vaccine Court had compensated autism cases. The formulation he preferred was that the Vaccine Court had compensated encephalopathy cases as a result of a vaccine injury, and that damages from that encephalopathy included autism.[8] I will let the reader determine whether this is substantially different from saying vaccines cause autism, at least in some children.

I attempted several times to get Special Master X's opinion as to how often doctors correctly identified an encephalopathy in a child as a result of a vaccine injury, but he did not give me a clear answer.

In his opinion, the difficulty was going back in time and seeing if records existed that hinted at an encephalopathy:

> What I'm looking for is the diagnosis of what this individual has. I was told early on that the record you should go to is the first neurology visit. The parents are going to give a history, in the last three or five days my child started doing this, and you may not even see a vaccination as part of that history. But you can go back and find when the vaccination took place. The parents can say these are the events that took place, and then in the differential by the doctor, there might be encephalopathy. So you piece it together. You very, very seldom find one record that is a smoking gun.[9]

Even if it is clear that something had gone wrong in a child after a vaccination, there might be issues at play other than encephalopathy:

> In the differential they may say there could be an encephalopathy or some form of brain damage. Or they'll have in there, check these viruses. The vast majority of the defenses in these cases are viruses. And then, Jesus Christ, you wonder how you even have a chance to be healthy when they start naming off all these viruses. Then you get into, well, there's no evidence of viruses. But fifty percent of these viruses are idiopathic.[10]

It was after this discussion of the difficulty in determining whether there was an encephalopathy or a viral infection (viral infections can also cause an encephalopathy) that Special Master X revealed some of his deepest frustrations with the program:

> The real problem with the Act is burden. The first problem is the causation standard because it's not defined. And second, it's burden. It puts the burden of proof on the petitioner, that's fifty percent and a penny, as they always used to say. And you will find decisions which say that the evidence is in equipoise. We don't know. Well, you lose. And that's a real kick in the ass for the petitioners. And I agree with them. The starting point is the policy. The policy was to give the manufacturers coverage in return for being over-inclusive and paying off the petitioners. But the Act was not written that way. Over the years people, including those on the Hill, said this was an administrative body. Well, they put it in a court and fed it through the court appellate

> system. They complained to me in the early days, why are you having court reporters? Because I have to have a record, I'd tell them. Why are you writing these long decisions? Because it's going up on appeal. Those judges handle all government cases and they're used to heavy-hitting lawyers.
>
> They write fifty, seventy, a hundred page opinions and the Federal Circuit is the Patent Review Court. If I don't do that, they send it back and say, why did you award or dent it? They wanted a down and dirty administrative process and then they put it in a court. There's a basic conflict in that. And people recognize that. But they're at a point where they can't change it because of the politics.[11]

Special Master X went on to compare the Vaccine Court to other federal programs: "You could actually write that it's fallen into the same black hole that we argue about with other government programs, like the milk subsidy, or the sugar subsidy. Once they get on the books, they stay on the books."[12]

Special Master X talked about how the Act might have been better and how simple changes might have allowed the Vaccine Court to continue without its current level of controversy:

> We're operating today with an Act designed to handle the DTP shortage, which we no longer even give. The original Act had I believe 6 vaccines, but now I think we're up to 18 or 19. So that original language is being applied to the HPV [Human Papilloma Virus] vaccine and also the flu vaccine, which is the number one source for work in the program. And it's not effective. And that's what's causing all the frustration from the parent's side, which is absolutely correct. The argument that the program takes too long, it's too litigious, and it's not quick justice. There's a tie-up in the courts right now. They're inundated with work. They expected a hundred and fifty cases a year and last year they hit over nine hundred. And this year they're on track to go over a thousand cases.[13]

I thought this would be a good point to ask Special Master X about the increase in the vaccine schedule. By his own admission the program had started with six vaccines, but was now up to eighteen or nineteen, with several of those requiring an initial shot and boosters. Perhaps I was unclear, but what I was attempting to convey to him is that if the usage of a certain product requires its removal from the traditional civil justice system because it is causing such damage that the entire system may collapse, using three times the amount of that product was probably at least three times as dangerous. He replied that he did not quite understand my question, and I sought to make myself clear.

In response, he seemed to somewhat get the point and acknowledged that whenever large sums of money were involved, there was bound to be a certain amount of shadiness, and that was simply human nature.[14] I must note, though, that he seemed to imply much of the shadiness inherent in the Vaccine Court came from the petitioner's side and their lawyers. He did not appear to acknowledge that a compensation program that removed manufacturers from liability might result in unintended mischief from the pharmaceutical companies that produced the vaccines.

In the opinion of Special Master X, the problems inherent in the Vaccine Program began in 1995 when the table of injuries was tightened up, which meant many of the cases would have to rely on proving causation in fact, but without the same rules of discovery commonly found in the traditional court system:

> I told people in 1995 when they were tightening down this table what the impact was going to be. And it did at that time, but it got exacerbated by the addition of all these new vaccines with no table of events, including the flu vaccine. The number one vaccine and injury right now is the flu vaccine with Guillain-Barre syndrome following the vaccine. It's the number one vaccine causing problems and it's got no table. So the number one injury that's being compensated in the program, has no table.[15]

Special Master X continued:

> Every problem that comes up or is talked about in the vaccine program, I can trace back to the table versus causation in fact. If you went in and could add a table of events for each vaccine, a generous table, as they talked about in the legislative history, you would not have the issues you are talking about now. All the issues that you're talking about, flow from causation in fact.[16]

Special Master X detailed how the tightening up of the table on the side of the petitioners, without any corresponding changes for the manufacturers, unbalanced the system:

> I can understand that what is so frustrating to the petitioner's side is that you continue to add new vaccines, so the manufacturers and administrators [physicians] get their protection from litigation, but you don't do anything

> on the petitioners' side of the equation. So the petitioner gets left out. And they've been left out ever since that table went away as well as the generous compensation as envisioned under the law. You keep adding to one side of the equation, the protection from litigation. But the petitioners are getting neglected on their side of the equation.[17]

An interview is always a delicate thing, dependent on the good will of both parties. But I found Special Master X to be such a friendly, open-minded person, willing to answer most of my questions, that I decided to bring up an issue that might derail the nearly hour and a half of productive conversation we'd had by that point. I prefaced my question by telling him it was clear to me the special masters had tried to do their very best. However, I was curious about how the allegations of government misconduct—including the cover-up of Generation Zero data from the Simpsonwood Conference on thimerosal in 2001; the claims of misconduct from Dr. William Thompson regarding the hiding of data in the MMR/Autism study from 2001 to 2004; and the 2009 embezzlement charges against Dr. Poul Thorsen, who had produced a large amount of the CDC's research that claimed to show no link between thimerosal and autism—affected the decisions of the Vaccine Court.

I held my breath as he started his answer.

I was surprised to have him tell me that he had been part of a group that met with Dr. Julie Gerberding, head of the CDC, and had told her that CDC needed to grant independent researchers like Dr. Geier access to the Vaccine Safety Datalink. Special Master X made it clear that he did not condone the hiding of information. He paused for a moment, as if considering a question he had known was coming. But to his credit, he did not shy away from what I'd asked:

> I would say this in answer to your question. The special masters, like any judge, rely upon information. If that information is not correct, it would obviously impact the information the judge is considering, and could potentially impact his decision. In that respect, the special masters are no different than any other judge. Your decision is only as good as the information you get.[18]

I exhaled upon his answer, and we continued with our interview.

I was intrigued that, in a long discussion of the pluses and minuses of the program, he talked about a parents' group that was appearing before the House Operations Committee and arguing that the Vaccine Act should be repealed in its entirety and vaccines should return to the traditional tort system. Special Master X believes it would be a mistake to repeal the 1986 National Childhood Vaccine Injury Act and let those cases return to the original civil justice system. As evidence he cites a longtime petitioner's lawyer who "has said publicly many times that the Act has its problems, but people have received a couple billion dollars in compensation that they would not have otherwise received."

The actual number paid out to vaccine-injured individuals as listed on the Health and Human Services website is $4.1 billion dollars,[19] a number that would probably come as a shock to most Americans.

But is that the real number?

In 1993, FDA Commissioner David Kessler suggested that only about 1 percent of serious adverse events to prescription drugs was reported.[20] If we assume a similar profile for vaccines, then the more accurate number for damages from vaccine injury would be around $330 billion dollars. But what if the age at which vaccines are given means that we are given another magnitude of difficulty in determining an adverse reaction? Kessler thought only about 1 percent of serious adverse drug reactions was reported for adults who received a medication that caused harm. Let's just assume that the added difficulty of determining an adverse reaction in an infant is 10 percent of that which we would observe in an adult. That means the damage from vaccines might be somewhere around $3.3 trillion dollars. I tried to get Special Master X to comment on this line of inquiry, but he deferred to it as being outside his scope of expertise. It was curious to me that a government employee who had spent more than twenty years of his professional life handling vaccine injury claims could not even give an estimate of the amount of damages vaccines had caused the country during that time. If one did not know this information, or could not even come up with a ballpark figure, how was it possible to conduct even the simplest risk vs. benefit analysis?

If the underreporting of vaccine injury mirrors what we observe with pharmaceutical drugs, the amount of damages is at least $330 billion dollars. If one considers the additional hurdles in determining whether a vaccine injury has been suffered in an infant, this probably suggests that the actual damages are far in excess of that number.

***

It was a pleasure to interview Stanford Law Professor Nora Freeman

Engstrom about the Vaccine Injury Compensation Program (VICP), also known as the "Vaccine Court." She has published two law review articles dealing with the VICP[21] as well as an op-ed in the *National Law Journal* titled "Heeding Vaccine Court's Failures."[22] Even after ten years as a science teacher, I find such familiarity in speaking with a fellow lawyer. I think some of it may have to do with our common training to fully understand our own point of view, as well as that of the other side. It is a skill I think we are losing as a society. A good lawyer does not dictate, but persuades with the logic of their argument. I remember one of my law professors saying that at the end of a well-presented case the jury should feel they have been educated and can now render a decision that is in accord with their own sense of justice.

In her article "A Dose of Reality for Specialized Courts: Lessons from the VICP," Engstrom sketched out the history of the Vaccine Court, which passed in the final hours of the 99th Congress as the "National Childhood Vaccine Injury Act of 1986." Engstrom wrote:

> Congress established the Vaccine Injury Compensation Program(VICP), a no-fault scheme run out of the U. S. Court of Federal Claims and jointly administered by the Department of Health and Human Services (HHS), which serves as the respondent and therefore represents the Fund's interest in all VICP proceedings) and the Department of Justice (DOJ, which represents HHS). Financed by a seventy-five cent excise tax on each vaccine administered (which creates the Fund upon which injury victims draw), the VICP is intended to provide adequate, though abridged, compensation to all individuals injured by covered vaccines via "less-adversarial, expeditious and informal proceedings[s]."[23]

Although many parents today may feel that the allegation that their child suffered a vaccine injury subjects them to societal ridicule, it is instructive to review the history of the issue prior to the passage of the National Childhood Vaccine Injury Act of 1986.

Engstrom reviewed the case of Anita Reyes, a young girl living near the Mexico border in the 1970s who contracted polio as a result of a dose of a Wyeth Laboratories polio vaccine. The case that resulted, *Reyes v. Wyeth*, heard in the Fifth Circuit, suggested that between victims and vaccine manufacturers, the manufacturers should bear the loss.[24] This was followed by the swine flu fiasco of 1976 in which forty-five million Americans subjected themselves to a flu shot based on the urging of President Gerald Ford, only to find that the flu was not particularly dangerous and in a small number of cases caused Guillain-Barré syndrome, a form of paralysis that was sometimes fatal.[25] Then on April 19, 1982, an Emmy-winning, hour-long television documentary titled *DTP: Vaccine Roulette* was aired on a NBC affiliate, which alleged that the DTP vaccine (diphtheria, tetanus, and pertussis) was causing seizures, mental retardation, and death in children.[26] Footage of dead and seizing children as a result of their vaccinations terrified Americans.

Engstrom wrote of the effects this combination of events had on the vaccine market in the 1980s:

> As the number of lawsuits ticked upward, so did manufacturers' dismay. In 1984, for example, Lederle's President went on record declaring that "[t]he present dollar demand of DTP lawsuits against Lederle is 200 times greater than our total sales of DTP vaccine in 1983." Then, the following year he complained the situation had deteriorated: All but two of the more than ninety DTP cases filed against Lederle—in more than forty years of distributing the vaccine—had been filed since 1982. Another vaccine manufacturer—Connaught Laboratories—faced a similar plight, as suits filed against it in 1985 and 1986 sought a combined billion dollars in damages.[27]

There can be little doubt that in 1985 and 1986 vaccine manufacturers found themselves in a perilous position. But so did American parents who wanted to keep their children healthy and were starting to believe a trip to the pediatrician's office might not be the best way to achieve that goal. The National Childhood Vaccine Injury Act was composed of two parts, the first of which was intended "to upgrade the nation's immunization program by perfecting vaccines and monitoring adverse reactions"[28] as well as to "provide simple justice to vaccine-injured children."[29]

Even though there was broad agreement between both parent groups and manufacturers, the Vaccine Injury Compensation Program soon ran into difficulties. Professor Engstrom has some opinions about the failure of that effort.

***

One of the surprising questions about the Vaccine Court, or any no-fault judicial system, is whether it is consistent with the United States Constitution. The Seventh Amendment, in a few simple and elegant sentences, provides that "in suits at common law, where the value in controversy shall exceed twenty dollars, the right of trial by jury shall be preserved, and no fact tried by a jury, shall be otherwise reexamined in any court of the United States, than according to the rules of the common law." How is it that a specialized court with significantly different rules of discovery and procedure is consistent with the justice system that Americans expect to rely on if they find themselves injured? Professor Engstrom's article looked specifically as to whether the Vaccine Court provided a good model for other, similar health courts but also gave some consideration to potential constitutional concerns:

> Constitutional questions loom large because if health courts are enacted, opponents are sure to challenge these tribunals. Opponents will allege that, in curtailing victims' compensation and denying them the right to a trial by jury, health courts violate victims' rights to due process and equal protection and run afoul of many states' open court, separation of powers, and right-to-jury trial guarantees. Evaluating these constitutional claims, many reviewing courts will presumably ask the same question they've asked and answered on other occasions: In abrogating victims' common law remedy, did the legislature accompany the abrogation with a sufficient tangible benefit? Was there, in other words, an adequate quid pro quo? So far, those defending health courts constitutionality have suggested that a tangible benefit justifying the withdrawal is "the system's promise to deliver faster, more reliable compensation decisions." Whether that "promise" is or is not realistic thus takes on weighty constitutional significance.[30]

While some might take a stricter view that the protections of the Constitution cannot under any circumstances be "curtailed," a different analysis has prevailed in most courts.

Under the "quid pro quo" theory mentioned by Professor Engstrom, the court will look to see if a "tangible" benefit has been given to the victim in return for the abrogation of their constitutional rights. The evidence from the Vaccine Court is not encouraging according to Professor Engstrom.

In her op-ed on the Vaccine Court for the *National Law Journal*, Engstrom laid out the expectations that had existed for the Vaccine Court and the failure to live up to those expectations:

> Despite predictions at enactment that it would "guarantee" equal treatment to similarly situated claimants, a lack of consistency has bedeviled the program.

> Even though Congress established that each petition would take at most 240 days to adjudicate, in reality, the average program adjudication takes more than five years. This is substantially longer than similar claims resolved by court judgment or trial verdict within the traditional tort system.[31]

The discrepancy between what was promised and what has been delivered by the Vaccine Court is nothing less than shocking. From a system that was expected to lead to a resolution within 240 days, one is left with a system with an average wait time of more than 1,825 days. If one of the main justifications for the National Childhood Vaccine Injury Act is that it would provide quicker settlements to desperate families, then it has clearly not delivered on its promise.

And as for the other significant justification for the Vaccine Court, that it would remove the contentiousness from the system, seems to have failed in that as well:

> And although claims within the system are supposed to be amicably resolved, in reality their resolution is frequently antagonistic. In the words of a medical expert who has long participated in the program: "What should be a quiet, civil, deliberative discussion of facts and medicine too frequently degenerates into a contentious, vituperative, decibelescalating exchange."
>
> The bottom line is that the Vaccine Injury Compensation Program was supposed to offer "simple justice" to vaccine-injured children. But it has largely failed to do so.[32]

Neither of the two main justifications for the Vaccine Court—speed of recovery and the dispelling of a confrontational system for parents who are unprepared to fight a pitched legal battle (at the same time they are dealing with the devastation of a formerly healthy child becoming vaccine-injured)—appears to have worked out the way its proponents had hoped.

While the battle lines between parent advocates and supporters of the current Vaccine Court have become sharply divided, it is perhaps surprising how similar people on both sides of this issue view what has taken place. One of the principal parent advocates for more than three decades has been Barbara Loe Fischer, founder of the National Vaccine Information Center, who had been instrumental in the passage of the National Childhood Vaccine Injury Act. When questioned about the Act in 2014, she was asked the question, "If you had to do it over again, would you support the VICP?"

Her answer was "If I knew then everything I know now, I would not support the enactment of the VICP compensation system. . . . It does not provide simple justice for children as we had hoped and been told that it would."[33] That bleak assessment was also shared by a report from the Government Accounting Office, including Chief Special Master Gary Golkewicz, as detailed by Engstrom:

> Indeed, the U.S. Government Accountability Office (GAO) has studied the Program and concluded: "While [the Program] was expected to provide compensation for vaccinerelated injuries quickly and easily, these expectations have often not been met." A leader in the parents' lobby, instrumental in the Act's passage, has concluded that the VICP's administration has constituted "a betrayal of the promise that was made to parents about how the compensation program would be implemented." And the man who served for over two decades as the VICP's chief special master has publicly lamented: "[L] itigating causation cases has proven the antithesis of Congress's desire for the program."[34]

When a program fails in the eyes of the group for whom it was meant to exist, in the eyes of those who administer it, and in the opinion of the Government Accountability Office, it is time to admit that a crisis exists.

* * *

Professor Engstrom's article "Exit, Adversarialism, and the Stubborn Persistence of Tort," published in the *Journal of Tort Law* in 2015, suggests that these problems are not unique to the Vaccine Court but exist in any forum that seeks to solve problems in a "no-fault" manner. Engstrom looked at four no-fault systems:

1. Worker's compensation.
2. No-fault auto insurance.
3. The Vaccine Court.
4. Forums in Florida and Virginia to compensate parents for injuries to their children suffered during delivery.

Engstrom writes first about those who seek to reform tort law by placing caps on recoveries to injured individuals but finds that does not address the underlying factors that led to the injury:

> Unlike those favoring discouragement mechanisms, those favoring no-fault or "replacement regimes" do not necessarily believe there is "too much" litigation. Rather, they believe that the litigation we *do* have, for at least particular kinds of claims, is misdirected, taking too long, costing too much, and compensating too few. As such, reformers seek to shuttle various categories of claims away from the tort system and into (typically) freestanding, newly minted administrative tribunals. There, it is assumed, with the fault obstacle gone, procedures simplified, and damages curtailed (and often paid not on an individualized basis, but pursuant to strict schedules), compensation can be more easily, cheaply, quickly, amicably, consistently, and predictably delivered.[35]

In theory, these aspirations are noble. Let's remove the anger and bureaucracy from the system and get people their money in the quickest manner possible. But does it work out that way in actual practice? The evidence suggests it does not.

Engstrom brings up previous criticisms of no-fault systems, such as the assertion it runs counter to American conceptions of individualized justice, permits guilty parties to "dodge responsibility for their misdeeds," and might run afoul of Seventh and Fourteenth Amendment guarantees, but Engstrom's contribution to legal scholarship posits two additional considerations:

> I call these the problems of *exit* and *adversarialism.* Across substantive areas, that is, nofault mechanisms have become plagued by the problem of exit, as claimants seeking full compensation make end-runs around no-fault, either to evade the regime entirely or to supplement no-fault's comparatively meager benefits with more generous payments, available only within traditional tort. Or, they become bogged down by adversarialism, marked by longer times to decision and increased combativeness, attorney involvement, and utilization of formal adjudicatory procedures. Some regimes, including auto no-fault and the VICP [Vaccine Injury Compensation Program], display just one of these afflictions. Others, like workers' compensation and neurological birth injury funds, display traces of both.[36]

It seems that dispensing with the traditional civil justice system in favor of trying to create something better often creates the opposite effect. If you limit the amount of a person's recovery based on some prearranged formula, many individuals will seek to escape from that system. And it appears that if you seek to take the adversarial nature out of a discussion about whether a product has harmed a person, possibly for the rest of their life, a little bit of contentiousness will inevitably creep back into the proceedings. Perhaps these are unavoidable qualities of human nature.

If you are defending a system, does your humanity allow you to fully understand the damage that the person standing before you has suffered? And on the side of one who suffers, isn't there an understandable desire to tell the entire truth of what they have gone through?

In her article on the Vaccine Court, Professor Engstrom ends with this warning:

> Yet it has been said, "Before the traditional tort system is abandoned . . . there must be substantial grounds to ensure confidence in an alternative institutional mechanism that would serve as its replacement." When it comes to resolving claims for medical injury, health court proponents seek to replace common law courts, in place for centuries, with a new and untested alternative. They have, in large measure, advocated their reform idea based on health courts' ability to offer a few concrete administrative advantages. The VICP [Vaccine Injury Compensation Program] casts significant doubt on health courts' ability to offer those advantages. That experience ought to shake public confidence in this new alternative mechanism—and inform future analysis.[37]

Professor Engstrom's analysis is not about the Vaccine Court in particular, but the broader question of whether these alternative courts function as well as their proponents claim. The evidence uncovered so far suggests that they are not.

***

When I first contacted Professor Engstrom in February 2016 and asked to interview her about her publications, she was reticent. Engstrom said she'd been contacted by various individuals like me, but instead of her legal opinions, they wanted to get her thoughts on vaccines. She said she did not feel informed enough on vaccines to state an opinion. I told her I understood the limits of her expertise and simply wanted to talk about whether the legal mechanism for answering the question of vaccine injury was working in the way it was intended. With the understanding I would not ask the broader question about the safety of vaccines in general, she consented to the interview.

I began by asking how she got interested in looking at the Vaccine

Court. She replied that in looking at the health courts issue she came across several mentions of the Vaccine Court and how well it was functioning:

> At that point my knowledge of the VICP [Vaccine Injury Compensation Program] was a little vague. But I talked to a colleague here and I asked, "Are we understanding the VICP wrong?" Because my understanding was that it hadn't actually been that successful at being non-adversarial or quick. That testing of my intuition is how I got interested in doing the piece. As I dug in, I saw that my intuition was right and things were worse in the VICP than I had initially thought.[38]

I then asked if there had been any surprises in her investigation. She answered:

> The time to adjudication surprised me as being worse than I thought it would be. I'm writing a book now called *Why No-Fault Fails*, or *The Fault with No-Fault* and I've written about how none of this stuff turns out the way we thought it would. I thought things would be bleak. But the time to adjudication, the fact that it's so much more than the tort system, and I don't think people have talked about that sufficiently. And it's not like it's a little more than the tort system. It's a lot more than the tort system. And not just for cases adjudicated to judgment in front of a judge. It's for all claims, even those resolved by settlement. The fact that folks are waiting five years is not something I expected to see. And I know some people can say, well, it's for this reason or that reason. And I'm sensitive to how this program has been hit by unforeseen or unforeseeable events. But still, even giving the benefit of the doubt, it's taking a very, very long time to get hurt kids their money.[39]

While Professor Engstrom was deeply concerned about the length of time it was taking to process cases, she was less concerned about the potential constitutional challenges to the Vaccine Court. The quid pro quo test she'd mentioned in her writings, although the subject of some commentary, would in all likelihood be satisfied by the congressional creation of the court. Engstrom felt this analysis would still remain, regardless of how the burden of proving causation had shifted so dramatically to the petitioners.

I asked about the mention in her article of fourteen thousand petitions filed in the program since its creation and whether that accurately reflected the number of vaccine injuries in our country since the court began functioning in 1989. "I have no idea," she replied. "It's an interesting question, but I have no idea."[40]

My next question revolved around the claim she'd made in her article that alternative courts work better in situations like automobile accidents, where the facts are often fairly clear. However, in a developing field, like the creation of new vaccines, I asked what effect such broad immunity might have on product safety. She gave a long answer, which I think is probably similar to how most legal scholars would answer:

> One answer is the VICP came along with extra scrutiny to insure the safety of the vaccine supply. On the face of it, that is very salutary, although the devil is in the details. How well is that side of the system working? I'm not an expert on that. Like all things, the devil is in the details. In general, do I believe tort liability provides a separate and important check on manufacturer safety? Yes, I believe that in general. I am against broad pre-emption, because I think in general the common law allowing individuals to bring suits provides a beneficial layer of protection in a country where we don't have strong bureaucracies and heavy regulation. In general, that was taken away here, but we got this extra thing instead, this FDA approval and oversight. How good is that really on the ground? One answer is to say the DTP vaccine sure got a lot safer even though this system's in place. So that's interesting. It would suggest that manufacturers still have some incentive in making stuff safer. But is it an adequate system? I don't know that. I don't know one way or the other."[41]

When I asked whether vaccine safety would have been better handled under traditional tort law, Engstrom was conflicted by the suggestion:

> Vaccine injuries are super-hard. Let's not forget how hard it is to get compensation for your vaccine-injured kid in court. You've got a whole lot of things working against you. You've got the fact that vaccines tend not to cause signature diseases. And so the causation questions are always going to be hard. You've got to show deviation from custom, and that gives doctors tons of latitude. You have to show the product is defective to argue against the manufacturer. You have to identify the manufacturer. You have to do all of this in the statute of limitations. It's really, really hard. And you have to find a lawyer who's willing to do this on a contingency fee basis when fewer lawyers are willing to do it.[42]

I reviewed my discussion with the former Special Master X who had told me of his concern that since the burden of proof had shifted so dramatically on the shoulders of the petitioners, without changing the rules of discovery against manufacturers, proving a case had become so much harder for the parents of vaccine-injured children. I asked Professor Engstrom what she thought of that observation. She said:

> I am totally sympathetic to that view. It sounds right to me. As an academic, I think there is something so interesting about the VICP. Some things are designed to fail. Auto no-fault in most states was designed to fail. It was a compromise and the trial lawyers' associations were able to embed poison pills into the legislation. So everybody knew it wasn't going to work out well, and it didn't surprise anybody. And you can say auto no-fault wasn't successful and that doesn't say anything bad about the no-fault idea, because of the way it had these poison pills in it. The VICP is so interesting because it was designed to succeed. On the face of it, it should have succeeded.[43]

***

In October 2010, I was convinced that the US Supreme Court would begin to limit the damage caused to our nation by the National Childhood Vaccine Injury Act of 1986.

The case was *Bruesewitz v. Wyeth, LLC* and concerned vaccine injuries suffered by Hannah Bruesewitz at the age of six months, when she developed seizures and developmental delays as a result of a DPT vaccination. The injuries had been compensable under the previous table of vaccine injuries, but not at the time her action was filed. In addition, the vaccine itself had been taken off the market. The section at issue, 42 U.S.C. section 300aa-22(b)(1)), read, "No vaccine manufacturer shall be liable for damages arising from a vaccine-related injury or death associated with the administration of a vaccine after October 1, 1988, if the injury or death resulted from side effects that were *unavoidable* [italics mine] even though the vaccine was properly prepared and was accompanied by proper directions and warnings."

I was thrilled when I read about the case because it seemed to finally offer the kind of public debate that was so critical to this issue. After all, couldn't just about everything, even dynamite, be made safer? Could we start to open the closed door of the pharmaceutical companies as we do with just about every other consumer product? Others were not so hopeful.

The reality was even worse than I had imagined.

On February 22, 2011, in a 6–2 decision, the Supreme Court ruled that not only did the National Childhood Vaccine Injury Act of 1986 prevent any claims against manufacturers of childhood vaccines, but it also closed off any claims against the manufacturers of adult vaccines.

Yes, that's right.

Under this decision from the Supreme Court, the National Childhood Vaccine Injury Act of 1986 now prevents every American from ever suing a vaccine manufacturer for a negligently designed vaccine. Apparently, I wasn't the only one stunned by the Supreme Court's decision. Justice Sonia Sotomayor wrote a blistering twenty-eight-page dissent on the case, joined by Justice Ruth Bader Ginsburg. Sotomayor wrote:

> Vaccine manufacturers have long been subject to a legal duty, rooted in basic principles of products liability law, to improve the designs of their vaccines in light of advances in science and technology. Until today, that duty was enforceable through traditional statelaw tort action for defective design. In holding that section 22(b)(1) of the National Childhood Vaccine Injury Act of 1986 (Vaccine Act or Act), 42 U.S.C. section 300aa22(b)(1), pre-empts all design defect claims for injuries stemming from vaccines covered under the Act, the Court imposes its own bare policy preference over the considered judgment of Congress. In doing so, the Court excises 13 words from the statutory text, misconstrues the Act's legislative history, and disturbs the careful balance Congress struck between compensating vaccine-injured children and stabilizing the childhood vaccine market. Its decision leaves a regulatory vacuum in which no one ensures that vaccine manufacturers adequately take account of scientific and technological advancements when designing or distributing their products. Because nothing in the text, structure, or legislative history of the Vaccine Act remotely suggests that Congress intended such a result, I respectfully dissent.[44]

Sotomayor's dissent could not have been any clearer. In her opinion, there was nothing in the text of the act, its structure, or legislative history of the National Childhood Vaccine Injury Act of 1986 that suggested it meant to ban claims of negligently designed or tested vaccines that harmed either children or adults.

The court just made up a new law. Now ALL vaccines were covered by the Vaccine Court, although there had not been a single congressional hearing or vote on the issue.

Parents of vaccine-injured children were stunned, but not surprised.

Vaccines were big business. The pharmaceutical companies had more than twenty years of immunity from claims involving devastating childhood vaccine injuries. Now they had complete immunity for the adult market, as well.

It looked like good times ahead.

The only possible threat to this legal and scientific juggernaut was insiders who turned against their former colleagues, and instead of bringing their claims in the legal courts of the land, which were now closed to them, took their grievances to the court of public opinion.

Perhaps never in the history of the modern world had so few stood against an army so powerful.

CHAPTER EIGHT

Those Who Would Oppose Goliath

When you are young, attractive, witty, and have a zest for life, it's easy to imagine these traits will propel you into a future of financial success and personal happiness. But perhaps you are being prepared for greater, more difficult tasks. Since graduating from the University of California, Santa Barbara, in 1999, Brandy Vaughn's life has undergone dramatic changes, taking her from working for one of the giants of the pharmaceutical industry, Merck, to leading the charge against their practices. Like many before her, it is the story of an insider who became an outsider.

At UC Santa Barbara, Brandy took three years of biology and biochemistry, planning to go to medical school, but changed her mind after working in a doctor's office.[1] She ended up graduating with a political science degree, thinking she might one day go to law school. After college, she worked for a few years in media planning and sales for a promotional item company, when a friend suggested she might enjoy pharmaceutical sales.

Brandy interviewed with several companies, including Pfizer and Aventis, before deciding to accept a position with Merck. While one part of her was not thrilled to be working for a pharmaceutical company, she considered herself "ethically-minded" and researched which nonprofit projects each company was supporting. Merck had a program to combat river blindness in Africa, and she was impressed with the effort. In their marketing materials, it seemed Merck was always portraying itself as one of the "good guys of pharma." Even when she was working for the company and "on the inside," that feeling was still present.[2]

When she signed the contract with Merck, Brandy knew she would need to go for two months of training on the product before being allowed to go out and sell. Smaller pharmaceutical companies normally conduct their two- or three-month training for their sales representatives at the home office, but large companies like Merck usually rent out a suite of rooms at a fancy hotel in different regions of the country. For Brandy's group of hires, Merck rented out a suite of rooms at the Hyatt in Irvine, California, for the approximately sixty newly hired sales representatives from California, Oregon, Nevada, Washington, Nevada, and Hawaii who arrived for the two-month training. Most of the hires were fresh out of college, or perhaps had a few years of work experience like Brandy, but it was a heady experience to find yourself as a new recruit for one of the world's most successful pharmaceutical companies. Each representative got a nice room at the Hyatt (the corporate hotel for Merck) and was given fifty dollars a day for food and incidentals. The company organized trips on the weekends for the new sales reps in an effort to build an esprit de'corps. They were being groomed for a fabulous life.

As Brandy recalled:

> They usually take people right out of college, or a few years out of college, so they can be influenced, right? They've gone through the public system. They've gone through college. They've been taught not to question. They've been fed lines and they come out fresh, thinking this is how it works. They usually take people with a scientific background, so that we have some kind of credibility, and we can talk about the science of it.
>
> And also because there's an indoctrination in this country about science. It's like a religion. You can't question science. Peer-reviewed studies or peer-reviewed journals, no matter how bought they are at this point, and no matter how many editors come out and say, "Hey, research is now totally biased," are sacred. There's still this sense that you don't question science. Science is what it is. It's fact. It's not theory.[3]

Brandy continued her explanation of how pharmaceutical companies make themselves attractive to potential sales representatives:

> When they can get you young and you don't question, you just really love the money they're throwing at you and the lifestyle you have compared to your other friends just out of college. And the wining and dining makes you feel powerful and they give you a flexible schedule so you kind of run your own sales business within the territory. You only meet with your manager once a

month. All of these things are very, very attractive to those who are just out of college.[4]

At first, the training resembled an intensive science class, as you studied various diseases and the drugs the company had developed to combat them. You made yourself familiar with the inserts, and you studied the drugs sold by your competitors, trying to fully grasp their strengths and weaknesses. At this stage, it was easy to consider yourself as not a salesman, but an educator for the overworked doctors who didn't have time to do their own research. Brandy was put into group one, and the drugs she would sell were Vioxx, Merck's blockbuster pain relieving drug which was especially good for those with arthritis; Singulair, an asthma medication; and Zocor, which lowered cholesterol levels.

Brandy recalls everybody at Merck being very excited over Vioxx:

> It was used by everyone for a lot of things. The reason it got us in the door is that it was so powerful that doctors were using it themselves. We were constantly running out of samples for Vioxx. Everybody wanted Vioxx samples. The nurses in the offices were taking it. The doctors were taking it for tennis elbow, for any kind of pain, for hangovers or menstrual cramps. It numbs your body. It was so powerful that it was also an extremely dangerous drug. It worked. That's why it was so popular. And they [Merck] put a huge promotion behind it. All of the sales reps were promoting it. Doctors were getting talked to about it three to four times a week. And that was because Celebrex [another competing pain reliever marketed to those with arthritis] had a huge market share because a lot of people on Medicare with arthritis were taking it. It was a huge money maker and had a near-monopoly on the market. So when Vioxx was launched, Merck went at it a hundred percent. All the sales reps, all the doctors, Vioxx was in their face constantly. That's the way it became a blockbuster. I think they were doing a hundred million dollars a year in advertising. I think at the time it was the most marketed drug ever.[5]

Merck would even use skating legend Dorothy Hamill and former Olympian Caitlyn Jenner in their television commercials.

Once you were secure in your knowledge of the product and the diseases, it was time for you to be polished. It was one thing to educate a physician; it was quite another to close a deal. And the company only made money when product got moved: "You learn how to speak subtly and plant seeds of doubt about competitive drugs. You learn how to deflect questions you don't want to answer, and you learn messaging."[6] The strategy even goes into areas that most people would consider highly intrusive, if not downright manipulative:

> They teach you how to deal with different personality types. How to determine what doctors fall into which categories. What speaks to them most. You had trainings on all of that. You do your own personality testing and then they classify you. They put you on certain doctors that they determine are a specific class. Then you are kind of their detail. The psychology and strategy behind it is so sophisticated. It's quite amazing. But when you make that kind of money, more than many small countries, you can do those types of things.[7]

Brandy was a good salesperson. She won sales awards for Vioxx and Zocor. As a sales representative, she was given a food budget and she'd buy muffins and coffee and deliver them to a doctor's office to see if she could get a meeting. Usually it worked. "I knew my stuff," she said. "It wasn't like I was just a machine. I was funny. I traveled all over the world. I had diving trips to Hawaii so I could talk to people about that. I guess I'm not bad to look at, either. Some of the doctors had crushes on me. I'm clever. I could talk the talk."[8] And how might she respond if one of those doctors brought up a competitive product? She replied:

> You never speak directly against them. You always say something like, Pfizer, they're a great company and they have some really nice people working for them. But I've heard from my Pfizer friends that they are struggling with Celebrex. They are really pissed off that Vioxx is taking all of their market share. And they're attacking us because they're in a corner because Vioxx is clearly a better drug.[9]

It was a nice life. Brandy got a car, a phone, Internet set up at her house, and a food budget that she was supposed to take into the offices, but there was usually some tasty stuff left over. And with all the prestige that accompanies working for a successful company and interacting daily with doctors and medical people, it was easy to overlook some of the more troubling aspects of the job.

Brandy said:

> They gave you a cell phone, but you knew they were monitoring it. And what about the car? That was tagged with a GPS as well, so they knew where you

> were at all times. And the home computer with the fastest Internet possible? They could monitor every page you looked at. It was easy to justify. They claimed they just wanted to know you were doing your job. They didn't want you going out to the competition and giving away secret marketing plans or studies.[10]

Surely you could understand that? Merck was a multibillion-dollar company. Of course they had the most advanced surveillance program money could buy. It was probably better than the National Security Agency or Homeland Security. No pesky congressman or senator ever looking over your shoulder and asking what you're doing. This wasn't just capitalism. It was the jungle where might and money made everything right.

Brandy had been working for Merck for over a year when she first had suspicions that something might be wrong with Vioxx:

> There was a study that we were already talking about that was showing preliminary results. And all of a sudden it was stopped. And promotional materials that talked about the study were replaced with new materials that didn't. And there were rumors. You heard rumors. Some of the senior reps in my territory had friends at corporate who used to be reps. Pfizer reps, too, because the reps all talked to each other. Rumors that the FDA was investigating Vioxx, and rumors that the study was stopped because it was showing an increased risk of death. You'd be at a regional sales meeting and you'd ask what happened to that study and they'd say something like, "Well, let's forget about that study and focus on this." It was "don't ask questions." It was a very shady kind of thing. Pfizer is saying Vioxx kills people and causes heart attacks. And you're being told, "Oh, despite what I said earlier [about the study], this is what you say when a doctor tells you Pfizer is saying Vioxx kills people." You say, "Oh, they're upset because have you seen the latest numbers? Vioxx works so well that we've taken away so much market share from them. And Pfizer is struggling to come up with rumors that are totally unfounded, because they're losing money." You just twist it around. It became a shady scenario.[11]

Brandy was fortunate in that around that time she was taken off Vioxx and placed on the launch of Merck's new drug, Zetia, a cholesterol-lowering drug. Also around that time she had a boyfriend who was living in Portland, Oregon, and she had become disillusioned by what she was seeing in the pharmaceutical industry. The boyfriend gave her an excuse to move to Oregon, and although she was supposed to get in contact with the Merck manager in Portland, she never did. She remained on the health plan as something of an employee on hiatus, but then an opportunity came to move to Europe, and she left the United States in early 2004.

How many people did Vioxx kill when it was on the market from 1999 until September 2004? It depends on which set of numbers you believe. The most conservative estimate would probably come from Merck itself, which in 2010 created a $4.85 billion dollar settlement fund to make payments on 3,468 deaths, as well as 20,591 heart attacks.[12] As a point of comparison, the greatest terrorist attack on US soil, the attacks of September 11, 2011, by Al Qaeda, killed 2,977 people.[13] By their own admission, Merck killed more people than the terrorists sent by Osama Bin Laden.

An investigator for the Food and Drug Administration came up with a different number. He estimated that Vioxx had "probably been responsible for at least 55,000 American deaths during the five years it had been on the market."[14] Most estimates for the Vietnam War list somewhere around 58,000 US soldier deaths. Our country tore itself apart over Vietnam. If this was the actual death toll for Vioxx, where were the protests and demonstrations? Why didn't the television networks cover the funerals of the victims and spend their precious broadcast time bringing us the grieving families left behind?

One commentator, Ron Unz, has suggested that the true death toll from Vioxx was far higher, somewhere in the range of 500,000 deaths. Could it be that this FDA-approved drug actually killed a half million people, a number comparable to the soldiers killed in our own Civil War? Unz wrote:

> A cursory examination of the most recent 15 years worth of national mortality data provided on the Centers for Disease Control and Prevention website offers some intriguing clues to this mystery. We find the largest rise in American mortality rates occurred in 1999, the year Vioxx was introduced, while the largest drop occurred in 2004, the year it was withdrawn. Vioxx was almost entirely marketed to the elderly, and these substantial changes in national death-rate were completely concentrated in the 65plus population. The FDA studies had proven that use of Vioxx led to deaths from cardiovascular diseases such as heart attacks and strokes, and these were exactly the factors driving the changes in national mortality rates.
>
> The impact of these shifts was not small. After a decade of remaining roughly constant, the overall American death rate began a substantial decline in 2004, soon falling by approximately 5 percent, despite the continued aging

> of the population. This drop corresponds to roughly 100,000 fewer deaths per year.[15]

In his conclusion, Unz admits that proving cause and effect between the recall of Vioxx and the staggering drop in US deaths would be difficult but laments that nobody in the political, media, or medical class seems to have the slightest interest in investigating the question. It is as if the famous quote attributed to the Soviet dictator, Josef Stalin, that "the death of one man is a tragedy, but the death of millions [or in this case a half million], is a statistic," has come to pass in modern-day America.

Probably the most damning example of Merck's actions in the Vioxx matter came from evidence revealed in a 2009 class action suit in Australia. One email circulated among Merck employees contained not only the names of doctors who had complained about Vioxx, with each doctor tagged with the label "neutralize," "neutralized," or "discredit," but also the suggestion, "We may need to seek them out and destroy them where they live . . ."[16] CBS News reported that other shenanigans included having a doctor sign his name to an entirely ghost-written article and creating a fake "peer-reviewed" journal to publicize pro-Vioxx articles (the *Australasian Journal of Bone and Joint Medicine*). The company also created a Ricky Martin-style song to get Merck sales representatives excited about selling Vioxx. Merck hired an aggressive team of public relations consultants to monitor the Australian trial and even followed reporters into the bathroom to make sure they got the story "right." There was also a plan (never executed) to seed seminars with speakers who were "sympathetic to Vioxx but under instructions not to mention the brand names too often."[17]

Brandy Vaughn was done with Merck and Vioxx in late 2003, and the recall of Vioxx would not take place until September 2004, a time period of more than ten months. Like many who have worked in pharma and later learned the products they were selling caused harm, Brandy could have genuinely considered herself blameless of any moral culpability. She had believed what she was doing. Others had lied. It was their responsibility and their crimes, although not a single executive would ever spend a single day in jail for causing the death of somewhere between 3,468 and 500,000 Americans. She could have washed her hands of the entire messy affair and considered it all over. But the soul of Brandy Vaughn would not rest when she saw the pharmaceutical industry making plans for a catastrophe that was potentially far worse than the Vioxx debacle.

***

When Brandy left for Europe, it was an opportunity to make a break from the American pharmaceutical industry and healthcare, as well as to observe how other areas of the world viewed the practice of medicine. Prior to leaving, Brandy had taken several writing and editing classes and was anxious to merge her interests in science and political action. She started working as a volunteer for Friends of the Earth in Europe and had her eyes opened to the issues of genetically modified organisms (GMOs), overfishing, chemicals in body care products, and issues in the agriculture and meat industry. She also started working for the Global Reporting Initiative, a group that certified corporate sustainability reports, and saw that some companies were more concerned about their impact on the planet than others. Brandy was a freelance consultant, so she also worked with a number of corporate clients, such as ING, Nike, and Phillips Health Care. She said:

> I was there for seven years and it really opened my eyes. I became really detached from the US mentality, especially regarding healthcare. In [the Netherlands] where I was living, people do things very, very differently. It's very natural. They don't like medicine. In Europe, medicine is used when you're sick. You don't use medicine when you're not sick. They don't have pharma reps. Pharma is not on TV. They don't have lobbyists. There are no pharma reps that visit doctors. At the same time as I was doing work for Phillips and Friends of the Earth I became an activist against things like GMOs. I'd also been researching chemicals on the cosmetics database through the Environmental Working Group for at least a decade. So in all these kinds of ways I was trying to minimize my chemical exposure and taking some serious steps. I don't wear perfume. I use an all-natural deodorant. At the time my grandmother was diagnosed with breast cancer. They biopsied her breast tissue and found aluminum in it. She was a big fan of Secret deodorant, and I put two and two together, because I was researching things like chemical toxicity in toiletries on the environmental working group database. And I started researching the links between breast cancer and traditional deodorants.[18]

Secret deodorant has aluminum zirconium trichlorohydrex as one of its active ingredients.[19]

While she was in Europe, Brandy became pregnant and decided she wanted to keep the child but not continue a relationship with the child's father. It had been a new relationship, and he had wanted her to get an abortion, something she felt she could not do.[20] She knew she wanted to raise her child, a son, in the United States, where she felt more comfortable with the culture. Even though there was much she appreciated about the Netherlands and the greater European community, she felt America was her home. Although she decided to have her son in Europe, and keep him there for the first six months of his life, she would spend four months of her pregnancy back in California, preparing for her eventual return. She recalled being surprised at how "medicalized" birth was in the United States and how there were such concerns expressed by medical people that her baby would likely die if he wasn't born in a hospital.

The European perspective on birth made much more sense to her, viewing it as a natural event, rather than a health emergency: "Pregnancy is not a sickness. There's no disease. Hospitals are for sick people. I would rather not have my baby in a hospital because there's disease there."[21] And in the United States she quickly became aware of all the medical interventions the professionals wanted to foist upon her son:

> I remember going to a childbirth prep class and the woman was a nurse. She was touting the flu shot and how every pregnant woman should get the flu shot, as well as their husbands. And that is where my first real red flags came up. I was like, what the heck? Because I was also being followed in [the Netherlands], where they don't give the Rho-gam shot. They don't give a vitamin K shot. They don't give vaccines at all in the first few months and really don't prefer to give them. It was a much more holistic and natural environment. I was doing some childbirth stuff here and I was so disillusioned by how many drugs and shots were pushed and how medicalized it all was.[22]

She had her son in Holland, by a natural birth at home with a doula (birth coach) and was very satisfied with the experience.

Brandy returned with her son to the United States when he was about six months old, then went to a midwife appointment to have him checked for weight and height. He was off the charts in terms of his growth, and very healthy. The midwife asked if Brandy wanted to talk about vaccines. Her response was no, she didn't. He was healthy. Why would he need vaccines? The answer seemed to relieve the midwife. She replied, "Good, I don't want to talk about them, either." They laughed about it, and the appointment ended soon after that.

The situation was much different a month or two later when she took her son to his first American pediatric appointment:

> The doctor was like, "Oh, we have to catch him up. He's really behind on his vaccines. We need to give him," oh I don't remember, something like sixteen or eighteen vaccines. And I was like, what? What are you talking about? He's healthy. Why does he need vaccines? I said, "I'm not putting anything into him. He's healthy. I'd like to see the inserts." And the doctor said, "I'm the doctor here. I've read the inserts. You can just trust me. Here's some information, [Author's Note: He handed her some brochures.] And I was like, "No, I've read my fair share of inserts, too. I used to call on pediatricians. And I don't always trust you guys. I used to be a pharma rep. Ha! Ha!" He didn't like that. I tried to say it jokingly, but he stormed out of the room. And then the nurse was like, we'll see you in two months. And then I left. He didn't get any vaccines. Then I started researching vaccines because that was a huge red flag to me. Somebody's pushing vaccines on my perfectly healthy son? Healthcare is about staying healthy. You don't need shots to be healthy.[23]

Brandy kept her son vaccine-free and would talk freely about the issue on natural mother forums on the Internet. The vaccine question had not been a high priority, as she was also concerned about other issues surrounding birth and childcare, such as urging women to consider a natural birth experience, avoiding drugs and pills, and asking them to investigate chemicals in their consumer products. Then, in 2015, she learned about California Senate Bill 277, which would deny a public education to any child who was not fully immunized according to the CDC schedule. For the first time in California history, in defiance of the Nuremberg Code passed after World War II, the right to an education was dependent on submission to a proscribed medical procedure. The right to decide which medical interventions were appropriate for your child was now being usurped by the state. Brandy said:

> I was like, "Holy shit! This is what's going on?" I had already spoken out among friends about Obamacare and how Obama would never have gotten Obamacare passed if pharma hadn't had a big seat at the table. And how interesting it was that pharma funneled him a lot of money, and funneled all the money to the Democrats because he was going to be elected on a universal health care platform. How better to mandate certain medical procedures, like ADHD meds and psychotropic drugs, and anti-depressants, and these types of things? The best way is to have a mandated, one-size-fits-all, health care. I already knew that was in the background, but it was like somebody hitting me over the head with a brick when SB 277 came on the docket. I said, "Holy shit! I totally know what's going on here!" I see what you guys are doing.[24]

* * *

It is said that tyrannies advance by slowly taking away a people's freedom, and cloaking their actions in soothing words of how this will be a benefit to the community. We have seen many examples of this in history. The tyrant does not come dressed in the clothes of a wolf, but in those of a shepherd. I want to share a story of how, two years before Brandy became aware of what was at stake, I had a very similar awakening.

In 2013, it was Assembly Bill 2109, sponsored by Assemblyman Richard Pan, who would later sponsor Senate Bill 277 in 2015. The law provided that if a parent declined a vaccination for their child, they would have to meet with a medical professional and have a conversation about the decision. One could argue it was a small and reasonable step, but others saw it as the beginning of an effort to mandate vaccinations. I freely confess that while I went to the state capitol in Sacramento to express my opposition to the bill, I did not really have a fire in the belly about it. If this was where the efforts stopped, I could live with it. Heck, I was looking forward to bending the ear of my pediatrician about the National Childhood Vaccine Injury Act of 1986, the rules of the Vaccine Court, and the Simpsonwood Conference. I actually thought it would be nice if medical professionals were required to listen to parent concerns. It brought to mind all those medical commercials where patients and doctors have genuine, caring, and respectful conversations about health choices.

But as I stood outside the committee hearing room, waiting for my turn to speak, I saw something that chilled me to the bone. A battalion of medical students walked into the capitol. (I heard they had been driven up from Southern California in a bus caravan. I wondered, *Who paid for that*?) I watched as each one of them was handed a new, white lab coat, prior to walking into the committee room. They all looked so crisp and professional as they waited to testify, the best and brightest of their generation. Our group spoke before them, and then I sat in the committee room and watched as these cute twentysomething medical students testified that they would never intrude upon a parent's right to choose the medical interventions they thought appropriate for their child. They just wanted to make sure we were fully informed. I cannot tell you how adorable and earnest they all looked, and how much I wanted to go up and pinch each one of them on the cheek.

And yet I was absolutely terrified by these students. I realized then that California Assembly Bill 2109 was not the final solution of pharma. I watched as Assemblyman Pan, later to be Senator Pan, reassured the committee that he had no intention of ever taking away a parent's right to choose the appropriate medical intervention for their child. I knew without a shadow of a doubt that at some point in the future, they would try to do exactly what they had promised, in the committee hearing in 2013, that they would never do.

In his signing letter for Assembly Bill 2109 in 2013, the governor of California, Edmund G. Brown Jr., wrote, "I am signing AB 2109 and am directing the Department of Public Health to oversee this policy so parents are not overly burdened in its implementation. Additionally, I will direct the department to allow for a separate religious exemption on the form. In this way, people whose religious beliefs preclude vaccination will not be required to seek a health care practitioner's signature."[25]

All of these promises and assurances would be broken two years later by California Senate Bill 277. I cannot help but wonder if any of those cute medical students would now like to withdraw their testimony.

* * *

"I cried for two weeks, then I started speaking out," said Brandy Vaughn. She continued:

> I knew on a spiritual level that this was it. This was my new canoe and the new river I was headed down. And I knew I was going to take a lot of heat. When you're a pharma rep your car is tagged. Your phone is tagged. Every screen on your computer is tagged. They track everything you do. They record all your phone calls. They have internal departments that track all of this. That's how they keep people from speaking out, from going out and selling Merck's secret formulations to Pfizer. They have a very advanced surveillance program. It was clear what kind of things could happen, but I didn't expect it to pop so quick. And I didn't expect it to feel so disturbing. The reality of it and hearing about it happen to somebody else are two different things.[26]

Brandy started going to the hearings for SB 277, and although she initially joined some of the groups opposing the legislation, such as Californians for Health Choice, she became suspicious that the groups had been infiltrated by pharma operatives:

> There are moles and controlled opposition in every movement. This was worth

> billions of dollars to pharma if they could get child mandates for vaccines. And if they could get a similar federal mandate, in addition to a federal mandate for adult vaccinations, the amount of money on the table is absolutely incredible. Why would they spend so much money on inner surveillance and not try to get into our movement? They've been in our movement for decades. And why would they not try and control it? To think otherwise would be absolutely naïve.[27]

There were five hearings on Senate Bill 227, and about halfway through them Brandy started to become suspicious. She observed a group that was controlling the testimony. They were present at the capitol forty to fifty hours a week and wouldn't let the parents speak to lawmakers without them present. This group had packets of information, and when Brandy looked at them, she realized that their setup was the complete opposite of how she had learned in pharma to persuade doctors. It seemed to Brandy that the informational packets were designed to fail.

As a pharmaceutical sales representative, Brandy had learned that the first minute or two when speaking to a new person were the most important. That's when you had to strike with your critical messaging. It was startling then to see the so-called "opposition" to this law spending the first couple minutes talking about how they weren't "antivaccine" but were instead "prochoice" or "pro-parental rights." Or they'd talk about the never-studied and, in her mind, mythical concept of "herd immunity" and saying that even with the relatively small number of parents who were not vaccinating their children (about 1–3 percent), this supposed herd immunity was still being preserved. And she saw other "controlled opposition" tactics, such as rallies that were cancelled at the last minute, permits that were never pulled, and a planned demonstration at the California Democratic convention that never materialized. There were supposedly "broke" autism moms who were flying in on private jets and staying at the local Hyatt hotel, which Brandy recalled from her past was the preferred corporate hotel for Merck. It wasn't adding up.

I was supposed to be one of the speakers at a Senate hearing but shortly before was told I had been replaced. To this day, I have no idea who made that decision. I had thought I would be an ideal person to testify, given my background as an attorney, science teacher, autism parent, and author, with the recent publication of my book *PLAGUE*, cowritten with a twenty-year government scientist. When I told this story to Brandy during the course of our interview, she replied:

> Yeah, they pulled anybody they couldn't control. They put up people who would be coached, who would say they were pro-vax, even if they weren't. They had this whole, you need to go with this messaging, or you don't get to testify. And the ones they couldn't control, they either tried to block or not invite. So you got uninvited, or banned.[28]

After Brandy said this, I realized she had given me one of the greatest compliments of my life. I have decided that on my tombstone I want written, "KENT HECKENLIVELY—He Could Not Be Controlled. He Would Not Go with the 'Messaging.' He Was the Uninvited and the Banned." My soul will rest easily.

The last hearing was on June 9, 2015, and Brandy coorganized a rally. Dr. Brian Hooker and Dr. Toni Bark were scheduled to speak at the hearing, but at the last minute they were prevented from speaking by this self-appointed "committee." Another person who was scheduled to testify was Allison Folmar, an attorney from the organization Parental Rights, but just as she went to speak, her microphone was turned off. Brandy was not alone in believing the hearing had been controlled and sabotaged. Some people seemed to back off as the sophistication of the pharma opposition became clear to them. Two weeks before the last hearing, Brandy noticed that some of her mail was missing. Was somebody gathering information on her? Then somebody broke into her car but did not steal anything. Had her car been tagged?

After the hearing on June 9, 2015, and the rally she had cosponsored at the capitol, she returned home to find her hide-a-key container open and, along with the key, placed conspicuously at the entry to her house. The next day she had her house alarmed, but because it was a 1950s home, there was difficulty installing a video camera. On June 11, 2015, in a Facebook post, Brandy called out the actions of this "committee" as being a part of a controlled opposition group of the pharmaceutical industry.

The next night, when Brandy was away from her home, somebody entered her house at 3:45 a.m. The intruder came in through the front door, punched the security code into the alarm pad, and stayed in the house for about five minutes. The hallway motion sensor went off, which meant the intruder had probably walked to the very threshold of the bedroom where she slept. The motion sensor in the kitchen went off. The intruder opened and closed the dining room window, which some corporate security guys later told her meant they were considering it as a spot for future entry. Her backyard was very private, so it would be much easier to gain access to the home unobserved. When the intruder left, he rekeyed the alarm and locked the door.

Brandy called the police and had them enter and clear the home before she went back inside. She gathered her things and made plans to go to Europe for three months with her son. She was not going to stay another night in that house. She would later rent it out for the summer, and then in November she rented it out to a more long-term tenant.

With her son safely in Europe, Brandy still had the obligation to appear at a rally in San Francisco at the Golden Gate Bridge two weeks later. She also thought that would give her the opportunity to do a more complete job of moving out her things so she could rent it. A friend picked her up at the Santa Barbara Airport. She had not let anybody else know she was returning. Prior to their flight from America, Brandy and her five-year-old son had adopted two puppies. When they left, Brandy placed them with a neighbor, realizing they might eventually have to find the dogs a new home. She arrived at the neighbor's house to retrieve the puppies, when the neighbor pointed to Brandy's house and said, "I didn't do that."

Somebody had gone into Brandy's garage, taken out her small stepladder, opened it, and placed it right beneath Brandy's bedroom window. They knew when she was returning home.

Brandy worked on the house during the day but then stayed at a friend's place at night. Every other night, for three or four nights, the intruder returned, setting off the alarm. Each time Brandy would call the police, and they would arrive and clear the place. The police would always ask her to check and see if anything was stolen, which meant a report would need to be filled out. But by this time, Brandy had heard enough stories to know that whoever was harassing her knew enough not to take anything. Theft of property put these incidents into an entirely different category. A friend who accompanied her one time asked, "Is your computer still there?"

Brandy answered that it was, and the police officer asked, "Where do you keep your computer?"

The friend answered, "She keeps it above the microwave."

Two nights later when the intruder came in for at least ten minutes, he went to the cupboard above her microwave, took out her computer, and left it open on the kitchen floor. A few nights later the intruder knocked over some paintings she had in the garage and took a Buddha statue off a high shelf and left it on the garage floor. The message was clear. *We are here and we know how to mess with you.*

Brandy later got a burner phone, but each time she did, a few days later she would notice calls being dropped, and sometimes it would sound like the person she was speaking to was inside a cave, a sign that somebody was listening in on the call. Twice during my interview with Brandy, the call was dropped, and three times the sound quality changed dramatically, as if she were talking to me from a cave. One of the times the cave sound appeared, it was just as she was talking about her phone problems, as if somebody were intentionally messing with us.

I must confess I have been on the other side of a telephone wiretap. As a law student, I spent a summer working for the United States Attorney's Office, Drug Task Force Division, in San Francisco. My job was to assemble evidence from wiretaps of an Oakland drug lord, Rudolph Henderson. Henderson had a six-million-dollar home in the Sonoma wine country called "The Skycastle" and a collection of expensive automobiles. They suspected their phones might be tapped but mistakenly thought the recording only started when they dialed the phone, not when they picked it up. And their code word for drugs was "cars," but they often forgot the code word. My job was to find those few instances in hundreds of hours of recordings. It was such a boring job, even though Henderson was eventually sentenced to twenty-five years in federal prison. So to the person who may have been listening in on my interview with Brandy Vaughn, I understand why it might have been fun to make the cave sound, just as we were talking about the cave sound. You were just trying to amuse yourself and stay awake. My advice? Get a better job!

Another time, Brandy was on her phone and a friend asked if she was going to stay at her house. Brandy replied, "No way. I feel like I'm a sitting duck." Two days later when she went back to the house, somebody had placed a metal duck on her patio table, looking directly into the kitchen. Brandy was later told by people who experienced similar situations that the safest thing for her to do was shout what was happening to her from the highest rooftop. Brandy even made a ten-minute video called "The Overt and Covert Intimidation of Brandy Vaughn," detailing the intimidation, and put it on YouTube.[29] She said, "I was told that the louder I am and the more people know my name, like Erin Brockovich, the—" and the phone call dropped.[30] I assume she meant to say "the safer I would be" but cannot be certain.

> When we reestablished the call, she started in and I wasn't sure if I'd missed anything: And I was told that was the best way to stay safe. Because that way it would be more suspicious if an accident happened. I'm not accident-prone.

> I'm not suicidal. And I just got a large life insurance policy and went through all the medical panels. My agent called me and said in twenty years "I've only seen a handful of people get the highest health rating." And I got it. I am not going to die of a heart attack. I am not going to mysteriously die. And I am not going to kill myself. I would never kill myself. I would never leave my son. I just want to put that on the record.[31]

Although she has made herself hard to find, the harassment has not stopped:

> My website is under constant attack. Wells Fargo, with whom I've had accounts for fifteen years, calls and tells me they've never seen the kind of hacking attempts that they've seen on my account. They suspended all my online banking and suggested I have no online banking. So I don't have any online banking. They keep giving my credit card to people in Nigeria, and I keep getting fraud alerts. They've ruined my credit. I can't tell you how many times, when I get a new credit card, there's suddenly thousands and thousands of dollars of fraud on it. So they keep me busy with things like that. So I don't have the energy for things like an interview with you.[32]

It's easy to understand why Brandy feels like a fugitive in her native California. The state of sunshine and unlimited dreams is not supposed to be a landscape of fear and nightmares:

> I've made it very difficult to find me, physically. It's very hard to have a social life because I don't tell anyone where I'm going to be. I just randomly pop into places here and there. And I live out of hotels now. My poor son tells people we live in a hotel now. They'll ask him, "Where do you live?" And he'll say, "We live out of a hotel. Mommy, which hotel are we in tonight?"[33]

But Brandy is undeterred. She has started her own nonprofit organization, the Council for Vaccine Safety, and has begun an ambitious billboard campaign titled "Learn the Risks" to inform the public about issues of vaccine safety.[34] One of these billboards was even located in San Jose, California, right on one of the main thoroughfares leading to the 2016 Super Bowl:

> This is how we fight pharma. We fight them at their own level. We have people on the ground, getting them unbiased information. Grassroots. And we need a legitimate, easy-to-manage website, which I think I've done, for people to go to. I get tens of thousands of hits on the website every week. We have to do

> marketing and massive advertising campaigns. That's why pharma spends so many hundreds of millions of dollars marketing their products, repeating and repeating the same message. We need to plant those seeds of truth. We need to get this information in front of people who don't even know to ask the question.[35]

Brandy is concerned, not just for what the pharmaceutical industry has done in the past, but also the plans they have for the future:

> They started with mandated medical procedures, vaccines, because they're the most accepted medical procedure. And new laws will come down the line after they've mandated vaccines. Either through Obamacare or Congress. They're shopping a bill around Congress right now. I say it's three to five years before we have a federal mandate on childhood vaccines and then another five to seven years before there's a similar law for adults.[36]

Even with all of these daunting challenges, Brandy is optimistic as to the eventual success of these efforts:

> If you think about it, at least seventy-five percent of the people already agree with us. Because what do you do when a needle comes close to you? Do you roll up your sleeves and say, here, I want more? No. You instinctively fear it. Our natural instinct is to keep that out of our system. So I say, most people already agree with us, they just don't know why. We're here to connect the dots for them.[37]

And the struggle to change people's minds will not be won in a single conversation, as Brandy learned when she was a sales representative for Merck:

> Pharma used to tell us that when we were going in and pitching a new drug, and maybe there's another drug the doctor likes, it takes eight to ten times to hear something that goes against their already ingrained mindset, for them to question it. But on that eleventh or twelfth time, you're going to hear them turn around and say what you've said. And I swear to you, it's true. So on that third, fourth, fifth time when you're telling somebody something, they're going to have a big wall up and they're going to bash you. But as soon as you get up around number ten, they'll say something like, "Oh my friend asked if I got the flu shot because you get a 20 percent off coupon and she bought some candy with it." And they'll turn to you and say something like, "I decided to hold off this year." They're going to start opening their minds.[38]

* * *

I warmly embraced Dr. Jeff Bradstreet when I saw him in the hallway just outside of the presentation rooms at the Omni Hotel in Chicago on May 30, 2015, during the Autism One Conference. Dr. Bradstreet was one of my daughter's doctors, and I respected him as one of the best thinkers in the autism field, even though nobody had been able to bring my daughter close to anything resembling recovery. We were both presenters at the conference. My talk was titled "AIDS—Autism Immune Deficiency Syndrome" and recounted the story I told in my book *PLAGUE*, with Dr. Judy Mikovits. Mikovits had come to believe that a retrovirus (specifically a mouse retrovirus, XMRV or xenotropic murine leukemia virus-related virus, which had somehow jumped into the human population) was implicated in autism, chronic fatigue syndrome (ME/CFS), and many types of cancer, helped along by chemicals like mercury and aluminum that tended to skew the response of the immune system. Bradstreet's talk was titled "How Close Are We to an Autism Cure?" His answer? Pretty damn close.

Bradstreet had come to believe one of the most important clues in autism was the elevated presence of a substance known as nagalese, given off by cancer cells and viruses. Can you guess why I was so interested in this line of inquiry? In an article I had written on Dr. Bradstreet in 2011, I reviewed what the good doctor had written about the subject:

> In the past months Dr. Bradstreet has become interested in nagalese, which he describes as an enzyme "produced by cancer cells and viruses." He thinks it unlikely that children with autism have undiagnosed cancers, and thus suspicion falls on a viral etiology. Dr. Bradstreet writes, "Viruses make the enzyme as part of their attachment proteins. It serves to get the virus into the cell and also decreases the body's immune reaction to the virus-thereby increasing the odds of viral survival."
>
> Further on Dr. Bradstreet writes, "It is reasonable and likely that the nature of the immune dysfunction and the frequently observed autoimmune problems in autism are mediated by persistent, unresolved viral infections." He claims to have tested approximately 400 children with autism and found that nearly 80% have significantly elevated levels.[39]

Bradstreet was investigating whether nagalese might also be contained in vaccines, which would not only lay the foundation for autism, but many cancers, as well. Whether the presence of this nagalese was intentional, or an inevitable byproduct of growing viruses in culture, then weakening or killing them before placing them in a vaccine was unclear. If this were true, it was easy to understand how many would conclude that all vaccines were, in effect, Trojan horses promising to prime the immune system for a viral onslaught, but in fact weakening it.

Since I had written about Bradstreet in 2011, he had become interested in a protein called GcMAF, which stands for Gc protein-derived macrophage activating factor. Macrophages are the soldiers of the immune system that will destroy cancer cells or virally infected cells, theoretically allowing the body to recover. Bradstreet was claiming about 85 percent of the children who received this treatment showed substantial benefit, with about 20 percent of those children experiencing a full recovery. I tried it on Jacqueline but saw no benefit. Like I said, she's a tough one.

Bradstreet was touting GcMAF (as part of a home-grown yogurt culture) and photobiomodulation (low intensity laser therapy to stimulate the mitochondria) as ways to help those children like my daughter who had not responded to other therapies. I talked to him about his presentation that I attended and started to pester him about what such a protocol might look like, when he lifted a hand to stop me. "When I get back to the office, why don't you give me a call, and we'll set up a time to talk?"

"Okay," I replied, giving him a wan half smile, as I could see he looked tired and didn't want to overburden him. I embraced him again, said I would call, and told him good-bye.

I had a number of other things to be concerned about at the Autism One Conference of 2015 in Chicago. First, I was giving a speech about the book I'd coauthored with Dr. Judy Mikovits, which suggested a retroviral factor in autism. The retrovirus theory was important because retroviruses tend to hide out in the B and T cells of the immune system, where any immune stimulation, such as a vaccine, might cause the retrovirus to replicate out of control. For example, babies born to HIV-positive mothers are routinely put on antiretrovirals prior to immunization because of the fear that the immunization will cause the baby to develop AIDS (Acquired Immune Deficiency Syndrome). I'd even been able to confirm this fact with the University of California, San Francisco, Pediatric AIDS unit, one of the world's leading facilities on HIV/AIDS.

My coauthor, Dr. Judy Mikovits, was scheduled to give three talks. In 2010, I'd met her at Autism One and told her that her research was going to get her into a lot of trouble. She didn't believe me. Five years later, she looked upon me as a prophet. But my skills don't extend that far. I'm just good at pattern recognition. I'd seen what happened to people who went up against the medical mafia. It's difficult to convey the sense of dislocation one experiences when you believe you live in a rational world that protects children and discover you do not live in that world. We can see it so easily in other countries when a good person believes the lies told by those in authority, but that sense of having been so completely wrong, and realizing that so few people in your own supposedly free and open-minded society will have the courage to look honestly at these issues, can be a bitter pill to swallow. I had worried greatly about Dr. Mikovits over the years. She has a fiery temperament and does not handle injustice well, but she seemed to be upbeat and cheerful at the conference.

Also at the conference I scheduled time to have dinner with Dr. Brian Hooker and discuss his participation in this book. The CDC whistleblower story had broken in our community the previous summer, got a few mentions in the mainstream press, and then went silent. I thought a book might help get the word out. I was fortunate that Brian's wife was a big fan of my writing on *Age of Autism*, and although I had originally thought we'd work on the book together, Brian thought it better that he not participate financially in any way in telling the story. As I sat with Brian and he spun out the story of his relationship with Dr. William Thompson, it was clear that despite the terrible crimes Thompson had participated in, Brian felt deeply for him as a fellow human being and wished him no harm. Brian Hooker is an amazing man.

I also wanted to secure the approval of Dr. Andy Wakefield, as more than any other researcher in this area he was the one who had spilled the most blood on the autism battlefield. Andy gave me a big hug when he saw me, recalling that I'd written positively about his almost superhuman forgiveness of William Thompson. Andy enthusiastically agreed to be interviewed by me and told me he was working on a documentary, as well. The book would make a nice accompaniment.

Autism One 2015 was a time for me to renew old friendships, plan new projects, and see if I could discover any different directions to improve my daughter's life. I have to admit that when I left the conference on Sunday afternoon, I was excited about what I had done, and the project I was planning to do, but the conviction of Dr. Bradstreet that we were close to an autism cure really made the trip worthwhile to me. More than anything, I wanted to make the life of my family easier, and Dr. Bradstreet might be the key.

I never got to have that conversation.

***

On June 17, 2015, the office of Dr. Bradstreet in Buford, Georgia, was raided by federal agents under the direction of Special Agent Marc Hogan and authorized to seize "for the time period of January 1, 2011, through the present, the following records, documents, and items listed below," which included "all Globulin component Macrophage Activating Factor (GcMAF), GC globulin, and/or any other products or component substances therof that constitute misbranded drugs under the Federal Food Drug, and Cosmetic Act."[40] Agents from the Food and Drug Administration and the Drug Enforcement Agency were on the scene for several hours.

On June 19, 2015, the Rutherford County Sheriff's Office received a report from a fisherman of a body floating in the Rocky Broad River in Chimney Rock, North Carolina. It was Dr. Bradstreet, dead of a single gunshot wound to the chest. For those who knew Dr. Bradstreet, it was difficult, if not impossible, to believe he had taken his own life, which was the initial conclusion of the local sheriff. As one blogger put it in language that would have found wide acceptance among those who knew Dr. Bradstreet:

> Let me see if I get this right: a working doctor taking care of patients with autism, which his son also suffers from, decided to kill himself so he travels a hundred miles to some obscure little river in North Carolina where he somehow manages to shoot himself in the chest in a deep enough part of the river that it requires divers to locate the gun that drops out of his hand when he pulls the trigger. Did he swim out there and shoot himself? Was he in a canoe which disappeared after the fact? Did he wade out in the river and shoot himself and linger long enough to toss the weapon out in the deep water as he slumped to his death?[41]

Bradstreet's death was even reported by CBS News.[42] But it seemed to be just the beginning of a suspicious pattern of deaths among similarly outspoken doctors. Erin Elizabeth, who runs a popular health website, *Health Nut News,* and knew Dr. Bradstreet personally, published an article on March 12, 2016, listing the suspicious deaths of at least thirteen different holistic doctors on the East Coast since the death of Jeff Bradstreet on June 19, 2015, and concluding on February 1, 2016.[43] Some died from apparent heart attacks although they were apparently healthy, others from gunshot wounds, and some just went missing.

On March 27, 2016, Erin Elizabeth published an interview with Tom and Candace Bradstreet, the brother and sister-in-law, who, after the death of Jeff Bradstreet, hired a private investigator to look more closely at the case. Although the investigation was still continuing, Tom Bradstreet felt there were some misconceptions about his brother that he wanted to address. Because of the federal raid there was much about the case they could not say, but they wanted to knock down any notions that his brother was estranged from his family or had left a suicide note. The death of Jeff had come as a complete shock to his family. Thom said:

> The Jeff Bradstreet we know would not do this. He fought the FDA in 2003 and won. He was always fighting for what he believed in. He always thought outside the box. The family is doing their own investigation because they love him. Jeff and I as brothers were close. He was one of the top generals of autism. He was caring and generous and he understood the parents because he was one. To think that this field general would walk away from his army and the largest fight of his life is absurd. Just because the FDA and DEA walked into his office and asked a few questions about his life? He was not afraid. He cared about his employees and the parents. He would not leave parents with treatments not done or prescriptions that were needed. He would not walk away from the harassment by them. He did not commit suicide.[44]

* * *

I spoke to Thomas and Candace Bradstreet in May 2016, almost a year after the death of Dr. Jeff Bradstreet, to get an update on their investigation. They were actually on their way to the Autism One Conference in Chicago, where they would be participating in a tribute to Dr. Bradstreet, an event that would leave them both overwhelmed by the number of families that Jeff had helped and confident that his work was continuing.

I began by telling them that I considered Jeff to be one of the leading thinkers of the movement, and while I often implemented his suggestions, they had not helped my daughter. She was not one of the success stories. Still, nobody in either the traditional or alternative health worlds had been able to make much impact on her condition, but I always appreciated those with innovative ideas on what to try next.

Thomas began by telling me that Jeff was his older brother, and that there was about a year-and-a-half gap between them. They grew up shooting guns, and Jeff eventually became an Air Force pilot and captain, in addition to a medical doctor. For a time he had even been an ER physician in East

St. Louis, an area known for its high crime rate. When I told them I had seen Jeff less than a month before his death, Thomas wanted to know if I had detected any stress or despondency in his brother. I told them no. If anything, he seemed very excited about what lay ahead, as if he were seeing the end of a twenty-year journey.

After I spoke in this way, it seemed to relax Thomas, as he felt that some of the news and Internet coverage about his brother's death was designed to make people feel Jeff had been "dirty," that he'd been "found out" and decided to take his life in response. Thomas said:

> It was common knowledge that they were going to be harassed. That's just the way it is. Jeff and Andy Wakefield worked together for a long time and they had numerous conversations about the harassment of doctors who are trying to cure the world of autism. It wasn't something new. Those were the parameters that he was used to operating in. The only reason I'm trying to establish this premise is that he wasn't under tremendous stress to the point where he lost it and shot himself. No investigation, public or private, has come to the conclusion that Jeff committed suicide.[45]

Thomas then spent a good deal of time talking about the "raid" on his brother's office by agents of the FDA and DEA: "I'm sensitive to the word 'raid' because it comes with so much guilt and weight attached to it. It's like a drug bust. They come in, armed with M-16s, they take the money, they take the drugs, and they take people off to jail. That's not what happened."[46]

Thomas then detailed what he had been able to learn about that day:

> They came into his office. They did look at financial records. They took some USB drives. They took some information out of his computers. They didn't lock him down. They didn't take his passport. They didn't sequester anything. He wasn't arrested. They didn't seize banks accounts or freeze them. At the time, it was just harassment. It wasn't some horrible thing, my life is over. This was just, we can't slow you up in any other way, so we're just going to do this. When you look at the investigation, you look at the report, you talk to the office staff, you see it's clearly just harassment. It wasn't, you're going out of business, you're going to jail.[47]

According to Thomas Bradstreet, after the visit by agents of the FDA and DEA, Dr. Bradstreet contacted some lawyers:

> He spoke to a couple attorneys and they said, "Yeah, it looks like you're going to get your hands slapped, at most. Maybe a fine. Maybe not." Yes, it was his second infraction with the FDA. But they were unrelated. One was IVIG (intravenous gamma-globulin) [more than a decade earlier] and the other was GcMAF (glycoprotein macrophage activating factor). They found no GcMAF. Everything that he did was legitimate.
>
> If it wasn't legitimate, they would have had legal precedent. They would have gone through this thing and you would see some sort of FDA stipulation that would prohibit doctors from using GcMAF. And there aren't. There are doctors that are still using it.[48]

The allegation by Thomas Bradstreet that GcMAF is not prohibited under current law is backed up by the language of the search warrant itself, which states the agents were looking for "misbranded" GcMAF, rather than GcMAF that was "prohibited."[49]

Thomas then went on to address the various theories about who might have been after his brother, and why: "Everybody thinks it's the GcMAF that Jeff was working on that was his demise. I don't believe that at all. I think it was one stone in a ten stone plate that eventually tipped the scales. But I think the biggest thing that was going on was the problems he'd found with vaccines. And what he had actually found in vaccines had really started to become significantly problematic."[50] Thomas spent some time detailing his brother's belief that many of the vaccine components, specifically the thimerosal, the aluminum, and other ingredients, shut down the body's ability to produce macrophages and allowed nagalese to rampage freely through the immune system, causing damage. Then he moved onto an issue I'd never even considered:

> The other thing he found came because he was looking at DNA markers from the kids with autism, and DNA from mom and dad. And when he was looking at the child, it should have been one plus one equals two. And that would have been the child. But he was finding one plus one plus another one. And that was making the child. He was finding DNA that wasn't from either the mom or the dad. It was actually in vaccines. The reason why is because a lot of times aborted fetuses are used in vaccines. [Specifically, aborted fetal tissue is often used to grow viruses in cell culture.] So we were getting this other problem from the vaccines and it was a third DNA. And the third DNA could be corrupt. It could be bad. It could be corrupt.[51]

Thomas also felt deeply troubled as a Christian on a spiritual level by the use of aborted fetal cell tissue:

> Knowing that the DNA was obtained in an environment of deep persecution and death, which is what happens in an abortion. You have a physical and emotional trauma involved in that. To me, that in its own right is problematic. Forget the rest of the stuff, just the spiritual and emotional stuff is bad enough. But then you add all the other stuff into it. There are serious problems.[52]

And now we come to June 19, 2015, the date of Dr. Jeff Bradstreet's death.

Jeff and his wife were going to the Lake Lure Inn in Lake Lure, North Carolina, for the Father's Day weekend. Lake Lure is a favored destination in the Blue Ridge Mountains, about a half hour away from Asheville, North Carolina, where Jeff had an office. Jeff's wife has an autistic son from a previous marriage, and she was dropping him off at her ex-husband's home for the weekend, meeting up with Jeff later at the Inn. Jeff drove to Lake Lure, stopping off at a grocery store to buy some supplies for the weekend. According to a later report, the food purchased was not of a "last meal" variety, often encountered in suicides, but typical fare. Jeff arrived at the hotel but was told his room was not ready. He informed the attendant he was going downtown and would be back in a few hours. A staff member took his cell-phone number and said they would call when his room was ready.

A few hours later, a fisherman found Jeff's body in a stream that fed into Lake Lure.

From the very beginning, Thomas was suspicious about the events surrounding his brother's death. The stream in which Jeff's body was found was about five miles away from Lake Lure, accessible by a two-lane highway, and there was a rest stop where Jeff's car was parked. Depending on the amount of rainfall, the stream can be more like a big creek, or a nice flowing river. "It's relatively remote, but there are houses around," Thomas said. "So if there had been a gunshot, especially a nine-millimeter gun shot, you would have been able to hear it. And there were fishermen, less than a thousand or two thousand yards away. And they heard nothing."[53]

The forensics report revealed several abnormalities. One would think that if you were a medical doctor and going to kill yourself by shooting yourself in the heart with a nine-millimeter Glock pistol, you would place the barrel at that approximate place on your chest and fire. But the bullet was fired from above, as if he held the gun at the extreme length of his arm and pointed it down at his heart. When I mentioned it sounded like an execution shot, such as a man on his knees, Thomas told me that was how it sounded to him, as well. He acknowledged that it was possible his brother had fired the shot, but it was an unusual way to shoot yourself, which was acknowledged by the investigators. There was also no stippling on his chest, which is the pattern of abrasions one gets from being shot at point-blank range. The pistol was found ten yards away from his body in the creek. Normally, when a person shoots themselves, the body will naturally have a death grip on the weapon, but it can release when the hand hits the ground. However, in most instances, the gun is found within six inches of the body.

When I asked Thomas what his best theory was about his brother's death, he replied:

> Well, we have several. I'll talk about this medical examiner who we hired to go in and look at the autopsy and understand the situation, look at the environment, go out to the site, and all that. And his professional opinion is that Jeff did not commit suicide. If we were to just take the body as is and as somebody would look at it and say, "is it possible this person could have committed suicide?" The general consensus from medical people who have looked at the report has been, "Yeah, there is a possibility. But a very rare possibility that would happen." It would almost look like it was an accident. Now add the environment, where it was at, and his credentials as a medical doctor, everybody we've talked to has said absolutely not. His history of guns, his understanding as a medical doctor, he was an ER doctor in East St. Louis, and he understood gunshot wounds. He knew how the body was going to react and how the bullet was going to travel. There's just no way he would take a chance of missing the heart and critically wounding himself.[54]

There was also one other issue that troubled Thomas about his brother's death:

> I'm going to say something I've never told anybody else in an interview. My brother was under the watchful eye of the FDA and DEA. He did not have a permit to carry a concealed weapon. But he was traveling, or allegedly traveling, with a weapon that he was not legally allowed to carry. At any moment he could get pulled over, checked, and be arrested for carrying a concealed weapon. Jeff would never take chances like that. Never, never, never.[55]

In conclusion, Thomas had the following to say about his brother: "Jeff was really outspoken. He had no fear. He didn't care about who he offended in the process. I think he was very close to standing on top of the highest mountain and saying, 'No, this is incredibly sinister. And we need to get to the bottom of it.'"[56]

* * *

And how do I, the author of this book, feel about the death of Dr. Bradstreet? He was one of my daughter's doctors, and I considered him a comrade-in-arms against the terrible disease that afflicts both our children and millions around the world. In my estimation, Dr. Bradstreet was one of the "great souls" of our movement. That is my bias, and I freely confess it. I am shattered by his death.

My years as an attorney also taught me that people can act in unexpected ways when under great stress. The hidden depths of another person will always remain something of a mystery, even for those whom you think you know well. I cannot claim to have been a close friend of Jeff Bradstreet. I do not know the truth of his death.

It is common for attorneys to look at a set of facts and generate multiple scenarios, all of which match the same set of facts. Every lawyer has had the experience of having been asked by a law professor to argue one side of a case, then flip over and argue the opposing side. It creates an agility of mind in a world where the truth can so often be hidden from us. We are taught to hold opinions, but not to cling to them too tightly.

On one hand, it seems obvious that if you are going to murder somebody and stage it as a suicide, a river is a great place to dispose of a body, as it will tend to wash away critical evidence. If one is inclined to believe Brandy Vaughn's account of pharma intimidation, it's easy to generate a plausible scenario. Bradstreet's car would have been geo-tagged so they knew where he was at all times, and they could have snuck into his house, found his pistol, and taken it, then waited for a moment of opportunity. Jeff did own a nine-millimeter Glock pistol, and it was not found in his home after his death. Did he bring it with him to Lake Lure, or did somebody take it from his house? I have no answer to this question.

On the other hand, if you're going to commit suicide, being visited by government agents is just the sort of thing that might tip you over the edge. Could there have been problems in his marriage? Might he have been crumbling under the weight of other problems and simply hiding it under an optimistic demeanor? Again, I have no answer to these questions.

But whoever had their finger on the gun that killed Dr. Jeff Bradstreet,

I must view it in a broader context. I must consider it in a world in which a pharmaceutical company admits it killed more people than died in the terrorist attacks of September 11, 2001, and it hardly registers a ripple in the public's consciousness. Maybe they killed a lot more. Maybe the casualties are as bad as those in the Vietnam War. Maybe they are as bad as those in the American Civil War, which ended more than 150 years ago. This is what happens for a medication that is under our traditional civil justice system. How many more unspeakable acts are possible with vaccines, for which the pharmaceutical companies have no liability and whose executives and scientists can never be brought into a courtroom?

I also look at Dr. Bradstreet's death in light of the story of Brandy Vaughn, living in hotel rooms with her young son as she battles the pharmaceutical industry. Is it possible that in today's America, a person can be subjected to such intimidation? Do we turn away from stories like hers because they are unbearable to consider? What world do we really live in?

The road for those who oppose Goliath appears to be dark and full of terrors.

***

On December 8, 2020, as this book was going to print, I received word that Brandy Vaughn, the activist profiled in this chapter, had been found unresponsive in the morning by her nine-year-old son, Bastien. By the end of the day, many activists who were close to her confirmed the reports. Brandy Vaughn was dead.

I reproduce this section of my interview with Brandy from March 14, 2016. We had been talking about how one remains safe in our world of activism taking on Big Pharma. I'd been saying one needs to get loud, because if something happens to you, it gets attention. Brandy agreed with that, but then talked about the difficulty of that kind of life, especially with a young son.

> I have made it very difficult to find me, physically. Because it's very hard to have a social life because I don't tell anyone where I'm going to be. And I just randomly pop into places here and there. And I live out of hotels. And my poor son tells people we live in a hotel now. They'll ask him, where do you live. And he'll say, we live out of a hotel. Mommy, which hotel are we in tonight? It's kind of funny, but it's really not. It's been harder to physically intimidate me now, but they've done other things.

My website is under constant attack. Wells Fargo, with whom I've had accounts with for fifteen years, calls me and tells me they've never seen the kind of hacking attempts that they've seen. They suspended all my on-line banking. And they suggested not to have any on-line banking at all. I don't have any on-line banking.

They keep giving my credit card to Nigeria and I keep getting fraud. They've ruined my credit. I can't tell you how many times, every time I get a new credit card, suddenly there's thousands and thousands of dollars of fraud on it. They keep me busy with things like that, so I don't have energy for things like an interview with you.

Billboards up around the world. Even just yesterday I got a message from Linked-In saying we've got multiple attempts to sign in, we think you should change your password. This thing is happening all the time with learntherisk.org, a lot of these security things, like webroot, now report this website as malicious. This is how they keep fucking with me on a very intense level, repeatedly.

But in the movement, people that I called out as controlled opposition, came after me. And I was like whoa, who are you? Now I know who you are.[57]

A few days later she called me up and told me how important the interviews were to her. She said if anything happened to her, she wanted me to make sure I gave her interviews to her son, so he would know who she was, when he was old enough to read and understand them. I told her they would always be available to him if anything ever happened to her.

But I told her she was going to be safe, because that's what we do when we're scared, and don't know what might be out there in the darkness.

I do not know what led to Brandy Vaughn's death. However, I know she was healthy, avoided all pharmaceutical drugs, and tried to live a life of integrity. She believed she was doing the most important work in the entire world. A few days later, Robert Kennedy, Jr. reported that early indications were that Brandy died from gallbladder complications, which she had been struggling with in 2020. My own initial research suggests up the seven thousand people a year die from these problems.

Perhaps that is true. As an activist, I would like to believe her death was unrelated to her activism. And yet it seems odd to me, that just as our country is gearing up to take the untested COVID-19 vaccine, that one of our most vocal activists has been taken from us so young.

CHAPTER NINE

Curious Alliances

Dr. Brian Hooker stood in the entryway of the Chicago home of Minister Louis Farrakhan, leader of the Nation of Islam, and watched as bodyguards conducted a security pat-down of Robert Kennedy Jr., to ensure he was carrying no weapons that might be used to assassinate the minister. Hooker looked around the room at the small and unlikely band that had traveled to Chicago to enlist the help of the Nation of Islam to get to the truth about vaccines and autism. There was Barry Segal, a wealthy Jewish philanthropist and founder of Focus for Health, who had bravely taken on the vaccine/autism issue, despite not having any family members afflicted with the disease. There was Michelle Ford, founder of a group called Vaccine Injury Awareness League in Southern California. Also in the band was documentary filmmaker Eric Gladen, who along with his partner, Shiloh Levine, had directed and produced the film *Trace Amounts*, about the possibility that mercury was contributing to the autism epidemic and other diseases. There was Robert Kennedy Jr., of course, but as Kennedy's security pat-down reached its end, Hooker's attention was drawn to the face of a man standing just outside their small circle, the man who had made all of this happen, Minister Tony Muhammad, leader of the Los Angeles mosque for the Nation of Islam.

***

Minister Tony Muhammad was born and raised in Atlanta, Georgia, to a single parent who raised ten children. He graduated from Morris Brown, a historically black college, with a degree in education. Tony was always an exceptional athlete and after college had tryouts with the Pittsburgh Steelers and Atlanta Falcons, but instead of football, he ended up playing baseball for two months in the farm system of the Atlanta Braves. After leaving sports, he got a job working for Eastern Airlines but eventually succumbed to the frustration of not getting the jobs he wanted because of his color and started to sell drugs on the side.[1]

When he was twenty-seven years old, Tony Muhammad went to hear a lecture by Minister Louis Farrakhan, and it changed his life. Tony was riveted by Minister Farrakhan's declaration that the true problem in life was not one's color, or even one's personality, but ignorance. "The biggest enemy to any human being is ignorance," said Minister Farrakhan. "It is ignorance that causes you to sell drugs to your own people." Tony felt as if the message were meant directly for him. Farrakhan proclaimed it was a person's duty to learn the true history of things, and when they did, their own individual path would become clear. That night Tony went home, flushed the drugs that were on his table down the toilet, and decided to join the Nation of Islam.

In his first two years in the Nation, Tony had such energy and enthusiasm, bringing in so many new members, that he came to the attention of Minister Farrakhan. When the two met, Farrakhan told the young man, "You are one of the best that I have. And I desire you to come into the ministry." Tony accepted the offer, and a few years later Minister Farrakhan tapped him to lead the Los Angeles Mosque, a position he has held for the past twenty-five years. Minister Tony is also now responsible for all of the Nation of Islam's mosques west of the Mississippi River.

***

Minister Tony would become involved in the vaccine-autism issue in May 2015 through one of his members, a man named Rizza, who was at a meeting of autism activists that included Robert Kennedy Jr., Brian Hooker, and Michelle Ford. In the course of the meeting, Kennedy expressed his frustration that he couldn't get any African American leaders to get involved in the issue and asked if anybody had suggestions for prominent individuals in the community who might consider taking a stand. Rizza suggested Minister Tony Muhammad.

Rizza called Minister Tony and said that Robert Kennedy Jr. wanted to meet with him.

"Really?" replied Minister Tony. "What does he want to talk to me about?"

"Well, we don't want to say too much, but it's real serious, and we want you to come. It has something to do with vaccines."

Minister Tony thought it took a lot of courage for somebody of Kennedy's stature to reach out to the Nation of Islam. The Nation is generally portrayed in the media as a bigoted and separatist black movement. Minister Tony agreed to meet with them and drove over to their location.

Brian Hooker led the presentation on the Thompson documents, including Thompson's long letter of confession to Congressman Bill Posey, in which he laid out exactly how scientists at the CDC had concealed the effect of earlier MMR administration on African American boys. Minister Tony was angered by the revelations, but not surprised. The founder of the Nation of Islam, Elijah Muhammad, had cautioned as early as 1941 against the use of vaccines. Elijah Muhammad had worried that medicines, and especially vaccines, would be used to target specific ethnic groups. Like the Amish community, the Nation of Islam has a mandate that their children not be vaccinated, although Minister Tony admitted he had not thought a great deal about the matter.

"I have to call Minister Farrakhan," said Minister Tony after the presentation had finished. He looked directly at Kennedy. "Would all of you be willing to travel to Chicago to meet with the minister?"

"Absolutely," replied Kennedy.

Tony called Minister Farrakhan on the spot, told him what he'd learned, and when Minister Farrakhan understood what they were talking about, he said, "Get to me immediately. I want to see this proof, and if it's what you say it is, we will support it."

A week later, the members of this unlikely band were all on planes from various parts of the country, headed for a fateful meeting at the Nation of Islam headquarters in Chicago with its controversial leader, Minister Louis Farrakhan.[2]

* * *

The home of Minister Louis Farrakhan, on Woodlawn Avenue in the historic Kenwood section of Hyde Park in Chicago, was originally owned by the founder of the Nation of Islam, the Honorable Elijah Muhammad. The residence has Mediterranean and modernist elements, beautiful stained glass windows with Muslim emblems, and a state-of-the-art security system in addition to guards. Less than a block down the street was the home of former heavyweight champion Muhammed Ali. A few blocks over from Ali's residence is the former home of President Obama and Michelle Obama when they lived in Chicago. If you continue down Woodlawn Avenue, you will eventually hit the University of Chicago campus. Taking a left on Hyde Park Boulevard from the Minister's home will bring you to the windy shores of Lake Michigan.

After going through the security check, the members of the group were brought to a living room with couches and introduced to the minister. Farrakhan was in his early eighties and did not look like the firebrand many of them expected, but projected more of a grandfatherly aura. When introduced to Robert Kennedy Jr., Farrakhan visibly brightened and said:

> You may not know this, but I grew up in Boston, around the Kennedy family. My wife was a member of the same Catholic Church as your uncle, President Kennedy, and we were married on the same day [September 12, 1953 at Saint Mary's Catholic Church in Newport, Rhode Island]. Your aunt and uncle were married in the morning, and my wife and I were married in the evening.

As they took their seats, Farrakhan continued with his discussion of the Kennedy family: "And we all know the part your uncle, Senator Ted Kennedy, played in ending the horrific Tuskegee syphilis experiments on African American men that the CDC was allowing to continue."

In 1932, the US Public Health Service initiated a study on the effects of syphilis in African American males, involving 399 men with the disease and 201 who did not have the disease. Even though penicillin came into use as a highly effective treatment against the disease in 1945, the participants were not informed of this change. In fact, they were not even informed they had syphilis but were told instead that they had "bad blood." The CDC, which came into existence in 1946, continued this study, not ending it until newspaper articles started appearing about the study in 1972.

In 1997, President Bill Clinton issued a formal apology, saying:

> What was done cannot be undone. But we can end the silence. We can stop turning our heads away. We can look you in the eye and finally say on behalf of the American people, what the United States government did was shameful, and I am sorry . . . To our African American citizens, I am sorry that your federal government orchestrated a study so clearly racist.

To the group gathered at Minister Farrakhan's home, the reference to Tuskegee was both powerful and appropriate. In his initial video announcement on the CDC whistleblower, Andy Wakefield had explicitly mentioned the Tuskegee experiment. But whereas the Tuskegee experiment had let 399 African American men with syphilis go untreated for decades, the MMR cover-up had in all likelihood affected hundreds of thousands of African American boys, as well as the rest of the population. The revelations of Dr. Thompson were far worse than what had happened at Tuskegee between 1932 and 1972.

Minister Farrakhan continued the account of his ties with the Kennedy family: "A few years before he died in that plane crash, John F. Kennedy Jr. came here and interviewed me for his political magazine, *George.* He had been planning to write a long article, essentially introducing me to white America and the things they may not know about me."[3] A short excerpt from that interview, conducted on July 31, 1996, conveys the flavor of their discussion, including Minister Farrakhan's interest in communicating to a wider audience:

> **John F. Kennedy, Jr.:** What sort of connotation does becoming more mainstream have for you? You are making more overtures to be received by that mainstream, yet, certainly in your own case that has some pitfalls potentially within your devoted following, does it not?
>
> **Minister Farrakhan:** I cannot act in a way that violates the mission of the Honorable Elijah Muhammad, but I must act responsibly by what the time demands . . . Farrakhan sees that the American democratic scene has made a provision for voices of discontent and dissent. I represent a people who are mainly democratic, but have not gotten from the Democratic Party that which satisfies our needs, our interests, and our rights. You have between 30 and 40 million Black people, you have a fast growing Hispanic community, you have an Asian and an Arab and an Indian community, and given the dissatisfaction in the country, we need to establish a national agenda that all of our groups can stand on so that in a united way we can leverage our vote to influence the direction of this nation toward the best interest of the poor and the weak rather than a nation held hostage by the rich and powerful. This is leading, in our humble judgment, to the destruction of this democracy. I felt and feel that it is time now for me as a spiritual teacher to broaden my own understanding of the message of Islam so that the message does not become exclusive, but inclusive.
>
> **John F. Kennedy, Jr.:** Who is your constituency?

> **Minister Farrakhan:** That is a wonderful question . . . I have a duty to Black people, but I think I also have a duty to the whites of this nation because of the interaction between Blacks and whites that started from a very negative position. Even at this moment there is a great divide between Black and white largely because we have dealt with each other falsely and hypocritically.
>
> I want to deal truthfully but not in a manner that would be considered an enemy to the goal. If my goal is to reach you, how best can I communicate with you without being so vicious in the manner of my speaking of truth that I turn you off or others off from what might be a truth that could save the country from its fall?[4]

Minister Farrakhan finished up by saying that Robert Kennedy Jr. was the second Kennedy he had welcomed into his home, and he hoped that it would happen more often than every couple of decades.

It was after about ten or fifteen minutes into the discussion that Minister Tony saw that some of the group were dealing with a very different Farrakhan then they'd been expecting. Barry Segal was first to voice this opinion and said, "I've met a lot of people, but I came in here with a predetermined idea that you would be a negative person. But I'm not seeing that. I'm seeing a warm and gracious person."

The minister started chuckling. "Aw, I know that our enemy has built a wall between us and you only know me through sound bites. If I gave you some of the words of Jesus in a sound bite, I could make you think he hates people."

The group found the comment funny and spoke for a little longer about the prejudices in the mainstream media about the Nation of Islam, as well as the prejudices against the autism parents. The minister started talking about those in power and said:

> They hate the very fact that we are meeting now. Because they started attacking the babies, they don't realize that walls are being torn down, and people of good will are making journeys they wouldn't normally make. I have never had any desire to meet or be with any white movement or white people in general. I didn't hate them. I just wasn't interested in them. I was interested in helping my own people. But now the attack on our babies has torn down walls and we all got to unite to fight a common enemy. This is a dragon.[5]

Robert Kennedy began his presentation to the minister, laying out some of the concerns about thimerosal and the other ingredients in the vaccines.

Minister Farrakhan was absolutely still as Kennedy spoke for several minutes, and it seemed to unnerve Kennedy after a while. He finally asked, "Minister, are you okay?"

Minister Farrakhan gave a small smile and said in a voice of gratitude, "I'm a student and I love knowledge. So I get quiet because you all are teaching me. But as you are talking, prophecy is popping up in my head, matching what you're telling me." He turned to Barry Segal and said:

> In the Old Testament there was a similar story. A pharaoh who thought the Children of Israel were multiplying and forgetting they were slaves. And he sent out a decree to kill all the male children under two years old, showing everybody who's in control, and what kind of nonsense was going to be tolerated. And as you were telling me about the vaccine schedule, a child at twelve or eighteen months that has to get this MMR shot, that's before they're two years old.
>
> That's in Scripture. It's what the satanic forces do to keep people slaves. That's why they hate social media. Because now CNN, NBC, CBS, they can't spin the message no more, by themselves. Social media is more powerful than CNN because people can do their own reporting. They are losing control.

The minister finished up by saying that these powerful forces get you under their control by telling you what you're going to put into your child's body, and then when that child gets injured, they've got a pharmaceutical customer for life.

Michelle Ford was following this information on the edge of her seat, and the tears slipping down her face were clear to everybody. Farrakhan asked if she was okay. She replied:

> Minister, I'm blown away. Many of my friends told me not to come here. I'm so happy because I went against them. My own husband was afraid for me to come here. And now I've got to go back and defend you and tell them, "We've got it wrong." I am not a fool. I came here looking for a hater. And I didn't care if you hated me. Even if you showed me hate, I just wanted the Nation to do something about black boys, about black children. But I'm listening to you and I'm messed up. I have great discernment and I know a hater when I'm in the presence of one.

The minister just laughed.

Kennedy finished his presentation, and then Hooker kept the dialogue going by talking about the Thompson information. When Hooker was reviewing the letter Thompson had written to Congressman Bill Posey, revealing how he and his coauthors had covered up the MMR data, Minister Farrakhan's eyes welled with tears. As Hooker was wrapping up his discussion, he said:

> Many people will tell you it's the mercury, Minister Farrakhan, but it's more than mercury. It's formaldehyde, aluminum salts, glyphosate, polysorbate 40, aborted human fetal tissues, maybe viruses from the animal tissues used to weaken the human viruses, or even the fact that with shots like the MMR we're making the body respond to multiple immune challenges. We're not really sure what's in those ingredients, or what's really causing the harm.[6]

"I see," said the minister, "but I am even more determined to get this out to the community. I thank you," he said, to both Brian Hooker and Robert Kennedy Jr. "You have brought this to the right man."

Barry Segal gave a warning to the minister: "You should know, minister, that the pharmaceutical companies are very powerful. They're worth billions of dollars. They're huge."

The minister chuckled. "I thank you for your concern, Mr. Segal. But I'm gonna be honest with you. I'm a man of God and I see the pharmaceutical industry as a gnat. They don't bother me." The minister then asked a question: "Are you sure you all want to stand with me on this issue? Because now that you have given me this information, you don't have to, because you have brought it to the right place."

The members of the group unanimously stated they wanted to stand with the Nation of Islam and would do so publicly. The minister was pleased. Then he had another question. "Mr. Kennedy, why haven't any other black leaders or black politicians stood up with you?"

"Many of the black politicians, as well as the white, are bought out by the pharmaceutical companies. They are the number one lobbyist group in the United States, donating twice as much as the oil companies. And they have no liability for their products."

"Oh, so now they have become pharaoh's magicians, have they?" asked the minister. "Oh well, I guess we'll have to expose them, as well. We will have to have community tribunals and we will bring up all of our black politicians on charges of treason, if they were told about this and did nothing."

When the minister looked around, he saw that there was support in the eyes of the group, as well as appreciation for the strength of his words.

He said, "This is meant to be. It's time for the white community to get to know me. You will get to know us and see we're not the monsters your community has made us out to be. We will fight this fight together, and as we fight, we'll get to know each other better. There's no better way for two groups to come together than to be fighting the same war."

The meeting lasted for more than four hours, and at the end, Minister Farrakhan asked if he could have his picture taken with the group.

"Doesn't the Nation of Islam have a publication, Minister?" Robert Kennedy Jr. asked. "Yes, it does," said the minister.

"Would it be okay to put the picture in your publication? Let everybody know we were here?" Kennedy looked around at the group, and they were all nodding in agreement.

"Are you all serious?" the minister asked. "You want a picture to go out with all of us?" Farrakhan was overwhelmed. "I've met with some of the top people in the country, Jewish rabbis, but none of them wanted nobody to know they was at my house. You all are the first of your caliber who has given us an okay to take a picture and allowed it to be put out."[7]

They took the picture, and Minister Tony Muhammad posted it on his Facebook page for all the world to see. It had been a small token, a courtesy extended by Minister Farrakhan to the man who had played such a vital role in their historic meeting.

* * *

After the meeting, Minister Farrakhan discussed the issue with Tony Muhammad and suggested he call up his old friend, Elijah Cummings, the powerful African American congressman from Baltimore, Maryland, and ranking member of the Committee on Oversight and Government Reform, which had authority over the CDC.

Tony made the call, and Cummings was happy to hear from him and asked how he could help. Tony told the congressman about the Thompson situation, what it meant specifically to black children, and asked for his help.

"Brother, we just had some of those CDC people up here and I kinda felt like they was lying to us," said Congressman Cummings. "I'm gonna look into this. I give you my word, Brother Mohammed."

Three days later, Congressman Cummings called back. "Brother Mohammed?" he said.

"Yes, sir?" replied Minister Tony.

"Look, I'm for vaccines, man. And I don't think I can deal with this."

Minister Tony immediately felt his blood pressure rise. "Wait a minute. I didn't tell you to be against vaccines, sir. That's not what we talked about. Good, I want you to be for vaccines. But don't you want them to be safe?"

"Look, we dealt with this. That man has been debunked."

"Who's been debunked?"

"Dr. William Thompson," declared Congressman Cummings.

"No, he hasn't! He hasn't spoken to nobody. He wants to be subpoenaed by you! Elijah, what happened?"

"I'm for vaccines, and I'm not going to deal with this issue. I don't mean you no harm. You just have to tell the minister that I'm not dealing with this."

Tony knew his anger was being felt on the other end of the line. "You sure? You sure you want me to tell THAT to the minister?"

The next thing Tony heard was the sound of Congressman Elijah Cummings, his old friend, hanging up on him.

Minister Tony immediately contacted Minister Farrakhan and told him what had happened. Farrakhan was disappointed, but not surprised:

> There's no telling what they've got on him. This is what they do to our black politicians and our black leaders who can't see they're being used. They invite many of them to high-level parties, get them into these compromising positions, whether it's drugs or infidelity, get all kinds of data and information, and use it against them. Or they give them so much money, and record that, so if they ever turn on them, they can threaten them with exposing it.[8]

Both of them knew that the FBI had done the same thing with Martin Luther King Jr., tape-recording his infidelities with other women, then sending the recordings to his wife, Coretta, and even going so far as sending him a letter telling King that he should kill himself. King hadn't let himself be bullied by the FBI, but he was the rare black leader who didn't succumb to the intimidation.

Farrakhan's remarks became a little more philosophical, believing that justice would eventually arrive:

> This act is so criminal. When the American people find out the truth about what's going on at the CDC, I feel sorry for all the white and black politicians. Brother, you are dealing with people's children. You ain't seen Caucasians get angry. They will go into the statehouses, grab these politicians, tie them to the back of a truck, and drag them down the street. The American people are at a

> boiling point. They'd better tell the truth while they have a chance. Because if the truth comes out in any other way, it could tear up this country. This truly is something that could tear up this country.

Minister Tony Muhammad felt he finally understood the enormity of this problem and the power of the pharmaceutical industry. He'd known Congressman Cummings for decades, considered him a friend and a fellow warrior for African Americans, and they'd turned his head around in the space of a few days. This really was a fight against the dragon. This was a monstrous evil, and it was harming children on a daily basis. Was there ever a more worthy fight than defending the babies?

***

Minister Tony Muhammad took the lead for the Nation of Islam in California to fight Senate Bill 227, authored by Senator Richard Pan, which would require all California children to follow the CDC's recommended vaccination schedule as a condition for going to school. He coordinated with the various groups that were protesting at the state capitol in Sacramento, primarily vaccine safety groups, but also members of libertarian, Tea Party and parental rights groups who were all concerned about various aspects of the proposed law. These groups generally didn't look at the issue in spiritual terms, as did Minister Farrakhan, but as an issue of uncontrolled government power. If the government could tell you which medicines and chemicals you had to put in your child's body and possibly damage them for life, what couldn't they do?

For Minister Tony, it was an interesting experience working so closely with groups that were predominately white. He recalled that when they asked him to speak on the steps of the state capitol, he gave a brief, five-to-seven-minute speech. When he finished and looked at the crowd:

> It was as if these people thought they'd heard Jesus himself. I said, "You all ain't used to this kind of talk?" This ain't nothing. In the Nation we confront evil. We stare down police officers. So we don't run from anything. We don't hurt people. In the Nation it's against our law. We don't even carry weapons. If a member of the Nation owns a weapon, he'll get put out of the Nation. The only weapon we believe in is truth.[9]

Minister Tony believes the Nation of Islam can play a vital role in the vaccine/autism fight, as sometimes an outsider can see something that the longtime members miss and can also bridge differences between various groups. When a dispute arose between some who wanted to focus more on thimerosal and others who wanted the emphasis to be on the MMR shot, Tony was quick to jump into the fray:

> It's boring if everybody brings the same weapon to fight the enemy. I said to some in the vaccine community, when America goes to war she calls on her European allies. They don't all bring the same weapon. One has better intelligence. One has a better navy. One has a better air force. One has a better army. But damn it, when they got a common enemy, they go to war. I said, the same thing here. If one is working on thimerosal, one is working on MMR, who gives a damn? We are fighting the same enemy and you all need to stop it. I'm a black man who is now in the vaccine movement and I'm gonna call you all together and I'm gonna call you all out. I'm having to use the bully pulpit, man. Because we all got to stand together. And that's what another Caucasian lady said to me. She said, "We needed you because we don't fight. We just live like Americans. We never really had to fight for anything. We established this country." I said, "I know. And that's why you're taking it lying down. You're saying, bite me one bite at a time. You all need the salt of the Earth. You need black people. You know how black people get when we think something is wrong? We get all emotional and we get loud!" If all of us are standing up against the pharmaceutical industry, asking our government to do right by its people, what's wrong with that?[10]

* * *

On June 17, 2015, Dr. Brian Hooker sat in a pew at a United Methodist Church in Los Angeles, CA, listening to Minister Louis Farrakhan talk to an interfaith audience of approximately fourteen hundred people about CDC whistleblower William Thompson. Hooker was pleasantly surprised to find that Minister Farrakhan actually seemed to have a good handle on the science. Most people had trouble with science. Near the end of his talk, Minister Farrakhan urged the audience to oppose California Senate Bill 227, which would forbid children from attending school unless they had been injected with all of the CDC's recommended vaccines. He also urged them to contact their congressional representatives and demand that Dr. William Thompson be subpoenaed to appear before Congress to testify about the MMR vaccine/autism cover-up.

Later that night, Hooker attended a meeting hosted by Minister Farrakhan at a private home in Tarzana, California, for about a hundred leading black entertainment figures so they could discuss the matter in greater depth. Again, Hooker was impressed with Farrakhan's understanding of the science, and he also found many people coming up to him to ask questions.

The next day, June 18, 2015, opened with an unexpected piece of good news. An article in the *Sacramento Bee* by Jim Miller listed the politicians who had received the most money from the pharmaceutical industry.[11] Topping the list was Senator Richard Pan, the author of SB 277, with $95,150. Close behind in the money race was Speaker of the Assembly Toni Atkins, with $90,250. The article reported pharmaceutical companies and their associated trade groups had donated more than two million dollars to the members of the California legislature in the 2013–2014 legislative session.[12] For two million dollars, the politicians of California were willing to take away the rights of parents to determine which chemicals and viruses they would allow to be injected into their children's bodies.

That night, Robert Kennedy Jr., Minister Tony Muhammad, and Brian Hooker took the stage for a town hall meeting at the Church of Scientology Community Center in Los Angeles to talk about the CDC whistleblower allegations, as well as Senate Bill 277. The founder of the Church of Scientology, L. Ron Hubbard, had been wary of pharmaceutical drugs and believed the industry would eventually try to wrap itself around the politicians and deliver people into their hands. Also at that meeting were representatives of the Weston Price Foundation, who believed in the principles of natural health.

Minister Tony opened the meeting, then was followed by Robert Kennedy Jr., who compared what was happening in the African American community to a new Tuskegee experiment, and then it was time for Hooker. Brian took to the stage a bit nervously, looking out at the sea of white, black, and brown faces of more than a thousand people who wanted to hear his words. He took a deep breath, whispered a quick prayer, and began:

> Autism and neurological injury due to vaccinations are extremely important problems specific to the African American community. There are strong evidences in the scientific literature that African Americans may be more susceptible to vaccine injury and may also have increased susceptibility to neurological disorders such as autism. The most reliable studies show that autism incidence is higher in African Americans as compared to Caucasians.

Durkin et al. (published in 2010 in the journal *PLOS One*) applied a correction to autism incidence to account for under-reporting at lower socio-economic status and found that autism incidence was about 25 percent higher in African Americans as compared to Caucasians. This was determined in a nationwide study using the CDC's Autism and Developmental Disability Monitoring Network. Further, in a study by Becerra et al. (published in 2014 in the journal *Pediatrics*), it was shown that the incidence of autism among African Americans in Los Angeles County was higher than that of Caucasians. The effect was most profound in foreign-born blacks (living in the US) with a 76 percent greater risk of autism as compared to US born whites. The effect was also seen to a lesser extent (14 percent greater risk) in US born blacks. However, when considering children with severe autism (autism with mental retardation), Becerra et al. found that the incidence was much higher in foreign-born blacks (163 percent greater) as well as US born blacks (52 percent greater) as compared to US born whites. This pronounced effect was not observed in any other race category considered.

In terms of vaccine injury, let me be clear—I am not antivaccine. I want safer vaccines that protect and not harm children. I want populations vulnerable to vaccine injury to be identified and protected as well. You don't call someone who wants safer automobiles, "anticar." Similarly, it is ridiculous to refer to vaccine safety advocates as "antivaccine."

In terms of vaccine injury, the study by Gallagher et al. (published in 2010 in the *Journal of Toxicology and Environmental Health*, Part A) showed that blacks were at significantly greater risk of regressing into autism after receiving the thimerosal-containing Hepatitis B vaccination series as infants. Thimerosal is a mercury-based preservative that is used in some vaccines in multidose vials and is still used in the flu shot, the tetanus vaccine and meningococcal pneumonia vaccine and is also in trace amounts (sufficient to cause harm) in the Hepatitis B, Hemophilus influenza B (HiB) and DTaP vaccines. The data show a 5.53 times greater risk of autism for black boys receiving the thimerosal-containing HepB vaccine series versus those black boys not receiving any HepB shot. White boys did not show a statistically significant risk in this instance.

Further, background information released by the CDC whistleblower, Dr. William Thompson, showed that the CDC found higher risks of autism in black children who received the MMR vaccine on time versus those that received the vaccine after 3 years of age. Unpublished data released by the CDC whistle blower show that black boys were up to 3.36 times greater risk of receiving an autism diagnosis when they received their first MMR vaccine prior to thirty-six months of age versus those black boys receiving their first MMR vaccine at or after thirty-six months of age. This effect was not observed in any other race category considered.

Although the CDC attempted to hide this information (which was discovered by Dr. Thompson on November 7, 2001), Dr. Thompson ultimately issued an August 27, 2014 press release through his attorney stating, "I regret that my coauthors and I omitted statistically significant information in our 2004 article published in the journal Pediatrics. The omitted data suggested that African American males who received the MMR vaccine before age 36 months were at increased risk for autism." Dr. Thompson further stated in his press release, "My concern has been the decision to omit relevant findings in a particular study for a particular sub- group for a particular vaccine. There have always been recognized risks for vaccination and I believe it is the responsibility of the CDC to properly convey the risks associated with receipt of those vaccines."

Over the period of November 2013 to August 2014, I had over thirty separate phone conversations with Dr. Thompson. He initially reached out to me in an unsolicited phone conversation to my cell phone. Dr. Thompson and I had talked on the phone and exchanged email correspondences much earlier, between 2002 and 2004, back when I was trying to advise the CDC on their vaccine safety studies related to childhood neurodevelopmental disorders. However, the CDC curtailed my conversations with him in 2004 due to my family's participation in the National Vaccine Injury Compensation Program where we were seeking remuneration for my own son's vaccine injuries. The phone calls from November 2013 to August 2014 were secret and Thompson did not let CDC officials know that he and I were talking as that could have cost him his employment.

I made the decision to record four of the last phone conversations I had with Dr. Thompson, without his knowledge, based on the revelation of harm to children, caused by the CDC's very dysfunctional and even criminal vaccine safety program. These recordings were obtained legally and involved advice from legal counsel in each instance.

In my phone conversations with Thompson, he also discussed thimerosal- containing vaccines. Dr. Thompson revealed adverse neurological outcomes specifically in boys exposed to thimerosal in vaccines within their first seven months of life. This consisted of motor and phonic tics present in "neurotypical boys" tested in standardized tests.

Although Dr. Thompson did not comment regarding the relationship between thimerosal and autism, he did note that tics were about five times more prevalent in autistic boys compared to the general population.

Dr. Thompson also described a culture of fraud in the CDC, an institution with a built-in conflict of interest regarding vaccine update versus vaccine safety. The CDC buys over four billion dollars of vaccines each year from the pharmaceutical industry to distribute to the states' public health departments. Vaccine uptake in the US must be high for the CDC to get reimbursed for that purchase. Thus, vaccine safety scientists are under tremendous pressure not to find associations between vaccines and neurological adverse events, among others. He has been specifically told "point blank" from his superiors in multiple instances to not report such findings and to find ways using fraudulent statistical methods to obviate the results and falsely give vaccines a clean bill of health. Dr. Thompson stepped forward due to the agony of over ten years of lying and covering up the real truth regarding vaccine injury.

I also wanted to talk about another specific whistle blower lawsuit, regarding MMR's effectiveness. There is a False Claims Act lawsuit pending against Merck in the U.S. District Court for the Eastern District of Pennsylvania, Case No. 10–4373 (CDJ). This case was brought by two former Merck virologists who were involved in the efficacy testing of the mumps portion of Merck's MMR vaccine. According to these scientists, Merck engaged in fraudulent testing and data falsification to conceal the vaccine's diminished efficacy.

As a result of Merck's fraudulent scheme, the scientists allege, American children are being injected with a vaccine that does not provide the efficacy Merck claims it provides and does not provide the public with adequate immunization. According to the scientists, Merck's MMR vaccine contributed to the recent mumps outbreaks in the US. Late last year, the Court denied Merck's motion to dismiss the case and the case is in the discovery phase.

SB 277 removes the last "check and balance" in preventing vaccine injury in children, parental consent rights. In the past, parents have been able to opt out of vaccines for their children based on personal beliefs, without jeopardizing school attendance. SB277 will change all that whereas the only children that will be able to attend school will be either fully vaccinated or receive a very "difficult to obtain" medical exemption based on some condition that would increase her/his susceptibility to vaccine injury. These exemptions are rare and extremely difficult to obtain. Based on CDC guidelines, even if an earlier vaccine leads to seizures or the death of a sibling, the child is still not exempt and this is being widely misrepresented by the proponents of SB277. Homeschool children will be exempt from the law but this is just not an

> option considering the large number of two income families in our underserved communities. I urge you to contact your state Assembly members and tell them to vote NO on this bill. I urge you to reach out to the legislative black caucus members and educate them about the CDC whistle blower and other issues regarding vaccine injury that make this bill nothing but medical tyranny.
>
> We want Congress to subpoena Dr. William Thompson. In fact, Dr. Thompson himself wants to be subpoenaed so the entire truth about the CDC can become public record. I urge you to contact key Congressional offices to ask that Dr. Thompson be subpoenaed in an open Congressional hearing. The truth needs to come out, period, and this is one way to bring the truth to light.[13]

Hooker finished and took a deep breath. There was a thunderous applause in the room, and Hooker felt he had done well with his speech. Many people asked questions, and Hooker did his best to answer them, knowing even then that they would be unlikely to block the legislature from passing the bill. There was some hope they might be able to get Governor Jerry Brown to veto the bill, but that also turned out to be a vain hope.

But there was no doubt that something new had been created. An alliance was forming that could not be denied. There would certainly be battles lost in the near future, but even so it seemed like the course of the war was changing.

***

On July 29, 2015, Congressman Bill Posey took to the floor of the House of Representatives to discuss the CDC whistleblower Dr. William Thompson and urge his congressional colleagues to subpoena Thompson. I have made several attempts to interview Congressman Posey, but his staff tells me he wants to pursue this matter quietly. I am disappointed in this response, as I think the issue should be shouted from the rooftops. But there is no question in my mind that Congressman Posey is the greatest advocate we have in Congress for pursuing the corruption at the CDC and getting the truth to the American public.

When I watch the video, it seems to me that Congressman Posey is nervous as he begins to speak, and his voice shakes, but even through his fear of what attacks may follow, he will not be turned away. His five-minute speech is nothing less than a profile in courage:

I rise today on matters of science and research integrity.

To begin with, I am absolutely, resolutely, pro-vaccine. Advancements in medical immunization have saved countless lives and greatly benefitted public health. That being said, it's troubling to me that in a recent Senate hearing on childhood vaccinations, it was never mentioned that our government has paid out over 3 billion dollars through a Vaccine Injury Compensation Program for children who have been injured by vaccinations.

Regardless of the subject matter, parents making decisions about their children's health deserve to have the best information available to them. They should be able to count on federal agencies to tell them the truth. For these reasons I bring the following matter to the House floor.

In August 2014, Dr. William Thompson, a senior scientist at the Centers for Disease Control and Prevention, worked with a whistleblower attorney to provide my office with documents related to a CDC study that examined the possibility of a relationship between mumps, measles, rubella vaccines, and autism. In a statement released in August 2014, Dr. Thompson stated, "I regret that my co-authors and I omitted statistically-significant information in our 2004 article published in the *Journal of Pediatrics*."

Mr. Speaker, I respectfully request the following excerpts from the statement written by Dr. Thompson be entered into the record. Now, quoting Dr. Thompson:

> My primary job duties while working in the Immunization Safety Branch from 2000 to 2006 were to lead or co-lead three major vaccine safety studies. The MADDSP-MMR Autism Cases-Control study was being carried out in response to the Wakefield Lancet Study that suggested an association between the MMR and an autism-like health outcome. There were several major concerns among scientists and consumer advocates outside the CDC in the fall of 2000 regarding the execution of the Verstraeten Study. One of the important goals that was determined up front in the spring of 2001 before any of these studies started, was to have all three protocols vetted outside the CDC prior to the start of analyses, so that consumer advocates could not claim that we were presenting analyses that suited our own goals and biases. We hypothesized that if we found statistically significant effects, at either 18 or 36 month thresholds, we would consider that vaccinating children early with MMR vaccine could lead to autism-like characteristics or features.
>
> We all met and finalized the study protocol and analysis plan. The goal was to not deviate from the analysis plan to avoid the debacle that occurred with the Verstraeten Thimerosal Study, published in *Pediatrics*

> in 2003. At the September 5th meeting we discussed in detail how to code race for both a sample and the birth certificate sample. At the bottom of Table 7 it also shows that for the non-birth certificate sample, the adjusted race-effect, statistical significance was HUGE.
>
> All the authors and I met and decided sometime between August and September 2002 not to report any race effects for the paper. Sometime soon after the meeting we decided to exclude reporting any race effects, the co-authors scheduled a meeting to destroy documents related to the study. The remaining four co-authors all met and brought a big garbage can into the meeting room and reviewed and went through all the hard copy documents that we had thought we should discard and put them in a big garbage can. However, because I assumed it was illegal and would violate both FOIA and DOJ requests, I kept hard copies of all documents in my office, and I retained all associated computer files. I believe we intentionally withheld controversial findings from the final draft of the *Pediatrics* paper.
>
> Mr. Speaker, I believe it's our duty to ensure that the documents Dr. Thompson provided are not ignored. Therefore, I will provide them to members of Congress, and the House Committee upon request.
>
> Considering the nature of the whistleblower's documents, as well as the involvement of the CDC, a hearing and a thorough investigation is warranted.
>
> So I ask, Mr. Speaker . . . I beg . . . I implore my colleagues on the appropriations committee, to please take such action. Thank you, Mr. Speaker. I yield back.[14]

It had been an amazing period of time, from the whistleblower revelations of William Thompson to Brian Hooker, the part played by Andy Wakefield, the previous efforts of Congressman Dan Burton and Dave Weldon, the passage of the documents to Congressman Posey, as well as the alliance with the Nation of Islam, the Church of Scientology, the Weston Price Foundation, and various libertarian, Tea Party, and other organizations. At the twentieth anniversary of the Million Man March on Washington, DC, Minister Tony Muhammad gave a short speech on the MMR vaccine/autism allegations to the hundreds of thousands gathered on the National Mall. On October 24, 2015, the Nation of Islam led a protest at the CDC and insisted that Dr. Thompson be subpoenaed by Congress about his allegations. In May 2016, Minister Tony Muhammad went on a forty-city tour, speaking at black churches across the country about the Thompson allegations.

CHAPTER TEN

DeNiro, Tribeca, and the Real *Goodfellas*

When I interviewed Dr. Andy Wakefield on February 25, 2016, he mentioned he was working on a documentary about the CDC whistleblower, William Thompson. Wakefield was joined in this effort by Dr. Brian Hooker, and I would later find out that Brandy Vaughn also had a good deal of screen time, sharing an insider's perspective on the pharmaceutical industry. I was pleased to hear about this effort, though privately I wondered if they would be able to make much of an impact. Other fine documentary filmmakers had attempted to cover this issue, notably *The Greater Good* by Leslie Manookian, *Bought* by Jeff Hayes and Bobby Sheehan, and *Trace Amounts* by Eric Gladen and Shiloh Levine. I had enthusiastically written about all of these films on the *Age of Autism* website and even hosted a screening in San Francisco for *Trace Amounts.* But it seemed to me that for the most part we were just talking to our own community. We were asking the real-life equivalent of asking the age-old philosophical question, *If a tree falls in a forest and nobody hears it, did it make a sound?* If you release a controversial movie and the press doesn't talk about it, did you even make it?

I expected there would be a media blackout of the film. They would simply act as if it didn't exist and it would die of loneliness.

I could not have been more wrong.

***

On Monday, March 21, 2016, Dr. Wakefield announced that his documentary film, *VAXXED: From Cover-up to Catastrophe*, had been made an Official Selection at the Tribeca Film Festival, founded by the actor Robert De Niro in the wake of the September 11, 2001, terror attacks, to bring commerce back to the southern part of Manhattan. The film was scheduled to have its screening on Sunday, April 24, 2016, the closing day of the festival. Suddenly, Robert De Niro was added to the litany of dangerous antiscience zealots like Andy Wakefield, Robert Kennedy Jr., Congressman Dan Burton, Congressman Dave Weldon, Congressman Darrel Issa, Congressman William Posey, Jenny McCarthy, and Jim Carrey. Could it be that the no-nonsense DeNiro, famed for his role in movies like *Taxi Driver, Raging Bull, The Godfather*, and *Goodfellas*, would join with the tinfoil hat crew?

And among us in the tinfoil hat brigade, we wondered whether the actor known for his tough-guy roles would stand up for a movie about injured children and a whistleblower scientist from the CDC who had turned over thousands of pages of documents to Congress and had been patiently waiting for nearly a year and a half to be subpoenaed by Congress to testify about the issue. (If Thompson wanted to retain his whistleblower status, he could not talk about the case to the press before he had talked to Congress.) Many suspected that the pharmaceutical industry was doing its best to exert influence among the politicians to whom it had so generously donated over the years to prevent Thompson from ever testifying to the American public.

Media outlets ranging from the *New York Times*, the *Washington Post*, *Forbes* magazine, *People*, and *Glamour* immediately jumped into the fray, speaking in a single, unified, amazingly creepy, anti-free-speech voice that the film should not be shown at Tribeca. An example from a March 25, 2016, *New York Times* article:

> The plan to show the film has unnerved and angered doctors, infectious disease experts and even other filmmakers.
>
> "Unless the Tribeca Film Festival plans to definitively unmask Andrew Wakefield, it will be yet another disheartening chapter where a scientific fraud continues to occupy a spotlight and overshadows the damage he has left behind in the important story of vaccine safety and success," Dr. Mary Anne Jackson, a professor of pediatrics at the University of Missouri-Kansas City, said in an email.
>
> The documentary filmmaker Penny Lane ("Our Nixon") published on Thursday an open letter to the festival's organizers in Filmmaker Magazine,

> suggesting that "Vaxxed" in the documentary section "threatens the credibility of not just the other filmmakers, but the field in general."[1]

A similar tone was taken by a writer for *Glamour* magazine on that very same day, March 25, 2016. We in the antivaccine movement learned long ago that this is just a "coincidence" and understand that just because the same line was used across many different media outlets on the same day does not mean there was any coordination of message by powerful and corrupt health organizations trying to hide the truth from the public. The article in *Glamour* began:

> In a true "WTF" moment, the Tribeca Film Festival will screen a documentary about a supposed conspiracy to cover up the dangers of vaccines film by discredited doctor and anti-vaccine movement leader Andrew Wakefield.
>
> For those who may be unfamiliar with Wakefield, here's a little background on the British ex-physician. Formerly a gastroenterologist, Wakefield rose to prominence in the late 1990s after publishing a paper linking vaccines to autism—essentially single-handedly inciting the anti-vaccine movement. In a case study of only twelve children, Wakefield suggested a connection between inflammatory bowel disease in children who had recently received the measles, mumps, rubella (MMR) vaccine and the development of autism. But in 2010, he was stripped of his medical license by the General Medical Council in the United Kingdom for conducting "unnecessary, invasive tests on children" specifically, lumbar punctures and colonoscopies intended to bolster his claims and forge a connection between MMR-induced gastroenterological disorders and autism-and behaving "dishonestly and irresponsibly."[2]

While I do not generally get my science and medical news from *Glamour* magazine, I have to admit that there was a good deal the writer got right in those first few paragraphs. Wakefield was stripped of his license by the General Medical Council in 2010 for conducting what they considered to be "unnecessary, invasive tests on children." It is apparently much better to not perform such tests (and a later investigation by Dr. David Lewis, mentioned earlier, shows Wakefield was allowed to perform such tests) and to forever declare those areas of inquiry off-limits. It's a little like Columbus sailing to the New World, returning to tell Queen Isabella of Spain about his discoveries, and then being told that in the future no ships may sail beyond the Azores Islands.

In case the allegation of "unnecessary, invasive tests on children" suffering from a disease whose cause is still unknown did not strike you as enough of a crime against humanity, *Glamour* magazine was ready to provide the public with the proper perspective:

> Unsurprisingly, the inclusion of Wakefield's "documentary" is controversial among both film critics and medical professionals alike have slammed the inclusion of the film, with *New Yorker* film critic Michael Specter ardently criticizing the decision telling the *Los Angeles Times*: "It's shocking. This is a criminal who is responsible for people dying. This isn't someone who has a 'point-of-view.' It's comparable to Leni Riefenstahl making a movie about the Third Reich, or Mike Tyson making a movie about violence towards women. The fact that a respectable organization like the Tribeca Film Festival is giving Wakefield a platform is a disgraceful thing to do."[3]

Do we understand the moral equivalence expressed by *Glamour* magazine? A doctor investigating the potential causes of autism is equivalent to a Nazi filmmaker or a domestic abuser. On the one hand you have those who are responsible for the death of millions in concentration camps and a war that left a great amount of the world in ruins. On the other, you have a scientist trying to determine what causes a disease that affects more than a million children in the United States alone. Both are going to end up in the same circle of hell, obviously. On Friday, March 25, 2016, the autism community held its breath as the onslaught against the whistleblower film seemed to reach its maximum force. The provaccine forces seemed to have been taken by surprise by the Monday announcement, but by Friday they were in full battle mode. What would be the outcome?

On Friday morning, Robert Kennedy Jr. sent out a quick email to a group of autism advocates, of which I was one. The email read: "De Niro just called me. We spoke for 30 min. He's going forward with film. Under huge pressure to not screen film."[4] The community was elated, and it seemed to be supported by a statement De Niro put out that day, as reported by the *New York Times*:

> Mr. De Niro's statement seemed to suggest that this was the first time he has expressed a preference that a particular film be shown at the festival.
>
> "Grace and I have a child with autism," he wrote, referring to his wife, Grace Hightower De Niro, "and we believe it is critical that all of the issues surrounding the causes of autism be openly discussed and examined. In the

> 15 years since the Tribeca Film Festival was founded, I have never asked for a film to be screened or gotten involved in the programming. However this is very personal to me and my family and I want there to be a discussion, which is why we will be screening VAXXED."[3]

The statement was important for a number of reasons. While many in the autism community had known that Robert De Niro had a son with autism, it was not public knowledge. The movie also had additional relevance to the De Niro family, since Grace Hightower De Niro is African American, exactly the group identified by the CDC whistleblower Dr. William Thompson as being at increased risk from earlier administration of the MMR vaccines. With this public acknowledgment, De Niro joined other well-known actors with autistic children, including John Travolta and Sylvester Stallone.

On Saturday morning, I called up my good friend J.B. Handley, cofounder of the group Generation Rescue, to express my delight over this remarkable show of support from De Niro: "They lied to us about Simpsonwood and got away with it, made the Congressional investigations into mercury look like they took place in a black hole, they went after Jenny McCarthy, Robert Kennedy Jr., and when Jim Carrey got on our side they were able to paint him as a wackadoodle. But what are they going to be able to do about this film that Robert De Niro is standing behind?"

"I know, dude. It's amazing. De Niro is not somebody you mess with," Handley replied.

"This is Robert De Niro. The ultimate tough guy." I proceeded to do a really bad impersonation of the Travis Bickle character De Niro played in *Taxi Driver*. "Are you talking to me? Are you talking to me? Because I don't see anybody else around. So you've got to be talking to me!"

"And he's got a kid with autism," said Handley. "He's one of us."

"De Niro is not going to back down," I said, as we ended our phone conversation.

I could not have been more wrong.

***

Just a few hours after my really bad Travis Bickle impersonation, word came that De Niro had reversed his stance and decided to pull *VAXXED: From Cover-Up to Catastrophe* from the Tribeca Film Festival. The *New York Times* wasted little time in trumpeting the news, putting out an article on March 26, 2016, just one day after De Niro had publicly defended the film:

> In a statement, Robert De Niro, a founder of the festival, writes: "My intent in screening this film was to provide an opportunity for conversation around an issue that is deeply personal to me and my family. But after reviewing it over the past few days with the Tribeca Film Festival team and others from the scientific community, we do not believe it contributes to or furthers the discussion I had hoped for."[6]

And for De Niro's capitulation, the seventy-two-year-old actor was given a pat on the head like a beloved, but rebellious, child who had finally seen the error of his ways:

> Dr. William Schaffner, a professor of preventive medicine at the Vanderbilt University School of Medicine, said on Saturday that he believed, "the entire board as well as Mr. De Niro have learned a lot in the last several days."
>
> "My hat is off to them for listening, thinking about it, discussing it, and responding," he said.
>
> Nevertheless, Dr. Schaffner said, it was troubling for scientists that a film promoting "discredited ideas" got so close to a forum as prestigious as the Tribeca Film Festival.
>
> "It gave these fraudulent ideas a face and a position and an energy that many of us thought they didn't deserve," he said. "We're all for ongoing reasonable debate and discussion, but these are ideas that have been proven to be incorrect many, many, many times over the past 15 years."[7]

Just in case you weren't really reading the last sentence of the previous paragraph, the respected Vanderbilt scientist used the word "many" three times, in describing how incorrect Wakefield's ideas are. That's a lot of times for a scientist to use the word "many" in a single sentence and is an indication that he really, really, *really* believes what he is saying. I guess it also means he is a very, very, very smart man.

An article from *Deadline* magazine on the same day, March 26, 2016 (Just because this article came out on the same day as the *New York Times* article and had a similar take, it's just a coincidence. I'm sure there was no collaboration or media "messaging" involved.), contained a statement from Andrew Wakefield, and his producer on the film, Del Bigtree:

> "Robert De Niro's original defense of the film happened Friday after a one-hour conversation between De Niro and Bill Posey, the congressman who has interacted directly and at length with the CDC Whistleblower (William Thompson)

> and whose team has scrutinized the documents that prove fraud at the CDC. It is our understanding that persons from an organization affiliated with the festival have made unspecified allegations against the film," the statement continued, "claims that we were given no opportunity to challenge or redress. We were denied due process. We have just witnessed yet another example of the power of corporate interests censoring free speech, art, and truth."[8]

And what was the "organization affiliated with the festival" that was interested in having the film withdrawn from the festival? One well-known blogger has suggested that it was the Alfred P. Sloan Foundation, a sponsor of the festival, which supposedly owns many shares of Merck stock, the sole supplier of the MMR vaccine to the American market.[9] However, at this point, the answer to exactly what arguments were made to De Niro to make him pull the film remains unknown. De Niro himself will need to further flesh out this question.

The allegation by Andrew Wakefield and Del Bigtree that prior to his initial defense of the film, Robert De Niro talked for an hour with Congressman Bill Posey is supported by an email that Robert Kennedy Jr. sent out to a group of autism advocates on Sunday, March 27, 2016. In the email, which gave some detail to his earlier and shorter explanation of his discussion with De Niro, Kennedy wrote:

> Just FYI. De Niro called me after talking to Posey. He asked if I would come to screening to defend film. I said yes. He and Grace asked for a copy of my book (Posey's suggestion) and a *Trace Amounts* screener. I told him he was walking into a typhoon. He said he intended to show the film no matter. I checked my schedule, realized I had a conflict and someone from my office informed them. They cancelled shortly thereafter. I'm not implying causation, only giving the chronicle. I know he was in a shitstorm beyond any experience and assume he got threats from sponsors.[10]

It may seem ironic that Kennedy, a liberal democrat with a legendary family name, would find himself working so closely with a Republican congressman, or that he would find himself so at odds with a medium that is overwhelmingly liberal, but such are the strange dynamics of this issue. One would expect the press to defend the right of free speech, following the view most commonly attributed to the eighteenth-century French philosopher, Voltaire, that "I may disapprove of what you say, but I will defend to the death your right to say it."

The situation becomes even more absurd when one considers that the press was working themselves into a lather over a film that they had not even seen. Is it too much to expect for the press to examine the evidence before writing an article on the subject? Remember, the documentary is about a whistleblower at the CDC making allegations that evidence on an important issue has been systematically suppressed. Would any scandal in history have ever been revealed if reporters simply accepted the bland denials of the organization or individuals implicated? Al Capone always claimed to be a businessman, rather than a mafia kingpin. President Richard Nixon said he was not a crook. The tobacco companies proclaimed smoking was safe and not linked to lung cancer. The oil industry said leaded gasoline was safe for the environment. The Catholic Church was indignant at the allegation that its priests were sexually abusing children. The financial institutions claimed the housing market was solid, right up until the 2008 economic meltdown. The National Football League claimed its players were not suffering from concussion injuries. Will I sound too conspiratorial if I ask the simple question: If you have not viewed the film, who is telling you what to think about it? Do these thoughts simply spring independently from the brains of multiple journalists at different publications and miraculously appear on the same day?

Many autism parents were so upset over De Niro's sudden turnaround that they ordered a bunch of t-shirts that read, TOUGHER THAN ROBERT DE NIRO, implying that they would never have folded in the face of such pressure. I ordered one immediately.

Emily Willingham, a contributor to *Forbes* magazine, seemed to be a day late to the party, publishing her article on March 27, 2016, but maybe that's because she was including the information about the interest of Congressman Bill Posey in the *VAXXED* documentary. Her article for *Forbes* magazine was titled "Why was Rep. Bill Posey Involved in Tribeca-De Nero-Wakefield Kerfuffle?" Here is her explanation:

> Some research suggests that people who are drawn to conspiracy theories tend to find themselves engaged in more than one web. Posey seems to have bought into the idea that CDC researcher William Thompson's Texas two-step around allegations of data destruction at the agency constitute just another such conspiracy. Swayed by helpfully annotated, repetitive documents provided by Thompson, Posey seems to be convinced that there be a conspiracy afoot here, too.[11]

If you are to follow Willingham's logic, if you believe that Al Capone was a mafia kingpin (still unproven because he was convicted of tax evasion, rather than murder and racketeering), Nixon was a crook (again, never proven because he was given a pardon), tobacco is linked to lung cancer, leaded gasoline is not safe for the environment, the Catholic church protected pedophile priests, the financial institutions lied about the stability of the housing market, and professional football players are at an increased risk for brain damage from concussions, then you are a conspiracy theorist.

About the only voices asserting those apparently antiquated American notions that issues should be debated in the marketplace of ideas and that citizens will eventually choose the ideas that made the most sense to them were voices most often associated with conservative right-wing politics. From an article by Jon Rappaport on the website Infowars about the removal of the film from the Tribeca Film Festival:

> You see, the parents themselves are children wandering in the wilderness, with no ability to analyze information. They must defer to the experts. They mustn't listen to other voices. They mustn't be allowed to think. Free speech? Never, ever heard of it. You see, this is Science. Only certain people know what science says or means. They are the chosen few in the palace. They decide for the rest of us. They are the little gods and censors. I don't know about you, but I am sick of this bullshit. On big screens all over this country, you can put up movies depicting people being torn limb from limb, drowning in their own blood, you can put up movies with panting soft-porn money shots, you can put up movies that blow up half the world; but you can't show a movie that questions the effects of vaccines.[12]

Personally, I find it difficult to disagree with the notion that people should be allowed to see a film that questions conventional wisdom. Perhaps it is my training as an attorney, because I was taught that the underpinning of any credible justice system is that every person is allowed to present their side, even if the majority is against them. The chant of many civil rights activists today is "No justice! No peace!" But this is an age-old idea. Preventing one side from presenting their side is a recipe for disaster in any society.

***

I'm a big fan of action-adventure movies, the kind where some plucky hero saves the world from a terrible fate, usually fighting off a deranged supervillain. I'm especially fond of the part in those movies where the villain thinks he is about to triumph over the hero and goes into an extended monologue about his evil plans. This is often referred to as "dialoging." This usually gives the hero time to escape, or gives him or her a critical piece of information needed to defeat the villain. If the narcissistic villain had simply kept his mouth shut and not wanted to impress the hero with his superior intellect, the villain could have succeeded. I had thought only villains in movies had this compulsion to reveal all of their dastardly schemes and plots.

I could not have been more wrong.

At this point I'm sure that many readers are probably on the fence about my assertion that the sudden tidal wave of opposition that arose in response to the screening of Dr. Wakefield's documentary film, *VAXXED*, was part of a coordinated plan. I submit for your consideration an article that published on March 29, 2016, in *The Guardian*, an English newspaper known for breaking the story of NSA whistleblower Edward Snowden and that has more recently have published articles about the inadequate Chinese response to COVID-19, and suggestions the virus may have been engineered in a lab:

> Within a half hour of Robert De Niro's Tribeca Film Festival posting on Facebook that it had scheduled an April viewing of Vaxxed, the highly controversial anti-vaccine documentary, a well-oiled network of scientists, autism experts, vaccine advocacy groups, film-makers and sponsors cranked into gear to oppose it.
>
> At the center of the network was a listserv group email list of more than 100 prominent individuals and science research bodies run out of the Immunization Action Coalition (IAC) based in St. Paul, Minnesota. The listserv acts an early warning system that sounds the alarm whenever the potent conspiracy theory that autism can be caused by vaccination surfaces.[13]

I guess you can't call it a conspiracy if the members of the conspiracy are willing to discuss their actions so openly. Let's call it what it is: a group that will defend against any accusation that vaccines are causing neurological harm to children, regardless of the evidence presented. One of the individuals who proudly proclaimed her membership in this group was Alison Singer, a former member of the Board of Directors of the group Autism Speaks, current president and cofounder of the Autism Science Foundation. Singer is quoted in the article as saying, "Today, we know that we have to respond to every incident however large or small, because if you leave any of these discredited theories unchallenged, it allows people to think that there's still something to be discussed."[14]

The listserv reveals its true ambitions. It doesn't want people to think about this issue. The members of the listserv do not trust the public to conduct their own investigations, ask questions, and determine their own beliefs. They must be "given" the truth and then told not to question what they have been given. We have seen this type of behavior from tyrants before in the past. We recognize it in every scandal that gets splashed across the papers, television, and the Internet.

I encourage readers to go to the webpage of the Immunization Action Coalition (IAC), and you will find that "Funding from the Centers for Disease Control and prevention (CDC) is provided for specific projects," and there is a list of the 2015 "current supporters and partners that share IAC's commitment to public health."[15] In 2020, the website listed more than a hundred Immunization Action Coalitions, with their many partners including the American Medical Association, the American Academy of pediatrics, and the American Association of School Nurses. Their listed supporters include the pharmaceutical companies AstraZeneca, GlaxoSmithKline, Merck Sharp & Dohme Corp., Pfizer Inc., Sanofi Pasteur, and Seriquis. I'm sure that in their defense the members will protest that this is a shining example of government and industry partnership. They just might not be able to answer so quickly if you ask who is representing the public.

When the writer of the *Guardian* article pressed Singer as to whether the pressure to pull *VAXXED* from the Tribeca Film Festival by the members of the listserv is censorship, this was her response: "This is not about free speech; this is about dangerous speech. The question of whether there is a link between autism and vaccines has been asked over and over again, and the answer is always the same—no. We don't discuss whether the world is flat or round anymore."[16]

When I read those words, I have to tell you it makes my skin crawl.

As an attorney, I was taught that I would be expected to make a passionate defense of my client but that the time would come when my argument would be at an end, and the members of the jury would talk privately amongst themselves. They would weigh the evidence presented by both sides, reflect on their own life experiences, and render a judgment.

I was an advocate, but the citizens made the decision.

As a science educator, I teach my students how to think, not what to think. If they want to assert a position that goes against current understanding, I will ask for their evidence. Perhaps they have some evidence, a new finding they've read about of which I am unaware, and I will acknowledge it. Perhaps they can't provide me with that evidence, but they feel it in their bones. Something doesn't make sense. I tell my students that most of the spectacular, world-changing discoveries were made by those who did not believe the current thinking.

Great scientists are bold and fearless.

In those action-adventure movies I so dearly love, there always comes a moment when the hero offers the villain a chance to surrender. I will make a similar offer to the members of the listserv. I will let any members of the listserv audit my sixth-grade science classes where I teach the scientific method and the expected conduct by members of a democratic society. We do not bully. We discuss and have a conversation. Maybe you never learned that in school, or you forgot it somewhere along the way.

I am willing to teach you that lesson again.

***

A good friend had mentioned to me that although the vaccine issue seems to get squelched in the media every time it gets brought up, the net of people who get identified as supposedly "antiscience" continues to grow. At some point, a tipping point is likely to be reached. A relative newcomer in the unfolding drama of autism is Del Bigtree, the producer of *VAXXED: From Cover-Up to Catastrophe.*

Del is a tall, athletic-looking man like Andrew Wakefield, and the two of them look like they could have been teammates on a championship rugby team. But whereas Wakefield has thin brown hair and looks every inch the English gentleman, Del looks more like a California surfer with ringlets of curly hair, streaked with gray, and a salt-and-pepper light beard. Del was a producer on the popular daytime television show *The Doctors* when he first became aware of Andy Wakefield. In a Facebook post on April 3, 2016, Del explained how he came to find himself in the middle of this media firestorm:

> My name is Del Bigtree. I am the producer of the film *Vaxxed: From Cover-Up To Catastrophe.* From the moment I started working on this movie people have asked why I would choose to leave my career as a respected producer on the medical talk show, *The Doctors,* to make a movie with Dr. Andrew Wakefield, arguably the most controversial figure in modern medicine. The answer is I had no choice.

Upon meeting Andrew (who I will refer to as Andy) I was haunted by all of the headlines that preceded him, "Baby Killer," "Father of the anti-vax movement," "The fraudulent Doctor who created a fake paper linking vaccines to autism," "The doctor who performed unnecessary experiments on innocent children," the list went on and on. But when Andy showed me the documentary film he was making about Dr. William Thompson, the CDC whistleblower, I was blown away. The evidence was undeniable. The CDC had lied to the world. It was the most important story of my life. As an Emmy Award-winning medical producer I knew I had the skill to help Andy deliver a documentary about complicated science, but before I could move forward I had to investigate Andy himself.

As soon as I started looking into the facts behind the case against Andy I realized that I had been repeating a lot of bumper sticker slogans about his story that weren't actually true. To begin with I was shocked to discover that Andy never came out against vaccines. WHAT?! What Andy had recommended was that parents vaccinate their children with the single Measles, single Mumps, and single Rubella vaccines instead of the triple MMR vaccine, which many parents were blaming for their child's regression into autism. Seems reasonable enough. More alarming was the realization that the Lancet paper in question clearly states that it does not prove a link between the MMR vaccine and autism. Though the majority of parents in the study and tens of thousands world-wide have made this claim, the paper admitted that more studies needed to be done before coming to a definitive conclusion.

I was surprised to find that the allegations against Andy were not initiated by a medical investigator or scientific institution, but by a freelance journalist named Brian Deer who wrote a Sunday Times article in the U.K. that was as scientifically accurate as a gossip column. The General Medical Council in the U.K. then used Brian's imaginative retelling of the story behind the Lancet paper as grounds for a medical trial that ultimately stripped Andy of his medical license.

I had always heard the Lancet paper described as Andrew Wakefield's fraud, but upon actually reading it the first thing one discovers is that there were twelve other co-authors on the paper. The claim that Andy used fake data to create a fraudulent paper is absurd when you realize that among his co-authors were top scientists in their fields who were responsible for performing the tests, outputting the data, and ultimately verifying that it was correctly represented in the paper before signing their names to it. If there was fraud all thirteen authors were implicated. So why is Andy the only one of the 13 co-authors currently barred from practicing medicine in the U.K.? Maybe because he was the only doctor brave enough to ignore pressure from the vaccine manufacturers and the U.K. Ministry of Health to begin larger, more in depth studies investigating the hypothesis that the MMR was causing autism in our children; studies that he never got to finish.

There is also the assertion that the Lancet study had been paid for by a biased outside source, which is easily refuted by following the paper trail that shows all financial contributions for the study were accepted by the Royal Free Hospital after the study had been completed.

Lastly the most disturbing accusation for me was the claim that Andy performed unnecessary procedures on innocent (mentally disabled) children. That sounds horrible until you discover that the parents had entered their children into the study because they were suffering from agonizing gastrointestinal pain and bowel issues in addition to their autism. The "unnecessary procedures" refers to the colonoscopies and intestinal biopsies that were performed by Andy's colleagues. I don't know how a gastroenterologist is supposed to investigate possible intestinal disease without performing these standard tests, but then again, I am not a doctor. And neither is the journalist who concocted this unfortunate obstruction of medical inquiry.

History has shown us time and time again that people like Galileo, who break from the scientific consensus to reveal discoveries like "the earth is not the center of the universe," are often persecuted. I suppose it's one of our great human flaws. Both of these men were essentially tried for heresy. The difference is that the imprisonment of Galileo just meant there would be a delay in the advancement of modern astronomy and physics. In Andy's case we may be responsible for a civilization-ending epidemic of autism that has skyrocketed from 1 in 10,000 to 1 in 45 children in less than forty years and is on a crash course for 1 in 2 by 2032. And still the CDC attempts to set our minds at ease with their official statement "we do not know what's causing autism."

In the Unites States of America we can never allow ourselves to believe that there is a case so sound that we avoid listening to a new witness or shirk our responsibility to investigate new forensic evidence. In the case of the possible link between vaccines and autism we have a new witness in the courageous CDC whistleblower, Dr. William Thompson, and we have new forensic evidence in the documents and data that he has provided us, much of which he claims was destroyed by the CDC.

Ultimately a documentary is not a court of law. It cannot prosecute a case. As journalists and filmmakers we can only reveal information as it is discovered and ask the questions we believe the citizens of the world should be asking. Much of the mainstream media have come out against this film

> warning people not to see it because it asks uncomfortable questions about the safety of the MMR vaccine. If we cannot ask important questions about the safety of a vaccine then I fear for more than the health of our children, I fear for the health of our democracy.
>
> We should not allow anyone to dictate what we say, what we do, and what we see. You still have the power to make up your own mind. We have made a very important film. It shines a spotlight on damning evidence that the pharmaceutical industry and the CDC do not want you to see. That is probably because they know if you do see it you will find yourself in the same predicament as me for as Einstein said "Those who have the privilege to know have the duty to act."
>
> Sincerely,
>
> Del Bigtree[17]

I will leave it to the reader to determine whether the course urged by Del Bigtree is more in keeping with our traditional democratic notions regarding new ideas, or whether we should let organizations such as *Forbes*, *Glamour* magazine, or even the *New York Times* do our thinking for us.

* * *

An old expression is that it's darkest just before the dawn.

After more than a decade in the autism wars, I have come to believe it's darkest just before it becomes pitch black. With De Niro folding on presenting the documentary at the Tribeca Film Festival and the massive media assault against people even viewing the movie, I figured that the film was essentially dead. We'd gotten to the starting gate, but just as the race was about to begin, somebody had snuck in and shot our horse.

I could not have been more wrong.

On Tuesday, March 29, 2016, *Variety* magazine announced that the distributor of *VAXXED*, Cinema Libre, had booked the film into the Angelika Film Center in New York City, for a two-week run starting on that Friday, April 1, and continuing until April 15.[18] Considering that the original plan was that there would be a single showing of the film on the closing night of the Tribeca Film Festival, Sunday, April 23, the opportunity to have the film run for two solid weeks in a theater right next to the Tribeca Film Festival was both a bold and potentially dangerous move. Would audiences stay away? Or were there enough rebels in the Big Apple to strike a blow for free speech? Richard Castro, head of distribution for Cinema Libre, put the issue in his own words:

> It's disturbing that an American Film Festival can succumb so easily to pressure to censor a film that it has already selected and announced. On Friday I received a call from Tribeca executives expressing concerns about showing the film, but no opportunity was afforded our filmmakers to even address those concerns. When I questioned the rationale, it was indicated that "sponsors" interest was a factor.[19]

The Cinema Libre team would be pulling out all the stops for their impromptu screening of *VAXXED*. After the eight o'clock showing there would be a panel discussion with Andy Wakefield, Brian Hooker, Del Bigtree, and Polly Toomey, an autism parent and founder of the magazine *The Autism File*.

It is said that no battle plan ever survives contact with the enemy, and this means that there can be both unexpected setbacks and unexpected victories.

* * *

And who were some of these members of the "scientific community" that Robert De Niro claims he communicated with and made the decision to pull the film from the Tribeca Film Festival? One might have thought that after winning such a signal victory as getting a documentary, which had already been approved as an "Official Selection" removed from a prestigious film festival, the perpetrators would slink off into anonymity.

But old habits seem to die hard.

On April 3, 2016, Dr. Ian Lipkin (is he a member of the listerv?) published an opinion piece in the *Wall Street Journal* titled "Anti-Vaccination Lunacy Won't Stop: Robert De Niro Made the Right Call in Pulling 'Vaxxed' from his Film Festival. But the Bogus Message Rolls On."[20] Like a creature from the Bad Science Lagoon, Lipkin just kept coming. Lipkin's stated concern was that the film "misrepresents what science knows about autism, undermines public confidence in the safety and efficacy of vaccines, and attacks the integrity of legitimate scientists and public-health officials." Apparently, we members of the public are unable to view a film about a whistleblower scientist at the CDC and come to our own conclusions. It's a little like somebody has been convicted of murder, but even if new evidence has come forward, the public will be advised not to consider it.

As somebody who extensively reviewed the scientific work of Ian Lipkin in my previous book, I have to say that the quality of his work performed in the Wakefield investigation was just as shoddy as the work he performed in the XMRV investigation of Dr. Judy Mikovits. If I had Dr. Lipkin on the witness stand, I would impeach the man with his own words. From his own article in the *Wall Street Journal*:

> We tested Mr. Wakefield's two major findings. First, whether MMR preceded gastrointestinal complaints (presumably leading to a breakdown of the gut wall, allowing molecules to enter the blood stream and travel to the brain to cause autism) and, second, whether we could find the measles virus in the gut of the majority of children with autism. Neither finding held up.
>
> In our peer-reviewed study, published in *PLOS One* in September 2008, we found that only 20% of children fit the Wakefield model in receiving MMR vaccines before onset of GI disturbances and autism. We found measles virus sequences in the gut of only one child with autism and one child with GI complaints.[21]

As a lawyer, you run across people who blatantly lie about important events. Those lies are usually easy to uncover through interviews with other parties or review of relevant documents. There is another type of liar, the one who lies by misdirection. There are usually a number of half-truths mixed in with the lies, and it can be daunting to pull apart this fictional web. These individuals are normally much smarter than the blatant liar.

You have to pay attention to what the second type of liar says, rather than the spin they want you to accept. I will give you my opinion as to why Dr. Lipkin is the second type of liar. It really is as simple as one, two, three. Just don't let the fact that Lipkin is a Columbia University professor blind you to the utter foolishness of what he proclaims.

First, Dr. Andrew Wakefield investigated the gastrointestinal complaints of twelve children, and the parents of nine of them who would link the beginning of these problems and their autism to the MMR shot. That was the group he studied. Lipkin studied a different group. Second, with a deft bit of misdirection, Lipkin informs the reader that he investigated whether the MMR shot preceded the development of autism and gastrointestinal complaints and found that this occurred in only about 20 percent of children with autism. Let's stop for a moment and consider that stunning admission. If we found that 20 percent of traffic fatalities were preceded by the downing of a gallon of alcohol, that percentage alone would probably be considered an important piece of information, and a cause for further investigation. But Lipkin acts like it's nothing.

Third, Dr. Wakefield stated that he found the measles virus only in a very specific section of the gastrointestinal tract and that Lipkin's collaborator, Tim Buie, was not very good at obtaining biopsies from that section of the colon. This is supported by Lipkin's own article, which does not identify the areas of the colon from which the samples were taken. And with all of that said, by Lipkin's own admission, he found the measles virus in the gut of at least one autistic child and one child with gastrointestinal problems. At the very least, that is some indication of an autoimmune issue in those children, as the virus should have cleared years earlier.

Lipkin's misdirections continue:

> In 2010, after concerns arose about his research, The U.K.'s General Medical Council revoked Mr. Wakefield's medical license based on cross-examination of physician's and evidence from 36 witnesses. The council found that he had done invasive research on children without ethical approval, acted against the clinical interests of each child, failed to disclose financial conflicts of interest, and misappropriated funds.[22]

Sometimes I find it very hard to be charitable. In 1998, Wakefield published his findings on the presence of the measles virus in the guts of children with autism. Their parents suspected their children were having severe gastrointestinal distress. The way to investigate that is to do a colonoscopy and take samples of suspicious tissue, which Dr. Wakefield's team did. Lipkin makes it sound like the children were subjected to Nazi experiments in some concentration camp. Are we truly to believe in 2016, eighteen years after Wakefield's publication, Lipkin's assertion that "If any good has come from it, the MMR controversy has sharpened the scientific community's focus on autism and may lead to insights into the biology of autism and new treatments?" Has science jumped right into curing autism? Are there any advances you can name? Is science so inept when studying autism, or is it simply not looking in the right place?

Another trick of the misdirecting liar is to ignore the main issue. Just for the sake of argument, let's say Wakefield's work doesn't hold up. Maybe it was well-intentioned, but wrong. I'm not saying it was, but any good attorney will always consider other possibilities.

The main focus of the film is Dr. William Thompson, a CDC scientist who claims the CDC Division of Vaccine Safety concealed information about the safety of vaccines, turned over thousands of pages of documents to Congressman Bill Posey in support of this charge, claimed federal whistleblower protection, engaged the services of one of the country's top whistleblower attorneys, and is waiting for Congress to subpoena him so he can talk about these issues without losing his federal whistleblower protection. Lipkin does not have one word in his article about Dr. William Thompson, even though Thompson was project manager for a time.

Does that influence your opinion of Dr. Ian Lipkin? What does it make you think of the *Wall Street Journal* for allowing him to publish his article? Do you expect the *Wall Street Journal* will extend a similar courtesy to Dr. Andy Wakefield? To Dr. Brian Hooker? To Emmy-winning producer Del Bigtree? To CDC scientist Dr. William Thompson?

Somehow I doubt the opinion articles of any of these individuals will ever appear in the august pages of the *Wall Street Journal*. You might want to ask yourself, Why? Who benefits when certain voices are silenced?

***

April is Autism Awareness Month, but for many it reveals the deep divides in the autism community. The group of parents to which I belong believes that autism is a man-made disease, and we find our natural home in shoestring organizations like Age of Autism, Generation Rescue, or Talk About Curing Autism (TACA). The better-funded groups, like Autism Speaks, believe autism is more of a genetic problem and have expended millions of dollars in that effort, with little to show for it.

However, there has been movement among the groups, and probably no family better exhibits this movement than the Wrights. Bob Wright was the chairman of NBC for many years and was well respected in the industry. Bob and his wife, Suzanne, moved in the very best social circles in New York. Their daughter, Katie, had a son named Christian, who developed autism. Bob and Suzanne then founded the group Autism Speaks, with a twenty-five million dollar endowment, to assist their grandson and other children like him. However, Katie eventually became dissatisfied with the scientific direction of Autism Speaks and its emphasis on genetics. In Katie's opinion, her son had not been born with autism. One might say that Katie "switched sides" and felt a greater allegiance to those groups like *Age of Autism* (for which she became a writer) who were looking at environmental factors.

On April 1, 2016, Bob Wright appeared on CNN with Alisyn Camerota to discuss his new book, *The Wright Stuff: From NBC to Autism Speaks*. The interview was startling, as it suggested that even among those considered most closely linked to the genetic view of autism, vaccines were also high on their list of suspects:

Camerota: I want to ask you about what's happening with autism today, because the new numbers are out. One in 68 children is diagnosed with autism. That number is staggeringly different than a generation ago when it was 1 in 10,000. And I couldn't help but notice your wording in the book when you talked about Christian—the timing of Christian's autism.

You say, "Right after he got the standard one-year vaccinations, he developed a very high fever and screamed for hours. Katie," your daughter, "was so frightened she called her husband to come home from work and they put the baby in an ice bath to bring down the fever. When they called the doctor they were told the reaction was completely normal."

Bob, I can't tell you how many parents, dozens, I have interviewed who had the exact same experience that you did. After the children got their standard vaccinations, that night the child had a high fever, they were clearly in distress, they were screaming in mortal pain, they called the doctor and the doctor said you're having a vaccine reaction. I know this is very controversial. Are you satisfied that enough research and studies have been done to prove that there is no link?

Wright: Well, I'm satisfied to date from what has been done, that we can't directly establish that link. And—but it's—you know, as we get smarter and we're able to do better research, it's very difficult to do research on vaccines when you're talking about vaccines that go to tens of millions of people, because you need a large sample to make conclusions about something like this. And that's part of the difficulty. I would also say, that you—that we all know without any controversy that a lot of children have very different reactions to vaccines, period.

Camerota: Yes.

Wright: And all vaccines are essentially the same, of the same type of vaccines, and the children are all different. And they all have different immune systems. So their responses are going to be like this and pediatricians are too quick to say, oh, you fall in a normal category. Well, that normal category is like this wide, and that's where vaccines safety comes in. And that's an area I did spend a lot of time in trying to understand the CDC's vaccine safety program.

And I can tell you conclusively in that one, that program can be significantly improved for very little money. And we tried. And I tried with two administrations, the Bush administration, Obama administration, and I failed to get it. It got stopped in the White House in both cases.

Camerota: My gosh.

Wright: And that's probably one of the most disappointing things that I didn't get done.[23]

The story Bob Wright tells of his grandson's vaccine reaction is familiar to many of those in the autism community, as well as their family members. Just as common is the response of doctors that the high fever and "screaming in mortal pain" of these infants is a normal reaction. Where Bob Wright differs from the rest of the community is that he was able to get high-level access in two presidential administrations to directly advocate for increased research but was denied by both Republican and Democratic presidents. How is it that what seems a sensible course whether you are a typical parent, or the former chairman of NBC, meets with such strong opposition across the political spectrum?

Probably no journalist working in America today has done a better job of trying to get to the bottom of the vaccine/autism story than Sharyl Attkisson, the former Emmy-winning CBS reporter and current host of the Sunday morning political show *Full Measure*, airing nationally on stations of the Sinclair Broadcast Network. In an article on her website posted on April 5, 2016, about the Wright book, she notes that the Bush White House was afraid of the political reaction to changes to the vaccine safety program, and that under Obama, the measure was killed by presidential adviser Valerie Jarrett. On the broader question of why such safety measures have not been implemented, she paints a picture of a corrupted scientific and industrial community that uses its financial resources to buy politicians and silence dissent:

> This political interference in the effort to produce the safest vaccines possible adds to my own first-hand knowledge of multiple Congressional hearings about vaccine safety and links to autism that have been scheduled, but then cancelled under pressure from the pharmaceutical industry. The pharmaceutical industry is able to wield inordinate pressure in the news media through its advertising relationships; in government and politics through its business relationships, donations and revolving door; and in universities and scientific research communities through its funding and contributions. Vaccine interests also use bloggers, Internet writers and sites, and social media—including Twitter and Facebook accounts under pseudonyms—to controversialize, bully and attempt to discredit scientists, journalists and others who discuss or report on vaccine safety issues.[24]

When one of the country's most well-respected investigative reporters describes what is going on today, most people will have difficulty recognizing that she is talking about the United States. We could more easily accept her words if she were talking about some corrupt dictatorship or communist country. Can the people really be so deceived in a country where we supposedly have a "free press"? It is an unusual arrangement, where those with more liberal sentiments are comfortable with a greater governmental role in people's lives, while those with more conservative beliefs are satisfied that industries can be trusted to police themselves. In many ways, the current system combines the very worst of liberal and conservative thought, government control over people lives, and industry control over the government.

In the concluding section of Attkisson's article, she wrote:

> It's a sad commentary on the state of our media that liberal reporters and commentators are so quick to take the side of Big Pharma against parents who have seen for themselves how vaccines have led to the dramatic increase in autism. The federal government recognizes the risks and dangers, having established the National Vaccine Injury Compensation Program to compensate victims of vaccines. This takes the vaccine makers off the hook for injuries and deaths caused by government-mandated vaccines. In short, Big Government is protecting Big Business from liability for their products. Wright notes that $100 million a year is paid in damages to victims of vaccines.[25]

When we speak of "freedom of the press," we harken back to American colonial times and look at a starkly different world. In those times, the primary threat was that one would insult a government official. But that government official was not kept in office by contributions from the local merchants. If you raised questions about the quality of a merchant's goods, the government was unlikely to act as an avenger for the merchant. Additionally, the exercise of any governmental power was likely to be out in the open, dramatically different from what Attkisson describes as pharmaceutical influence over politicians, media, and academic institutions.

For the parents of many children with autism, it has come as something of a shock to have friends, family members, and professional colleagues more likely to believe media messaging than their own experiences. *VAXXED* held out the possibility that the veil covering the eyes of so many of the public could be ripped away. At the end of the film, four simple suggestions were made:

1. That Congress subpoena Dr. William Thompson and investigate the CDC fraud.
2. That Congress repeal the 1986 National Childhood Vaccine Injury Act and hold manufacturers liable for injury caused by their vaccines.
3. That the single measles, mumps, and rubella vaccines be made available immediately.
4. That all vaccines be classified as pharmaceutical drugs and tested accordingly.

It would be interesting to interview a cross section of the American public and determine if a majority thought any of these steps were unreasonable. My guess is that these suggestions would meet with overwhelming approval.

***

The reviews from actual filmgoers started to come in when viewers posted their reviews on the website *Fandango*. As of April 5, 2016, there were 132 audience reviews, and the film had achieved a five-star rating. Here are some of the reviews from people who saw the documentary:

> **Such an Important Movie that Everybody Should See**
> By Corinne Brown
> This was an extremely well done documentary. It is heartbreaking as a parent to see how corrupt our government is and how much the pharmaceutical industry controls everything. It is so important for our society to see this eye opening documentary so that a change can finally be made. I so appreciate the bravery of all involved in putting this together.
>
> **Compelling**
> By JRose668
> The reviews you may have read in the established media bear no resemblance to the actual film, which offers compelling scientific evidence that the MMR is a major cause of autism. The CDC had that evidence and covered it up—which will not surprise anyone familiar with the CDC's failure to address the problem of lead in drinking water.
>
> **Shocking!**
> By Melissa Alfieri
> This is an absolute must see! I was able to view this documentary at the Anjelika theatre downtown. I also had the opportunity to meet Andrew Wakefield, Del Bigtree, Brian Hooker, and Polly Toomey prior to the showing. This was

> an extremely emotional and shocking film. This was not an anti-vax fil, but rather a call for stringent testing for all vaccines, especially separating the MMR back to single dose. Powerful testament to corruption at the CDC. How can data or facts be destroyed? How can they allow children to become sick? Why are they so adamant that the triple dose MMR is safe, when clearly it is not? As Wakefield said, they would only need to go back to the single mumps, measles, and rubella shots and the data would reflect the truth!
>
> **Not the Conspiracy Crap I was Expecting!**
> By Gagabear 3215
> I was very skeptical, but this turned out to be excellent. I highly recommend ANYONE to go see it, no matter which side of the coin you're on.
>
> **Excellent Movie!**
> By Dr. Steve Thiele
> Great job exposing the lack of true informed consent for parents to decide if the risks outweigh the benefits for vaccines. The fact that a vaccine injury court exists and that millions have been paid out to vaccine injured kids, and that the drug companies have immunity from suits as a result of adverse side effects is a true eye opener. Parents are entitled to review ALL of the potential side effects before making an informed decision. The fact that the CDC his data to push through a paper falsely showing there was no link between vaccines and autism, and the fact that the CDC whistleblower has not yet been subpoenaed before Congress smells of a rat. This movie is a must see![26]

The news from the box office was even more spectacular. One of the most important measurements of the potential success of a film is the per screen daily average on the first weekend a movie is shown. That was one of the critical metrics Cinema Libre was going to use in determining how large of a launch the film would receive in theaters across the country. At that single theater in New York City, in an auditorium that looked like it probably didn't seat more than 250 people, *VAXXED* pulled in $28,339 for that Friday, Saturday, and Sunday, for a daily average of $9,446.[27] As a point of comparison, the blockbuster film *Batman vs. Superman: Dawn of Justice*, during its second week of wide release, stood at number one at the box office with a per screen daily average of $4,021 for the same three days.

* * *

And yet surprisingly, the censorship continued. The controversy generated by the documentary's removal from the Tribeca Film Festival brought several other film festivals running, such as the Manhattan Film Festival (which would be showing the film on April 23, 2016, the same night it had been scheduled to show at the Tribeca Film Festival), as well as the Houston Film Festival and the Silver Springs Film Festival in Ocala, Florida.

On April 6, 2016, the head of the Houston Film Festival, Hunter Todd, sent out an email to Phillippe Diaz, chairman of Cinema Libre Studio. It read:

> Dear Phillippe:
>
> Good morning. I wanted you to know that just like De Niro and Tribeca, we must withdraw our invitation to screen VAXXED! It has been cancelled and there was no press release about the film . . . that was scheduled for today, but after very threatening calls late yesterday (Monday) from high Houston Government officials (the first and only time they have ever called in 49 years)—we had no choice but to drop the film. Heavy handed censorship, to say the least . . . they both threatened severe action against the festival if we showed, so it is out. The actions would have cost us more than $100,000 in grants.
>
> I do hope they did not call or threaten you. It is done, it is out and we have been censored . . . There are some very powerful forces against this project. It does seem a bit of overkill, as I am confident it will be released Online soon and millions of people can see it.
>
> My Thanks and Best Regards,
>
> Hunter Todd
>
> Chairman and Founding Director
>
> Team Worldfest
>
> The 49th Annual WorldFest-Houston[28]

One might have thought that the "high Houston Government officials" would realize that this action was in direct conflict with the First Amendment, which guarantees freedom of expression from government censorship, but they were apparently unfamiliar with the concept. In response to questions raised by the group Health Choice, the mayor's office released a statement:

> The mayor asked that it be removed from the lineup. I believe Judge Emmett did the same. The film festival is being funded in part through a grant from the City of Houston. The mayor felt it was inappropriate for the city to

endorse an event that would be screening a film that is counter to the city's efforts to ensure children receive vaccinations.

The film was also removed from the Tribeca Film Festival lineup so Houston is not alone. In fact, it was that move that raised the concerns locally.

Janice Evans—Chief Policy Officer & Director of Communications.[29]

The protection of free speech from government influence did not seem to weigh at all on the minds of the Houston government officials. Many advocates also commented on the irony of Mayor Sylvester Turner's censoring of the film, considering that he is African American, and one of the main allegations of the documentary is that earlier administration of the MMR vaccines dramatically increased the autism rates in African American males.

Phillipe Diaz put out a statement addressing the Tribeca and Houston situations:

> As I explained to the directors of Tribeca, their decision has created a huge precedent in the filmmaking world. Film festivals are no longer a space for free speech, but a place run by sponsors and corporate interests. It sends the message to filmmakers that they better make movies "approved" by corporate sponsors so they'll have more of a chance to be selected and therefore find a way to sell their films. We should never forget that the life of an independent film depends very often on festivals. If festivals can act as censors for corporate interests, it's a slap in the face to the First Amendment.[30]

It seemed that courage in the defense of the First Amendment would be left to Greg Thompson, director of the Silver Springs Film Festival, who, in response to an angry Facebook post from "John" (a member of the listserv?), wrote back a response that would have likely warmed the heart of any person who believes in the principles of free speech:

> Hi John,
> I trust you have seen the film. And I appreciate your concern for our festival.
> And I trust in the relationships that we have built over the last 4 years-with our community, its businesses and leaders, students and teachers and friends from around the world. But the most important relationship we have in our city and our festival is our relationship with the 1st Amendment, which protects our freedom of speech and expression.

> Based on your statement, I have to wonder if you and some of the folks writing about this film have indeed seen it—and not through the filter of personal agendas. Our only agenda is, "Good films deserve to be seen." And people have a right and need for access to every side of the story and the freedom to make their own decision, with their own physicians, regarding their children. If the film turns out to be misinformation then I trust we will have supported the bigger conversation that will bring that to light. If it is proven to be a fact, we will have been part of a conversation that could positively impact many lives.
>
> You may choose to paint our festival in any light you wish, but now you have the truth from the horse's mouth—we do not censor and we do not represent any one's agenda. We present good films and trust our audiences to use that information in a powerful way that will serve their best interests and inform their conversations!
>
> If you, or anyone you know, has not seen the film yet, I invite you to join us and bring your knowledge and expertise to the table and enrich our community. Thank you for caring!
>
> Best Regards,
>
> Greg Thompson, Festival Director
>
> Silver Springs International Film Festival

Could it be that the director of a four-year-old film festival seems to have such a better appreciation of the democratic process than the sitting mayor of a major American city like Houston? And why is it that the supposedly free press in our country isn't defending the right to a free press and open discussion of ideas? I encourage any reader to look at the coverage in the first few weeks of the *VAXXED*/Tribeca controversy and find any mainstream publication that defends the right of the filmmakers to show this film and the public to have a discussion about it.

How is it that the mainstream press does not believe that a documentary about a whistleblower at the CDC who has provided evidence to Congress about a massive cover-up relating to the health of our children is not an important subject for public discussion?

* * *

In a remarkable irony, as all of this was taking place, Dr. Wakefield's wife, Carmel, was on her own constitutional journey. In an interview with Celia Farber of the website *Truth Barrier*, Carmel was asked how she was feeling about the Tribeca and Houston Film Festival controversies and gave this response:

> Well, I'm just about to become an American citizen. I'm genuinely excited and proud. One of the massive tenets for me of this country has been the freedom of speech, which is the essence of America. The founding fathers would be horrified at these attempts to censor truth.
>
> Free speech will prevail. Everybody must see this film. Like it or loathe it, it must be seen. This is a statement of fact to which we all have to face up, and make our decisions going forward. The Houston revelation [that the mayor and a prominent judge had threatened the film festival with loss of funding if they chose to show the film], while massively disappointing, was a relief because the whispered, shadowed, nameless, faceless thing now had a name. And that name was suppression of freedom of speech. That name was cash dollars in the bank. That name was deceit and defilement of the American constitution.[31]

I think that Carmel Wakefield will become an extraordinary citizen of our democracy.

***

And just as I was ready to place Robert De Niro among those timid souls who could be bullied away from an honest investigation into vaccines and autism, even at the cost of sacrificing his own child, I was surprised again. Appearing on NBC's *TODAY* show on April 13, 2016, at the start of the Tribeca Film Festival with cofounder Jane Rosenthal, the most amazing conversation took place. De Niro and Rosenthal, an attractive looking older woman with straight brown hair, were interviewed by TV hosts Savannah Guthrie and Willie Geist. From the very start of the interview, it was clear that De Niro was back in the fight.

> **Savannah Guthrie:** There was a bit of controversy, some headlines at the beginning of this year's festival when it was announced that this film called *VAXXED* would be screened at the festival. Later, the festival pulled it. Was it because of the backlash? Were you surprised that people reacted the way they did?
>
> **Robert De Niro:** I was shooting a movie, in the middle of a lot of stuff. I think the movie is something people should see. There was a backlash that I haven't fully explored, that I will. And I didn't want it to start affecting the

festival in ways I couldn't see. But definitely there's something to that movie. And another movie called *Trace Amounts*. There's a lot of information about things that are happening with the CDC, the pharmaceutical companies, there's a lot of things that are not said. I, as a parent of a child who has autism, am concerned, and I want to know the truth. I'm not antivaccine.

I want safe vaccines. Some people can't get a certain kind of shot and they can die from it. Even penicillin. So why shouldn't that not be with vaccines? Which it isn't.

Willie Geist: You went public for the first time, saying your eighteen-year-old son does have autism. That had been a very private thing, part of the reason you wanted the film shown is to start that conversation.

Robert De Niro: Absolutely.

Willie Geist: Do you believe you'll have a role in that conversation, going forward?

Robert De Niro: Possibly yes. The thing is, they shut it down, there's no reason to. If you're scientists, let's see, let's hear. Everybody doesn't seem to want to hear much about it. It's shut down. [Motions toward the TV hosts] And you guys are the ones who should be doing the investigating.

Savannah Guthrie: I think the film was controversial because people felt the filmmaker had been discredited.

Robert De Niro: Even he, I'm not so sure about, at the end of the day. Even him.

Savannah Guthrie: Jane.

Jane Rosenthal: The one thing, it wasn't sponsors or donors that were threatening to pull out of the festival, it was our filmmakers. And we're known for having amazing documentary films. You can take a look at our lineup, whether it's what we're starting with tonight, or some other documentaries that we have at the festival. So it was our filmmakers that were pulling out—

Robert De Niro: [Interrupts] I find that amazing and we're going to talk about that.

Jane Rosenthal: There's another amazing film that was done by Roger Ross Williams that won the Audience Award at Sundance called *Life Animated*, about autism and it's a really beautiful film about the Susskind family—

Robert De Niro: [Interrupts again, looking a little angry] It's a beautiful film, but it's another thing. It's the result of, it's not questioning how some people got autism. How the vaccines are dangerous if given to certain people who are more susceptible. And nobody seems to want to address that. Or they say they've addressed it and it's a closed issue. But it doesn't seem to be. Because there are many people who will come out and say, "No, I saw my kid change overnight. I saw what happened and I should have done something, but I didn't do something." There's more to this than meets the eye, believe me.

Willie Geist: Is that the experience you had, Robert? Something changed overnight?

Robert De Niro: My wife says that. I don't remember. But my child is autistic and every kid is different. But there's something there. There's something there that people aren't addressing. And for me to get so upset today, on the TODAY show, with you guys, means there's something there. All I wanted was for the movie to be seen. People can make their own judgments. But you must see it. There are other films, other things that document and show, it's not such a simple thing.

Savannah Guthrie: Do you regret pulling it now, in some sense?

Robert De Niro: Part of me does. And part of me says, let it go for now. And I'll deal with it later, in another way. Because I didn't want the festival to be affected in a way. Because it was like a knee jerk reaction. Especially from the filmmakers, frankly that I, you know.

Savannah Guthrie: The other filmmakers who were in the festival.

Robert De Niro: Whoever they were. I really didn't want to ask. But now I will ask.

Willie Geist: Robert, it is nearly consensus in the scientific community that there's no link there. Do you believe that's not true?

Robert De Niro: I believe it's much more complicated than that. It's much more complicated than that. There is a link and they're saying there isn't. But there are certain things, the obvious one is thimerosal, which is a mercury-based preservative. But there are other things. I don't know, I'm not a scientist, but I know because I've seen so much reaction. About just let's find out the truth. Let's find out the truth. I'm not antivaccine, as I say, but I'm pro-safe vaccine. And there are some people who cannot take a vaccine and they have to be found out and warned. Don't just give a kid a bunch of shots and then something happens.

Some parents in this documentary say, "I knew I shouldn't have done it. I knew I shouldn't have done it. I talked to the doctor, he's the doctor, I should listen, I should listen, I did it, and the next day . . ." You imagine how that parent feels?

Savannah Guthrie: The worry is that people who hear those words and wonder about it will then not have their children vaccinated, which has led to a higher incidence of things like mumps and measles.

Robert De Niro: I don't know if those statistics are accurate. I'm not one to say, but I would question even that. There's kind of a hysteria, a knee-jerk reaction. Let's see. As I say, everyone should have the choice to take vaccines. Some places it's becoming mandatory. But it does benefit the drug companies. Funnily enough.

Jane Rosenthal: If we're going to start to take a look at facts and statistics one of the things we need to look at in this movie, VAXXED, is the contradiction with facts and statistics right off the top. What is state, with what the rise is, what the graph is, then people saying something different. So you need to read the reviews of that movie, make your own decision. Clearly the festival has about a hundred other movies that are in the festival. This was only going to be screened once. They certainly had their voice and their time. And there's amazing films, other films of social impact, whether it's the criminal justice system, an amazing film called *The Return*. Another film about Herbal Life, a whistleblower story. Another film about drone warfare. *National Bird*, which really asks a lot of questions about how we got to war. A film festival is about having conversations. And there's also some fun films, too. *Family Fang* with Jason Bateman and Nicole Kidman. So we've got a lot for you at the festival.

Willie Geist: It's always fun. [Turns attention to De Niro.] And also highlighted by the forty year anniversary of *Taxi Driver*.

Robert De Niro: I'm very curious and interested to go and meet with everybody who was in it. Marty and Jodie and Paul Schrader. It'll be very interesting.

Savannah Guthrie: Thank you very much.[32]

And the interview ended. What had just happened? Had NBC, a major television network, really let De Niro have that kind of screen time to discuss autism? Was it because Bob Wright, the former head of NBC and cofounder of Autism Speaks, had let it be known a few days earlier in his interview with CNN that he had tried to make changes in the vaccine program with both the Bush and Obama administrations but was shot down? Leslie Manookian, the documentary filmmaker of *The Greater Good*, offered what seemed to be inside information in an email she sent to a large group: "I just heard from a friend who has a friend who works on the Today show. She said that for some reason they just let the cameras keep rolling—almost seems like some people wanted this to come out."[33]

The aggrieved tone of those in the media who had so recently been praising De Niro for pulling the film from the Tribeca Film Festival was probably best expressed by the title of an article written for the website *This Week in Tomorrow* by a blogger, Richard Ford Burley, titled "Damnit

De Niro, I Thought We Were Good."[34] The website describes Burley as "a human, writer, and doctoral candidate at Boston College, as well as an editor at *Ledger*, the first academic journal devoted to Bitcoin and other cryptocurrencies. In his spare time he writes about science, skepticism, feminism, and futurism here at This Week in Tomorrow."

In the opinion of many autism parents, though, De Niro had finally figured out the tone he wanted to take, and a hashtag titled #DeNirotheHero started to appear on Twitter.

* * *

On April 30, 2016, at the end of the most amazing Autism Awareness month in memory, I went to the Opera Plaza Cinemas in San Francisco to watch *VAXXED: From Cover-up to Catastrophe* and listen to a panel discussion with Andrew Wakefield and producers Del Bigtree and Polly Tommey.

As I sat watching this calm, methodical, and powerful movie play out on the screen in front of me, I felt like an old boxing coach watching a new fighter in the ring. I anticipated the punches, the bobbing and weaving, and when it was all over, I was impressed. The film plays like an extended *60 Minutes* piece, putting the pieces of the puzzle together, interspersed with audio from the four conversations Brain Hooker recorded of CDC whistleblower William Thompson, the reactions of nationally known physicians to the MMR vaccine/autism fraud, and the personal stories of autism parents, backed up with home videos of their children before and after their adverse vaccine reactions.

The panel discussion was fantastic, and afterward I said hello to Andy Wakefield, who smiled when he saw me and gave me a warm embrace. I had not met Del Bigtree or Polly Tommey before, so I introduced myself to them, and after a few minutes we were talking like old friends. I'd shared a ride to the film with Terry Roark, the California State director of the National Vaccine Information Center who is widely referred to as the grandmother of the vaccine safety movement in California. For Terry, it was clear that the film represented a fundamental shift in the vaccine/autism debate and that, for once, we were poised to be on the winning side. As the group was talking in the courtyard of the Opera Plaza cinema, word came that a Chinese distributor was interested in showing the film in China and that they would devote significant financial resources to widening the US release. It did seem that the tide was turning.

I was a little more guarded. I have never been present at a fundamental shift in public consciousness. I may not recognize that we are in fact in the middle of one. As a society, we take a great interest in the health and safety of our children. We agonize over the right car seat, the food they will eat, how much television they should watch, the amount of video games they play, who their friends are, and many parents check their kids' progress in school daily through online teacher gradebooks. Parents want the best for their children. They want to make healthy choices for their sons and daughters. But will they consider that the most dangerous thing they may ever do is to walk their child into a pediatrician's office?

* * *

On Monday, May 9, 2016, the *VAXXED* production team, including Andy Wakefield, Del Bigtree, and Polly Tommey, went to the California state capitol in Sacramento. They wanted to talk to legislators about their film, and what it meant for SB 277, the recent law that provided that all California children had to be up to date with the CDC's vaccine schedule or else they could not attend school in the 2016–2017 academic year.

One office in particular they wanted to visit was that of Senator Richard Pan, the lead author and guiding soul of SB 277. After being told the senator was not in the office, Del Bigtree spotted him sneaking out through a side door and gave chase, telling Pan he wanted to talk about their documentary. As reported in one account:

> While most senators would be proud to talk about their legislative achievements, Senator Pan was filmed running away as the producers approached his office. Presently, Pan has still not disclosed—or addressed in any fashion—his ties to pharmaceutical companies and their lobbying groups seen on the house floor. The cardiovascular-minded Sacramento senator now has another permanent blemish on his record that will need to be confronted sooner or later. In the absence of any meaningful reason for his impromptu hallway scamper, people around the US are left scratching their heads while laughing at Pan's behavior. Social media memes comparing Pan to the Hollywood character Forrest Gump went viral in mere hours after Pan's gallop.[35]

"We were in his office, he saw us, and he literally ran out a back door," said producer Del Bigtree in a video interview right after the event, looking bemused at this turn of events. He continued:

I ran out into the hallway. He's actually pretty fast. I have to give Senator Pan that. I don't know if he did track and field when he was younger. He zipped down the hallway, and took two flights of stairs in about seven seconds. Very impressive. Obviously, all we wanted to do was have a conversation. I mean, this is the United States of America. This is a conversation about health. We have a senior scientist at the CDC who says that the MMR vaccine/autism study was a fraud.

I think Senator Pan should talk about that since he wants to pass bills mandating vaccines.[36]

Maybe things are finally starting to change. Maybe the bad guys are finally on the run. Will the American public let them escape?

***

On May 19, 2016, Mayor Aja L. Brown of Compton, California, a primarily African American community, offered a free showing of *VAXXED* to his constituents, followed by a question-and-answer session with Andy Wakefield, Polly Tommey, and Del Bigtree, as well as Minister Tony Muhammad. But perhaps the most chilling words came from Sheila Ealey, the African American mother of an autistic son depicted in the film. Her section was so emotional because her son has a twin sister who is completely normal, speaks three languages, is an honor student, and plays the piano with exceptional skill. At their eighteen-month checkup, the shots were lined up and the nurse accidentally gave the boy the girl's shots in addition to his own. The boy had an immediate reaction, and the mother left without getting the same shots for her daughter. Up until that point, both children were on the same developmental path. Ealey said:

What we have is a holocaust. Our children are being maimed and they are being killed. And you've got a governments sitting in Washington D.C. that doesn't think enough to subpoena Dr. Thompson who came out and said what they were doing. So what we have to do today is take back our communities and take back our children. And how do we do that? We walk out of the doctor's offices, we decide no, we're not going to take that shot in the dark. We take our children out of the school system because the only thing they understand is money.[37]

The walls of the room were lined with men from the Nation of Islam, dressed in sharp black suits and red bow ties, when Minister Tony Muhammad stepped to the podium and gave his account of what he believed was taking place and what needed to be done:

> What we are finding out is that the pharmaceutical industry is one of the richest lobbyist groups in the world. And they are now financing many pastors. They are financing black leaders—our politicians. I went personally to Sacramento. Minister Farrakhan said get to them quickly and I showed them this information before they took the vote [on SB 277] to [Senator] Isadore Hall, [Assemblyman] Mike Gibson, [Supervisor] Mark Ridley—I showed it to all of them—Sawyer, Mitchell—and they all pushed it back to me and said 'you got to do what you need to do.' Only two did not vote for it [SB 277] and it was black females. And then when we looked at the research, all of them took money from the pharmaceutical industry. We got some sellouts in our own community.[38]

As the question-and-answer session in the packed auditorium reached its conclusion, Ealey ended the night with a call to action: "You have to be willing to put your life on the line for this cause. Because if not, your children are not going to make it and you're not going to make it. The pharmaceutical industry has developed a client from the womb to the grave and it breaks you."[39]

* * *

On May 20, 2016, while Robert De Niro was at the Cannes Film Festival with the premiere of his new film, *Hands of Stone*, he was interviewed by a magazine, and he addressed the fallout over *VAXXED* in a way that probably won't win him any fans in the pharmaceutical industry:

> **And I have to ask, you have an autistic child and your own film festival, Tribeca, started out with a controversy over an antivaccination documentary you programmed. What do you think you learned from choosing it and then choosing not to show it?**
>
> **RD:** Well, what I learned, first of all, there was a big reaction, which I didn't see coming, and it was from filmmakers, supposedly. I have yet to find out who it was. I wanted to just know who they were, because to me, there was no reason *not* to see the movie. The movie is not hurting anybody. It says something. It said something to me that was valid. Maybe some things were

> inaccurate, but if the movie is 20% accurate, it was worth seeing. And they were saying it was because of the filmmaker and he was discredited, but *how* was he discredited? By the medical establishment? There's a lot going on that I still don't understand, but it makes me question the whole thing, and the whole vaccine issue is a real one. It's big money. So it did get attention. I was happy about that. And I talked about another movie, *Trace Amounts,* that I saw and spoke about a lot, that people should see it, and it's there. Something is there with the vaccines, because they're not tested in some ways the way other medicines are, and they're just taken for granted and mandated in some states. And people do get sick from it. Not everybody, but certain people are sensitive, like anything, penicillin.[40]

The other news from the article was the biggest shock of all. De Niro had paired with Hollywood producing legend Harvey Weinstein (not yet discredited by the #MeToo movement) of Miramax Films, to make his own vaccine documentary: "I don't want to talk much about it, because when I talk about it, something happens," said the Oscar-winning actor.[41]

On May 21, 2016, Robert Kennedy Jr. sent out an email to a small group of autism advocates in which he let them know he was assisting Weinstein and De Niro on their vaccine documentary.[42] De Niro had become a hero to the autism parents by deciding to show *VAXXED* at the Tribeca Film Festival, momentarily lost his footing when he gave into demands to pull the documentary, but regained the strength of character he displayed in the best of his film roles. Because of his actions, EVERYBODY was going to be looking at the issue of vaccines and autism.

***

I will give the final word to Dr. Andy Wakefield, one of the most courageous individuals of our time. This is what he had to say in March 2016 as the documentary began its improbable journey:

> For the last 20 years I have had to watch the suffering of those affected by autism as the problem multiples year on year. What started with hope for a new understanding, new and effective treatments, and even prevention, turned to despair as special interests exploited their influence over the media to crush the science and the scientists.
>
> And then, two decades and a million damaged children later, one man, Dr. William Thompson—a CDC insider—decided to tell the truth and the

embers of that early hope glow once more. Several years ago I decided that to take on the media, you had to become the media. The best medium for this story is film. Our aim with this movie was to take this complex, high-level fraud and to give it context, and weave through it the tragic street-level narratives of ordinary families affected by autism.

This film brings to the public a dark and uncomfortable truth. To ignore it would be most unwise.[43]

CHAPTER ELEVEN

The Battle for California, America, and My Hometown

The Golden State of California was a big prize for the pharmaceutical industry, the place where I believe they planned to lay the groundwork for a strategy they would seek to implement across the country. To steal an expression about New York, if they could make it in California, they could make it anywhere. Senate Bill 277 (SB 277) would set the precedent that a parent could not refuse a vaccination for their child or else lose their right to public and even most private education. If this goal could be achieved, they would move onto the next phase, mandating vaccines for adults, and the price of refusal would be the loss of employment. The pharmaceutical industry had already put such bills before the California legislature, mandating vaccines for those who worked in the childcare industry as a condition of employment.

With the passage of SB 277, the failure of a planned ballot referendum, and the looming date of July 1, 2016, when the bill was scheduled to become law, all appeared lost.

But on July 1, 2016, a lawsuit filed in the San Diego Federal court by attorney Carl Lewis took everybody involved in this fight by complete and total surprise. The lawsuit challenged SB 277 and had the potential to throw pharma's plans for California into disarray.

The initial filing consisted of ten plaintiffs and ten defendants, including the State of California Department of Health, Department of Education, three school districts, and the Santa Barbara County Public

Health Department. The complaint listed six causes of action, including infringement of rights protected by the California Constitution, infringement of rights protected by the US Constitution, violation of Federal Family Educational Rights and Privacy Act (FERPA), violation of California Confidentiality of Medical Information Act, violation of California Information Practices Act, and violation of California Health and Safety Code Section 120440.[1]

On July 15, 2016, an amended complaint was filed, adding additional defendants and causes of action. The hearing on the motion to temporarily halt the implementation of SB 277 until all of these issues could be properly litigated was set for August 12, 2016, in San Diego, a scant three days before many students and teachers, including myself, would be heading back to school. When a lawsuit of this nature is filed, the attorneys try to raise as many claims as possible in the hope that at least one of them will induce the presiding judge to issue a ruling to stop the law from being implemented. In an abbreviated form for an article in *Age of Autism*, I wrote about these claims:

> Violation of the California Constitution—Access to public education is a fundamental right under the California constitution, which means that any abridgement of that right must be viewed with great suspicion. The defendants must show there was an overwhelming need for the abridgment of this right and that the abridgment is narrowly tailored. Furthermore, this proposed law discriminates on the basis of wealth, as more affluent parents are more able to homeschool their children. It also discriminates on the basis of English language fluency as California law states that any parent who wants to homeschool their child must be fluent in English.
>
> Violation of Freedom of Religion—Many parents are against abortion and many of the vaccines are made using aborted fetal cells, a clear violation of their religious liberty. Even allowing children to go to religious schools which are consistent with this belief offers no relief, as religious schools are also bound by SB 277.
>
> Violation of the Right to Freedom of Assembly—By locking the schoolhouse doors to children whose parents alter the vaccine schedule, this law is depriving children of their right to attend a secular or religious school of their choosing.
>
> Violation of Parental Rights—Parents have the right to control the upbringing and education of their minor children and determine what medicines they will and will not take.

Violation of Equal Protection—14th Amendment—The 14th Amendment prohibits the state from treating people differently and SB 277 does this in many ways, such as depriving children of their fundamental right to an education, depriving children with religious objections to the use of aborted fetal tissue in their medical products, and institutionalizing a prejudice against a politically unpopular group.

Violation of Due Process, Fifth, and Fourteenth Amendment—This law impinges on fundamental liberties by denying certain children the right to enter school and subjects them to indignity and humiliation.

Violation of the Individuals with Disabilities Education Act—As a number of the parents with concerns about vaccines came to this belief after they watched their children apparently suffer a vaccine injury, this law will impact those children currently receiving services under current law.

Violation of Section 504 of the Rehabilitation Act of 1973—Federal law provides funds to the state of California and these funds will not go to many of their intended beneficiaries if this law is enacted.

Violation of the Americans with Disabilities Act—In California children are supposed to receive a "free and appropriate public education" and this law will prevent that from happening for a class of students.

Violation of Title VI of the Civil Rights Act of 1964—Because this law will disproportionately impact non-native English speakers who may not be able to qualify as competent to homeschool their children, this law violates the Civil Rights Act of 1964 which bars discrimination on the basis of race, color, or national origin.

Violation of Article IX, Sections 1 and 5 of the California Constitution—These constitutional provisions impose on the state a duty to educate all the children. These duties are non-delegable.

Violation of the Equal Protection Clauses of the California Constitution, Article I, Section 7(a) and Article IV, Section 16 (a))—This law prevents the state from denying basic educational opportunities to a certain class of students and SB 277 would violate that provision.

Violation of the Due Process Clauses of the California Constitution, Article 1, Sections 7(a) and 15—Students have a protected property interest in their education and the state is depriving them of that right without due process of law.

Violation of California Education Code Section 51004—This section provides that students have a right to obtain an education, regardless of race, color, national origin, or economic status.

Violation of California Confidentiality of Medical Information Act—There has been an attempt to gather information on medical exemptions provided by doctors, a clear violation of the California law.

Violation of California Information Practices Act—Personal information about students must be protected.
Violation of Health and Safety Code Section 120440—California law allows a parent to limit the sharing of information about their student.
Violation of Federal Family Educational Rights and Privacy Act—A similar federal law which deals with the privacy of student information.
Violation of California Code of Civil Procedure, Section 526a—Prevents the state of California from spending any money on activities which are in conflict with the law.[2]

The lawyers making up the team to stop SB 277 were well versed in the issues of consumer protection and rights. Leading the team was Jim Turner, a partner in the well-known Washington, DC, law firm of Swankin and Turner. Turner had come to Washington, DC, in 1968 to work with famed consumer activist Ralph Nader and had authored or coauthored several books dealing with environmental issues. Turner had also served as special counsel to the Senate Select Committee on Food, Nutrition, and Health, as well as the Senate Government Operations Subcommittee. In addition, Turner had successfully opposed a Federal Trade Commission effort that had sought to ban the words "organic," "natural," and "health food" from consumer products. Prior to becoming an attorney, Turner was a gunnery officer in the United States Navy.[3]

Attorney Betsy Lehrfeld graduated from the University of California at Berkeley's Boalt Hall School of Law and is a principal in the law firm of Swankin and Turner, as well as the executive director of the National Institute for Science, Law, and Public Policy. She has appeared before the Food and Drug Administration multiple times and focuses her practice on health care and corporate and tax matters for nonprofits.[4]

Attorney Carl Lewis has been practicing since 1985 and is admitted to practice in several courts, including the United States Supreme Court. He practices in the area of civil litigation and appeals, focusing on individuals in the areas of employment, discrimination, and civil rights.[5]

Rounding out the public team was Bob Moxley, who started practicing law in 1979, becoming a principal in the law firm of Gage and Moxley (dissolved in 2006), which specialized in vaccine compensation law. For several years, Moxley served as an assistant public defender for two counties in his native Wyoming and has worked in the National Childhood Vaccine Injury Compensation Program since it was established in 1988. Moxley is a member of the Wyoming Trial Lawyers Association, the American Association for Justice, and the American Civil Liberties Union and has worked on many religious freedom cases in association with the Rutherford Institute.[6]

According to Tim Bolen, a longtime health activist, there were many other attorneys who were offering support and advice, but they were not publicly identified at this time.[7] As the summer of 2016 waned and parents started thinking about the day when their children would return to their classes, a great uncertainty hung in the air of the Golden State as to whether many students would find themselves unwelcome at the gates of their local schools.

* * *

As the battle for California was continuing to be fought, an advance guard of activists—including Del Bigtree, the producer of *VAXXED*; Tami Canal, the founder and head of a group called March Against Monsanto, which boasted more than a million members worldwide; blogger Levi Quackenboss; Mark Blaxill, the editor of *Age of Autism*; activist Jen Larson; and a number of prominent attorneys—made its way to Washington, DC. On June 16, 2016, they met with Congressman Jason Chaffetz, the head of the House Oversight and Government Reform Committee. Prior to the meeting, Del Bigtree and Tami Canal filmed and posted a brief report on YouTube of how the remarkable meeting had been arranged.[8]

Del was dressed in a gray blazer, his salt-and-pepper hair falling in ringlets along his forehead, and a lopsided grin crept onto his bearded face as he stood in front of the marbled facade of the Rayburn House Office Building. Tami Canal was next to him, an attractive woman with straight blonde hair and that wholesome look that seems indigenous to her native state of Utah.

Del was quick to start talking. As a television producer, he was aware of the concept of dead air. Even though he had spent his life behind the camera, he certainly knew how to act in front of it: "Hey everyone! I'm here in Washington, DC, and we're about to go into a meeting with Jason Chaffetz, who is head of the House Oversight Committee, that is going to have to subpoena Dr. William Thompson, which is what this fight with *VAXXED* has been all about. This has been made possible by my friend Tami Canal. Say 'Hi,' Tami!"

The camera panned to Tami and she said hello, then the camera returned to Del.

"So how did you set this up?" Del asked Tami.

Tami gave a small nervous laugh. "I was very pushy and persistent. He hemmed and hawed and didn't want to grant me a meeting in Provo, which is forty-five minutes from my house. So instead of taking a little car ride I traveled twenty-two hours for a one hour meeting. And we're here to guarantee that whatever congressional meeting comes about it is not a sham. And we're demanding action today."

Del picked up the thread. "Tami is not just your average constituent. She is head and founder of March Against Monsanto, an awesome group. Tell us a little about that."

She nodded and began. "March Against Monsanto is 1.4 million strong across the globe, and today I'm here to speak on behalf of most of our supporters who realize the importance of uniting our movements. The vaccine transparency movement and the food safety movement go hand in hand. It's all about protecting our children and insuring their healthy futures."

Tami looked to Del, who seemed to almost bounce with excitement about the burgeoning collaboration between the two movements as he started to talk:

> This is huge for me. This has been one of my issues from the very beginning: this huge disconnect between my progressive liberal friends who were really into labeling GMOs, see the danger of a company like Monsanto, but have been supportive of mandatory vaccine laws. And they don't seem to be afraid of Merck, with all the issues and legal situations Merck has found itself in. All the lying they've been blamed for and the cases they've lost in court, it's just like Monsanto. They're very similar companies to me. So I think it's just so amazing to have Tami jumping onto this issue with us. Because I think the merging of a very strong progressive liberal issue like GMO foods and pesticides, things that we're all fighting for food and health in our communities, to have Tami helping us make the connection that the same thing is happening in vaccines, is huge. How did you get into the vaccine issue? Was it difficult? What made you jump over?

Tami said, "It was two years ago when I was pregnant with my third and final child. I began to hear the whisperings of a CDC whistleblower. And down the rabbit hole I fell and my eyes have been opened to the mass corruption and the horrific vaccine ingredients. Those syringes have no business being anywhere near our children."

The footage seemed to jump, and Tami did a plug for her March Against Monsanto website, and Del did a plug for the *VAXXED* website and then continued talking: "We're joining some other really heavy hitters, Mark Blaxill and Jen Larson, and some other friends who'd rather remain anonymous."

Tami jumped in with some additional information: "Some powerful lawyers who flew in from all over the country will be joining us today."

Del finished off with their closing statement as his eyes grew wide and his smile broadened. "Really, really big meeting. We'll let you know how it all went. Take care."[9]

* * *

A few hours later, Del had a new video for those who'd been following from home. This time the location was what looked like a hotel hallway, and he was alone. His message was significantly longer, but he wanted to convey to the thousands who were following his activities what had taken place in this issue, which for many had consumed decades of their lives.[10]

When the camera started rolling, Del was sucking on a mint, and he laughed, quickly taking it out of his mouth and apologizing with a sheepish grin. Then he got into the meat of the issue:

> I want to apologize for keeping everybody waiting. Obviously it's a big day. We were at Congress. We met with Congressman Jason Chaffetz, the head of the Oversight Committee that is going to be responsible for subpoenaing Dr. William Thompson. That meeting went an hour and a half. That is unprecedented in these situations. Mark Blaxill and Jen Larson have been doing this for a very long time. They were sitting with us, amongst many others, Tami Canal, who stood with me before we went in and a few other lawyers. It was a very important meeting. And a lot of what was talked about I'm not allowed to share. That was sort of the agreement that we all made that would allow Jason Chaffetz to speak freely.

Del paused for a moment, as if trying to determine how to present the information to the community:

> I want to report as best I can on what I can share with you. What I can share with you is this: There is an investigation taking place now at the CDC. That we know for sure. Of the CDC by Jason Chaffetz's office. Interviews are being done. And I would say that all of the work you have done—we sent 2,700 letters to Jason Chaffetz's office over the last couple weeks. His official statement

> I can give to you, he basically said that "You have our attention." We have the attention of Jason Chaffetz's office. And I want to report that I can honestly say that I enjoyed meeting Jason Chaffetz. This is a genuinely heartfelt individual. And I can report to you my own instincts of what happened in that room is that this is a man and a staff that recognize we have an issue with vaccines and the health of our children.

Del then shifted gears, telling the community what he hoped they would do next:

> What he asked is that we stop calling his office so much because the very people he's needing to do the investigation are getting hung up on answering your phone calls. So you've done an amazing job. I guess what I'd say, for now, and obviously this is a politician and he'd like to have his life a little easier. But I will say this, he's genuinely requesting we stop calling his office so much because he understands. They are investigating and they are going to try to get to the bottom of this. He said, "I can't guarantee you where this will end," but they are doing the investigation.

With that all said, Del seemed to visibly relax:

> So let's give them a little bit of time. What I think we need to start doing is contacting other people that are on this committee now, that are going to have to assist Jason in bringing this before the Congress. I will talk more specifically about that tomorrow. And in the next few days we're working on a strategy of who that might be and the best way to contact them. And right after speaking with Jason Chaffetz's office we went over to Congressman Posey's office and had a really great meeting there. I've been in meeting ever since, really trying to understand what just happened today.

Del took a deep breath, as if he could not believe how much had happened in the day. He had spent an hour and a half with one of the most powerful men in Congress and received a pledge that the claims of Dr. William Thompson would be thoroughly investigated. And after that meeting, he'd talked with another congressman about the same issue:

> I want to say this: This was a very positive day and there is a real recognition of the issue. And we have done an amazing job of bringing pressure upon this issue. I believe the movement of *VAXXED* is forwarding this issue on top of

> the amazing work that people like Mark Blaxill and Jen Larson have done through the years. *VAXXED* is obviously taking this thing to another level. I think the screening we had in Utah with all of the politicians is definitely affecting how Jason Chaffetz is looking at this issue. So, positive movement, but it is still Washington. I will never say I totally trust the system. I'm not sure the system works. But if it does, this is a guy who can do something. And I feel very good about what happened. I will tell you more, but I just wanted to check in and say, continue to sign-up to our mailing list. Continue to support all of your state groups that are fighting for vaccine freedom. It's a positive day in America for everybody fighting for vaccine freedom. Thank you for your work. I'll talk to you tomorrow. Bye.[11]

***

One of the people in the meeting with Congressman Chaffetz was blogger Levi Quackenboss, and he was so keyed up by the meeting he had to immediately write a post about it when he arrived home.[12] Quackenboss honored the agreement Del Bigtree had talked about, not reporting what Chaffetz had said, but believed he was under no obligation to withhold what they had said to the congressman. He also told his readers that he had first made contact with Tami Canal of March Against Monsanto and piqued her interest in the subject.

The meeting was scheduled to begin at 1:00 p.m., but Congressman Chaffetz was late because he had to vote on twenty different bills. While waiting for the meeting to begin, the group sat with Fred Ferguson, Chaffetz's chief of staff, and Sean Hayes, an attorney on the House Committee on Oversight and Government Reform (OGR). Sean told the group that the worst-case scenario was that the Committee would build a full record against the CDC through document requests and make criminal referrals. The investigation was complicated by the fact that the records are more than fourteen years old and much of the investigation depended on cooperation from the CDC.[13] The issue of the poisoned drinking water in Flint, Michigan, was brought up, which in the public's mind quickly resulted in congressional hearings. However, the Committee had been investigating the Flint situation for more than a year and a half before the public ever heard about it.

Chaffetz came in at 1:35, and as Quackenboss recounted in his article:

> [T]here was a wee bit of hostility from some of our members because we're sitting there with the impression that the OGR hasn't done jack for nearly

> two years now, so Chaffetz had to walk into that. Some of us introduced ourselves and when it got to Del, Del just started talking. He gave the history of Vaxxed, how he got pulled into it, and took Chaffetz through the timeline of it getting kicked out of Tribeca and going nuclear.
>
> Del said that in the fall of 2014 when the Whistleblower story was never reported that it made him question the state of democracy. It would have been one thing to have the media address it and dismiss it, but no one said a word. It's a sad statement as to who is running the news in the country.
>
> There was a huge smile from Chaffetz when Del commended him on how brave his staffer was to stand up in Utah, announce herself, and take questions at their Vaxxed Q&A. He knows how to make a boss feel proud.
>
> Del asked what he can say in the theaters when people ask about Chaffetz night after night. But Chaffetz made it clear that he can't broadcast what he's doing, where he's been, or where he's going next. It's not how investigations are done. He can't promise how this investigation will end. It may not end with a hearing. They may not even be able to construct the record. We can't hang everything we've got on one witness.
>
> Chaffetz asked what we wanted to see happen beyond a Thompson hearing and this was our wish list.
>
> One: that the power to police vaccine safety is taken away from the CDC.
>
> Two: That the 1986 National Childhood Vaccine Injury Act is repealed and pharmaceutical companies are responsible for the safety of their products.
>
> Three: That vaccinated vs. unvaccinated total health outcome study (Congressman Posey's HR 1636) is conducted.[14]

Quackenboss went over some of the same points Del had made in his video, such as Chaffetz's request to stop the phone calls so his staff could proceed with their investigation and possibly having the community focusing their efforts on some of the other members of the committee. Quackenboss finished his article with an account that transported his readers into what may be one of the pivotal historical meetings of the autism epidemic:

> I wish I'd taken a picture of Chaffetz leaning back in his chair biting a pen in his mouth and grinning nearly ear-to-ear while listening to Del. Del could thaw a cold war. He was endearing and disarming and changed the energy in the room when he talked about how he'd changed political parties since coming upon the story of the CDC Whistleblower.

> But, I did snap this picture in the beginning before Del turned on the charm and I'm risking Jason Chaffetz never hosting me in his office for the rest of his career, but I want to share this with you because you deserve to see it.
>
> This is the face of a man who cares, people. This is real. You are not crazy. The CDC switching their study criteria midstream is known by this man and his team. This meeting happened today. And our one hour meeting turned into a two hour meeting, 90 minutes of which was spent with the Congressman. That is simply unheard of.[15]

The picture taken by Quackenboss that accompanies his article shows three men in the congressional office. Chaffetz is on the far right, the largest figure in the picture and closest to Quackenboss, leaning forward in his chair, his chin resting on his left hand as he listens attentively. Chief of Staff Fred Ferguson sits to Chaffetz's right, leaning back in his chair, his legs crossed, with his left hand at his mouth, similar to his boss. It is a pose of intense concentration. Attorney for the House Committee on Oversight and Government Reform, Sean Hayes, sits on a couch in front of a window that almost seems to glow from the afternoon sun. His legs are folded, and he appears to be looking down at a document that has been provided to the group for the meeting. It is a picture that nearly makes me weep to look upon it. After a decade and a half as an activist, more than fifty thousand children a year coming down with this affliction, more than a million with the disease in our country and each day bringing the threat of yet another autistic child wandering away from their exhausted caregivers, only to be found floating in a lake or river or dead from exposure to elements they do not fully understand, somebody was finally paying attention.

Eventually, Congressman Jason Chaffetz mysteriously resigned in the middle of his term on June 30, 2017, and became a commentator for Fox News.

***

And so I figured the time had come for me to bring *VAXXED* to the local movie theater in my hometown. The producers had contracted with a company called GATHR, which had an agreement with many theater chains that if enough tickets could be presold to a planned screening (sixty-seven tickets in my case), the theater would guarantee a showing at a certain time. On Monday, July 18, 2016, I bit the bullet and registered myself as a film captain for a showing of *VAXXED* to take place on Thursday, September 8, at 7:30 p.m. at the Crow Canyon Cinemas in San Ramon. It was a convenient location, just off the main highway, and in the center of town.

San Ramon has a population of about fifty thousand people, and the two neighboring towns, Danville and Dublin, contain roughly similar numbers. Within about ten miles of the theater lived at least a hundred and fifty thousand people and I had about six weeks to sell the tickets. To increase interest in the showing, one of the people profiled in the movie, Brandy Vaughn, the former Merck pharmaceutical representative who was putting up billboards against vaccines, agreed to come and appear on a panel with me after the movie for a question-and-answer session.

My mother-in-law was an early and enthusiastic supporter, quickly buying ten tickets, then finding some other people interested in the film, who bought five tickets. My brother and sister-in-law, who live in Sacramento, were willing to make the two-hour drive and also agreed to come to the showing. I am aware that my opinions about vaccines are controversial, but I had thought that the showing of a documentary in a regular theater would cause people to give me the benefit of the doubt. I mean, haven't all of us heard about "banned" books and movies and the stupidity of censorship? You can imagine my surprise when some family and friends reacted almost violently to my invitation to view a documentary.

The first was a family member I'll simply call the Accountant. The Accountant is a first-rate family man, good to his wife and children, and an extreme athlete, qualifying twice for the Kona Ironman competition in Hawaii. He also volunteers at the local school, teaches a STEM class, and offers his accounting skills in whatever way is needed for the local PTA. The Accountant has been dismissive of the views of my wife and me about the dangers of vaccinations, but we have pretty much avoided the topic since my wife came home from one Christmas dinner in tears after what seemed like a particularly violent argument. I could not be at that Christmas dinner because our daughter was having a particularly rough day. She really doesn't like to be in places other than our home, so that means for most family events, my wife or I go solo with our son, while the other is at home with our daughter. Whenever I have steered the conversation in the direction of vaccines with the Accountant or his wife, I have been met with a brick wall. I can honestly say that, while I know the opinion of the Accountant and his wife, I have never had an actual discussion with them. However, if you cause my wife to come home in tears from a Christmas dinner, you're not really on my good side. After I booked the showing, I sent a Facebook message to the Accountant saying my expectation was that he attend the showing. (He lives five miles from the theater, and I was giving him six weeks notice!) This was his answer:

> Kent, you make me laugh. You shouldn't set expectations for me. You could ask nicely though. As for the family thing, no hall passes . . . As to VAXXED, I'm actually very well-read on the subject, believe it or not. I subscribe to lots of science blogs on a variety of topics. I even read your FB and blog posts! I'm certainly not a scientist and don't claim to be, but I do teach STEM to 500 kids a week so I must be an expert. I've taken courses on evaluating scientific information and think as a lay person I'm pretty open-minded. There is a reason that the scientific community picks on Dr. Oz, Foodbabe, Mercola, Wakefield, etc. VAXXED is in the same category. A lot of the mainstream science bloggers have already covered the topic. I always liked the very detailed Steven Novella site . . . I generally don't go to movies rated zero by Rotten Tomatoes (nor pay $11), but I'm happy to check in with family members and see what our calendar holds. Thanks for the invite!

Although the Accountant prides himself on his wide reading, he made an enormous mistake about the Rotten Tomatoes ranking for *VAXXED*. As of July 29, 2016, when I checked, VAXXED had a 36 percent rating from critics (four liked the film and seven disliked the film), and among 1,017 audience members who ranked the film, it had an 83 percent positive rating.[16] Facts can be stubborn things. This was more than an accounting error.

The second person is somebody I will refer to as the Economist. I first met the Economist when I was sixteen years old in the Brothers Residence of the Catholic high school we both attended, and he needed a ride home. Even at fifteen years old, he was clearly brilliant, and it wasn't a surprise to me when he graduated first in his class from high school, first in his class from UC Berkeley, and second in his graduate economics program from Stanford University. He teaches today at one of the best universities in the country and is widely expected to win a Nobel Prize in Economics at some point in the next ten years. We have enjoyed both a personal and intellectual friendship over the years, and I can still remember him giving me what I consider to be one of the best compliments of my life. He said, "Kent, I don't know a single person who has such a wide-ranging and consistent curiosity as you." But all of that seemed to come to an end when I asked him to come to a showing of *VAXXED*. He wrote:

> Dear Kent:
>
> I can't in good conscience support this documentary. From what I understand, it overstates the health risks of vaccines and understates their health benefits. I believe that misinformation on this topic has tragically put many children at risk, both in the U.S. and elsewhere.
>
> To convey how far apart we are on the subject, let me say this: I would rather spend time and money increasing access to vaccines.
>
> I have tried to engage skeptically in our many conversations on this topic. But judging from the 8 (!) junk science books you sent me recently, our views (standards of evidence and logic?) are just too far apart for us to talk productively about this topic any more.
>
> I am still happy to meet for movies and meals, and to talk about politics, sports, family—basically anything else.

At the end, he added what he called "a few self-serving quotes":

> "Friendship, to be real, must ever sustain the weight of honest differences, however sharp they may be."—Mahatma Ghandi
>
> "It takes a great deal of courage to stand up to your enemies, but a great deal more to stand up to your friends."—Albus Dumbledore to Neville Longbottom in *Harry Potter and the Sorcerer's Stone*
>
> "Only a true friend would be that truly honest."—Shrek[17]

Okay, so I don't think you'll be surprised to know that I considered that to be quite a blow. We are free to talk about whether God exists, but the question of whether the increase in vaccines is harming a generation of children is now off-limits. For the record, one of the eight books I had provided him was *Science for Sale*, by Dr. David Lewis, a longtime government scientist and the only EPA scientist to be the lead author for articles for *Nature* and *Lancet*. Another book was *Vaccine Epidemic*, written by attorney Mary Holland, a research scholar at the New York University School of Law, and Louise Habakus, who graduated Phi Beta Kappa from Stanford University, where she also received a graduate degree in international policy studies. The book was endorsed by Dr. Bernadine Healy, the former director of the

National Institutes of Health (NIH). You really can't credibly call these people "junk" scientists.

The third person I will call the Attorney, and it may be in him that I am most disappointed, as I consider the legal profession to be dedicated to a full and open hearing on any issue. I mean, isn't the wrongly convicted man the staple of so many legal dramas and historical cases? I also met the Attorney when I was in high school, and we were both English majors at the same college, reading the works of John Milton and taking on the longest book written in the English language, Samuel Richardson's eighteenth-century epistolary novel *Clarissa*. We also went to the same law school. The Attorney is a great husband and father and donates some of his time to charitable causes, serving on the board of his local hospital. But again, the viewing of a documentary film on alleged corruption at the CDC was too much for this fine man. He wrote to me:

> Dear Kent:
>
> Thanks for the kind words and invitation. I know how passionate you are about this issue, but . . . The premise of this film is completely at odds with my views on the risks/benefits of vaccination. I am by no means an expert on this topic, but all that I have read leads me to the inexorable conclusion that whatever risks may be associated with vaccines, those risks are so far outweighed by the health benefits of, in many cases, virtually eliminating horrible diseases that afflicted generations of families before the vaccines were discovered (i.e., polio, diphtheria, Hepatitis A and B, pertussis, etc.) So, my friend, I can't support this film or your endorsement of it, and I in fact believe that the film poses serious harm to the public if even a single person chooses not to vaccinate a child as a result of this film. I am communicating this so clearly Kent because, as your friend, I feel compelled to share my honest opinion (again, as informed or uninformed as it may be) with you on this subject. You can of course take it for what it is worth.
>
> Perhaps I should have shared this with you well before now. Regardless, please believe that it is not my intention to insult or harm you in any way, or to minimize the challenges you and Linda face daily in caring for your beautiful daughter Jacqueline. I care and feel deeply for you and truly value your friendship.

When I was in high school with the Economist and the Attorney, one of the topics we covered in our morality classes was the famed Milgram Experiments on obedience to authority conducted at Yale University. In the experiment, a person in a white lab jacket instructed the test subject to ask a series of questions to a person seated in another room who they did not know was an actor. If the person in the other room answered incorrectly, the test subject was to push a button to administer an electric shock. As the actor in the other room continued to answer incorrectly, the test subject was to increase the amount of the electric shock, even to levels clearly marked as dangerous. As the voltage was increased, the actor would begin to bang on the wall and complain about his heart condition, which the test subject could clearly hear. If the subject indicated he wanted to stop, the man in the white lab jacket would give him this series of verbal prods:

1. Please continue.
2. The experiment requires that you continue.
3. It is absolutely essential that you continue.
4. You have no other choice, you must go on.

No threats of any kind were made by the man in the white jacket. The stated purpose of the experiment was to determine how ordinary Germans could have participated so willingly in the Holocaust. As the voltage increased, the actor continued to scream, then went silent. The experiment only ended after the test subject had administered three 450-volt shocks to the actor in the other room who was nonresponsive.

The scientists who set up the experiment expected only 1–3 percent of the test subjects to continue to the end. However, 65 percent of the test subjects continued administering the dosages until the very end of the experiment. How different were Americans in their obedience to authority than ordinary Germans during the Nazi era? Unfortunately, the answer was, not much. Especially if that authority figure was dressed in a white lab jacket. In a 1974 article on his experiment titled "The Perils of Obedience," Milgram wrote:

> The extreme willingness of adults to go to almost any lengths on the command of an authority constitutes the chief finding of the study and the fact most urgently demanding explanation. Ordinary people, simply doing their jobs, and without any particular hostility on their part, can become agents in a terrible destructive process. Moreover, even when the destructive effects of their work become patently clear, and they are asked to carry out actions incompatible with fundamental standards of morality, relatively few people have the resources needed to resist authority.

I can clearly recall discussing this experiment with my friends the Economist and the Attorney. We were united in the belief that we would not be obedient to authority if it clashed with our values. Would our high school selves have refused to see a documentary? One may genuinely ask if I am any better than my friends, given that I have observed something differently in my family from what they have in their families. Would I think like them if I did not have a child with autism? I am incapable of making that judgment. But it is an important and critical question.

* * *

I wrote an article about the issue of family and friends not wanting to view the documentary *VAXXED*.[18] I compared those people not wanting to come and see the documentary to the Cardinals of the Catholic Church who, during the trial of Galileo, refused his pleas to simply come to his observatory and look through his telescope so they would know he was telling the truth. They refused. I found that some of the comments left on the article expressed my range of emotions on this issue:

> Excellent article. So sad. So cognitive dissonance even works on smart people? Whoa. See that's where I like being from the streets. If my friend said he wasn't going to my movie screening he'd get knocked out. Fricken asshole.[19]

I have to admit that one made me laugh. The idea of me, at the age of fifty-three, knocking out one of my fellow middle-aged friends, was very funny. I'm not sure I even still know how to throw a punch. My last fight was in the sixth grade.

Another comment from a reader made me sad:

> While you cannot actually divorce yourself from a member of your family, short of just cutting them out of your life, if I had such a relative or even a soon-to-be Nobel Peace Prize winner "friend" who placed censorship of speech conditions on my relationship with them, I would cut them right out of my life. I have no tolerance or use for such intolerant, ignorant, and controlling people who have no conception of or sympathy for another's intensely painful, directly witnessed experience. That said, I wish you the best, Kent. Permitting such people to be a part of your life will only cause you pain as you will be reminded, every time you look at them, of how unjustifiably and

> inexcusably disdainful they are of your personal pain. With friends/relatives like that—who needs enemies?[20]

I hate to admit it, but I'm not really built for the big emotional gestures where I cut somebody out of my life. Yes, I am half-Sicilian on my mother's side, and we do love those operatic emotions. But I've often found that they're more trouble than they're worth. Still, I wonder if it makes any sense for me to expend any energy on these family members or friends. There might be great wisdom in the comment that I am not yet ready to fully accept. I find that this comment best exemplifies what I feel about the issue at this point in my life:

> There is a very uncomfortable new reality that comes with such a change in a person's viewpoint on an issue such as this. I have had to do more than one 180-degree turn on a few subjects and it wasn't an easy psychological process. I still recall the thought process when a long held assumption is smashed by the evidence in front of your eyes.
>
> Okay, so my assumption that conventional medicine is real and honest and alternative medicine is for fruitcakes and suckers was proven wrong to me in dramatic fashion. First thought is: if I was so obviously wrong on this, what other important things am I completely wrong on? Not a good place to be mentally. You have to basically start over again. Question everything, even yourself.
>
> It turns your world upside down and trashes what you previously held as "your good judgment." It's easier to live in a world where things are much simpler and more pleasant. My doctor knows best. Pharmaceutical companies are working tirelessly to cure disease and the people that are ill. Politicians are looking out for my best interests.
>
> Basically, it all boils down to "don't change my comfortable world." I don't want to live in "the real, but muddy and unclear world." It is just a form of self-preservation for the average person. It's not rational, but it is important for them not to tear apart what is comfortable for them. It's hard to blame them for that.[21]

It is difficult to accept that, in the eyes of many family members and longtime friends, you have become a dangerous person. Still, I do not believe the problem is with me. When the Cardinals of the Catholic Church refused to visit Galileo's observatory and look through his telescope, they were guilty of the same kind of blindness as those who refuse to see a documentary about corruption at the CDC. I am certain those Cardinals considered themselves good men, and in all likelihood they were. They were simply wrong about Galileo and in their blindness were more than willing to burn him at the stake.

But even on his deathbed, Galileo was reportedly still defiant, whispering the words *Eppur si muove*: "And yet it moves." Autism is still with us, and it is rising. I am but the herald of the danger. I am not the danger. The disease is the danger.

***

And what might a reasonable response to a request to see a documentary about corruption at the CDC that endangers an entire generation of children look like? It often takes so little to be a decent person.

Consider my neighbor Fred, who has a grandson with autism. Fred was a marine in Vietnam, just retired from a long career selling computer systems, and spent twenty years coaching the local high school soccer and track teams. He even coached my son, Ben, during his freshman year in track. In Fred's front yard he has a large flagpole, and every day he flies an American flag.

I was in the car with Jacqueline (because she needs to take several car rides every day) when I saw Fred outside his house. I told him about the movie, gave him a flyer, and asked if he would be interested in coming.

"Sure," he said, "And I'll see if I can get my son to come. I know he'd be interested." Fred and I have talked about the vaccine issue before. Fred's son definitely thinks the vaccines were involved in the development of his son's autism.

"But you know I'm not sure if it's the vaccines," said Fred.

"That's fine," I replied. "That's why you should see the movie."

"But damn it, we've got to start talking about it. Because maybe it's not the vaccines, but somebody will see the film and they'll say, 'Hey, I don't think it's that, but it could be this.' Then we'll start to solve this thing."

"Fred, I often say, maybe I'm totally wrong, but if I am, somebody's got to come up with a better explanation. I've been waiting for that better explanation for fifteen years."

Fred smiled and nodded his head. "I'll be there."

And with that, Jacqueline and I drove away. Maybe it's the residual lawyer in me, but I expect people to keep an open mind. You don't have to believe me. Just listen to what I have to say, do your own investigation, and make up your own mind.

* * *

It took me just two weeks to sell seventy tickets, guaranteeing the showing. The theater holds 125 people, so that meant I had fifty-five tickets left to sell and five weeks to do it.

Starbucks was a great place to put up posters, as the only condition for their community board is that it has to be in support of a nonprofit. Andy Wakefield's Autism Media Channel, which produced *VAXXED*, is a 501(c)(3) organization, so I can put up flyers at Starbucks from coast to coast.

Chiropractors were also a sympathetic group, as they'd endured a challenge by the American Medical Association (AMA) to their very existence about twenty years ago. Chiropractors are generally leery of vaccines, as they believe in a different model of health from that of most medical doctors. But I was unprepared for how welcoming the chiropractor's offices would be when I showed up with a small poster for *VAXXED* and my daughter, Jacqueline, who usually looked immediately for a chair to sit in. She fatigues very easily.

I made my pitch to the receptionist at the first chiropractor's office, that this movie was being supported by some of the same people who had helped chiropractors twenty years ago when the AMA was trying to outlaw their profession. The chiropractor was in a treatment room near the front, heard my pitch, and actually came out of the room to give me a thumbs up. "We'll be there," he said.

Of the twenty chiropractor's offices I visited, I'd say five or six gave me similar responses. Sensing a new openness, I said to several of them, "How about having your office buy five or ten tickets and give them out to your best clients?"

To my surprise, many of them said that was a great idea and that they would do it.

I was even treated well when I went to the office of my local congressman, Mark Desaulnier (D), Eleventh District, California, and talked to his district scheduler, Jessi Bailey. "We should be able to have somebody there," she said.

"We're going to have a big crowd," I replied. "And I know how politicians love a crowd."

I put on a cheerful face as I was pitching the documentary, my autistic eighteen-year-old daughter walking slowly with me on the way to the various offices, as her balance is still not good, but I was being profoundly changed by the experience. I am aware most people have considered my family to be the victims of a cruel and capricious fate to have such an impaired child. They would say they have "sympathy" for our plight. But they have not seen us in the way I see us. As the victims of a monstrous evil. But those who worked at the chiropractors' offices did. I saw it in the faces of the receptionists as their brows furrowed and they said, "Oh, we know about this film." They'd look at me with compassion and say, "Count us in," and I knew they understood. I had not realized how much I hated being shut out from people's consideration until there were those who welcomed me in.

There's a Bible passage from 1 Corinthians that says, "Love does not rejoice in evil, but delights in the truth." The truth does not shatter you, and it will not destroy our society. It will set us free. And it is the only thing that can truly heal all of our wounds.

CHAPTER TWELVE

Can Science End the Autism Epidemic?

In one of his presentations, Dr. Robert Naviaux, professor of Medicine, Pediatrics, and Pathology at the University of California, San Diego (UCSD), since 1996, depicts the "cell danger response" as the One Ring from Tolkien's fantasy classic *Lord of the Rings*, hovering in space, the elvish writing visible as if it has just been pulled from the fire, but surrounded by words Tolkien probably never considered, words like phospholipids, pyrimidines, serotonin, autonomic imbalance, protein unfolding, epigenetics, flame retardants, antibiotics, dioxins, plasticizers, parabens, heavy metals, air pollution, systematic infections, and yes, vaccinations. Naviaux believes this "cell danger response" may underlie many diseases such as autism, and amid all the hue and cry, a simple solution a century old may hold the key to a goal that hopefully all sides share, an effective treatment for autism and other chronic diseases.

In person, Dr. Naviaux is a physically imposing person, with the shoulders of an NFL linebacker, the white hair you would associate with an august man of science, but his face appears almost boyish, as if he has always stayed curious about the world. It is easy to imagine him as a young man interested in marine biology, perhaps inspired by the writings of Jules Verne, or the *National Geographic* television documentaries of Jacques Cousteau, putting on his diving gear and slipping beneath the waters of the ocean to discover for himself a new world.

I have had dinner with Dr. Naviaux. I have tried to help him raise money. And I even interviewed him at length for a book on chemical exposures and their link to human health. But when I contacted him on Friday, April 9, 2016, asking if he would consent to an on-the-record interview about his work on autism, I thought the man might have a nervous breakdown. "I just want to do the work," he said, "I don't want to cause controversy."

"Bob, I understand the danger of this work," I replied. "I don't want to cause trouble, either. I just want to get answers."

We talked for a few minutes longer, and he gave his blessing for me to use any previous interview or comment on his published work.

***

When I had first read about Naviaux's work, I sent him a gift, a small piece of "moon rock" from a place called The Evolution Store in New York City. I also received some paperwork that explained it was from a meteorite with identical chemical properties to the rocks brought back from the moon by the Apollo astronauts. The implication was that a large meteorite explosion on the moon had sent pieces of moon rock hurtling into space, from where they eventually fell to Earth. In the card that accompanied the gift, I wrote something along the lines of "Like you, I believe in shooting for the moon to help children with autism." The gift deeply touched him, and he later told me he placed it in a spot in his office where he could always look at it and be inspired.

If Naviaux is correct, then the cause of autism is easy to understand. I could explain the rationale to my sixth-grade students in about five minutes. I am always reminding my students of the quote attributed to Albert Einstein that "If you cannot explain it to a six-year-old, you don't understand it yourself."

The most important idea is that it's all about "communication." If you ever find yourself getting lost in this discussion, I want you to run back to the idea of "communication." If you think that science is difficult to understand, just put it in the context of a relationship. It's all about communication, just like every woman has told every man at some point in their relationship since the dawn of time.

According to Naviaux, in 2008 he was asked to look into autism by Dan Wright, chairman of the board of the United Mitochondrial Disease Foundation.[1] Wright sent Naviaux to a meeting at the National Institutes of Health to start thinking about autism. Within a month, Naviaux had the germ of an idea. He started looking at what he called the "cell danger response" theory of autism, often referred to as the "purinergic" theory of autism, or signaling between cells. (Remember, it's all about communication.)

In April 2011, Naviaux was awarded one of just three international "Trailblazer" awards by Autism Speaks to test his theory of impaired cellular communication in a mouse model of autism. Now you might be asking yourself, "WHAT? There's a mouse model of autism? How do they induce something like autism in mice? I understand that chain-smoking monkeys will likely develop cancer, but we don't know what causes autism! How exactly do we create autism in mice?"

The answer is something called "maternal immune activation," which means the mother is given an infection or some sort of adjuvant to stimulate an immune response, as if she were responding to an infection. At the risk of sounding antiscience, I can't help but comment on the fact that stimulating the immune response of an infant through vaccinations is exactly what concerns many in the autism parent community today. If this process works in pregnant mice to give their offspring what looks like autism, how is this any different from giving a newborn baby a hepatitis B shot on the first day of life?

However, overstimulating the immune system of a pregnant mouse is exactly how you produce something similar to autism in her offspring. This is how Naviaux detailed the creation of the mice with "autism-like features":

> In the MIA [maternal immune activation] model of ASD [autism spectrum disorders], adult females are exposed to a simulated viral infection by injection of a synthetic, double-stranded RNA poly (Inosine:Cytosice) (poly(IC)) at vulnerable times during pregnancy. This produces offspring with neurodevelopmental abnormalities associated with ASD and schizophrenia. Injected poly (IC) RNA is not replicated, but is recognized by the antiviral response machinery within the cell.[2]

In plain language, the immune systems of these pregnant mice are induced to make them think they were suffering from a viral assault. The body responded as if it were under attack, and something about that attack made the offspring have conditions associated with either autism or schizophrenia. Isn't science wonderful? Kind of makes you wonder about those flu shots for pregnant women, doesn't it? And maybe something similar could happen in infants who do not yet have a fully functioning immune system? There are some, I'm sure, who would just say it's a "coincidence."

After a good deal of searching, Naviaux found a compound that he thought might reestablish cellular communication by telling the cells that the danger had passed. The medication was called suramin and had been developed by Bayer in Germany in 1916 to combat African sleeping sickness. African sleeping sickness was caused by a parasite, and as Naviaux considered how the parasite was able to maintain its presence in the body, the answer seemed so simple. Maybe it simply overloaded the cell danger response, essentially shutting down the system to allow the parasite to remain unmolested by the body's immune system.

In March 2013, Naviaux and his team published their findings in *PLOS One,* using suramin on the mice whose mothers had their immune systems overstimulated during pregnancy. As reported in an article on *Science Daily* on March 13, 2013:

> Describing a completely new theory for the origin and treatment of autism using APT, Naviaux and colleagues introduce the concept that a large majority of both genetic and environmental causes for autism act by producing a sustained cell danger response—the metabolic state underlying innate immunity and inflammation.
>
> "When cells are exposed to classical forms of dangers, such as a virus, infection, or toxic environmental substance, a defense mechanism is activated," Naviaux explained. "This results in changes to metabolism and gene expression, and reduces the communication between neighboring cells. Simply put, when cells stop talking to each other, children stop talking."[3]

The idea Naviaux was expressing is a simple one. Imagine a small village in Europe during medieval times. People move freely about performing their daily chores, raising food, making goods, and perhaps right next to the village is a castle with large walls, designed to shelter the population in case of trouble. Then war begins to trouble the land. Maybe there are not enemy soldiers present, but there are new people in the town, and they don't appear to be friendly. Maybe an enemy army is rumored to be nearby. The townspeople flee to the protection of the castle, pulling up the drawbridge, and placing soldiers along the wall. The people are safe inside, but they're not producing food or products. They may be surviving, but they are not prospering. If this is an accurate explanation for what is happening in autism, their brains may not be developing as they should, but they are not being damaged. The immune system may simply be protecting them, waiting for the proper signal to be triggered, so that they can develop in the way they were meant to develop.

Naviaux went looking for signs of the cell danger response being activated in the mice whose mothers had been exposed to an immune challenge and found them in signaling molecules related to both immunity and mitochondrial function. These were called "mitokines," signaling molecules directly associated with distressed mitochondria. Naviaux had his target (as he could measure these mitokines with a specially programmed mass spectrometer), and he had his potential solution, suramin.

Science News Daily reported:

> The drug restored 17 types of multi-symptom abnormalities including brain synapse structure, cell-to-cell signaling, social behavior, motor coordination and normalizing mitochondrial metabolism. "The striking effectiveness shown in this study using APT [anti-purinergic therapy] to 'reprogram' the cell danger response and reduce inflammation showcases an opportunity to develop a completely new class of antiinflammatory drugs to treat autism and several other disorders," Naviaux said.[4]

This is the type of change that has the potential to revolutionize the treatment of autism. From the Conclusions section of the paper: "Antipurinergic therapy with suramin ***corrected all of the core behavioral abnormalities and multisystem comorbidities*** that we observed in the MIA model of autism spectrum disorders." [Italics and bold added by the author.]

How is it that this finding did not immediately become the center of media and scientific attention? Is it because implicit in Naviaux's discovery is the unmistakable indication that "something" caused the cell danger response to be activated? There are so many potential triggers. Could it be air pollution? Antibiotics? Fertilizers? Pesticides? New kinds of plastics? Or maybe something in a medical product injected directly into the bloodstream of day-old infants, and at regular intervals after that?

Is it this possibility that makes Naviaux decline an interview with me, even given our association? Could inoculations in some people set off an abnormal immune response? Early in my association with Naviaux, I brought up the question of vaccinations. He told me he had investigated the question in collaboration with Dr. Judy Van De Water of the University of California, Davis. They found that the immune response of normally developing children to a vaccine was a few days. Children with autism had an immune response from a vaccine that lasted weeks and months. Had this abnormal immune response to a vaccine set off the cell danger response? Naviaux understood the very dangerous ground upon which he was treading. Naviaux and Van De Water decided not to publish the research but simply to focus on how to fix what had gone wrong. Somehow this does not strike me as science, but something more akin to politics, and not the amiable give-and-take of a Norman Rockwell painting, but something dark, primal, and fundamentally corrupt.

I hope that someday soon it will be safe again to be a scientist. We must end the silence.

* * *

Current estimates are that more than one million in America suffer from autism, with the vast majority of them under twenty-five years old. The more traditional thinking is that the brain goes through critical periods of development, and if things do not happen at that time, the brain is essentially set, like concrete drying on a hot day. Many experts say that if these critical developments do not take place by the age of five, then little hope remains. Others, however, believe that evidence indicates the brain is continually changing and that given the right conditions will heal itself and work as God intended our wonderful minds to function. For me, these differing views are not mere idle intellectual curiosity but produce the most important questions of my life.

My wife and I have two children, Jacqueline and Ben. At twenty-two years old, Jacqueline does not speak. I have never had a conversation with my daughter. She has some sign language, but it is rudimentary, and she often signs one word when I think she intends to sign another. Sometimes when she walks I worry she will topple over because she is unsteady on her feet. And yet, when I tuck her into bed at night and tell her I know she has thoughts in her head that she can't share and that one day we will have the greatest conversations, she giggles with an abandon I could not have imagined coming out of her.

When we share stories like this, my wife and I will ask each other, "How much is inside her?" The answer is, we do not know.

By contrast, Ben is a normal and healthy twenty-year-old. He is in college studying accounting and business, holds strong opinions, and can sometimes be a little bossy. He was on the track team in high school, got good times in the 100-meter and 200-meter races, and qualified for the league championship in both events. He set himself a goal to be one of the top ten fastest times in the hundred meter at his school, so that his name would appear on a large record board in the gymnasium. He achieved his goal. Currently, he is deep into weightlifting, can dead-lift five hundred pounds, and plans to start competing in a year or so in local weightlifting competitions. When he's older, he wants to be an entrepreneur. He reads books on finance and idolizes people like Elon Musk and Richard Branson. He tells me he wants to be rich and already has a fictional "dream garage" filled with the expensive sports cars he will someday purchase. And then he will surprise me by saying, "And after I get those cars, I'll make sure to give some money to research scientists so they can help kids, or to families in need." I do not usually expect a young adult to have such awareness or compassion. My two children are the north and south stars of my life, and I live both the American Nightmare and the American Dream. While my wife and I are in good health, what becomes of somebody like my daughter when we are too old to take care of her? She requires twenty-four-hour-a-day care and monitoring. Our situation is not unique, as it is estimated that 1–2 percent of children in our country have an autism spectrum disorder.

This concern is why Dr. Naviaux's next publication filled me with even more excitement than his first study. In this study, he took mice that had been given autism through the maternal immune activation model, waited until they were the human age equivalent of thirty years old, and gave them a dose of suramin. Again, the suramin dramatically affected the mice. As recorded in the article:

> In the present study, we tested the hypothesis that the behavioral manifestations of the MIA [maternal immune activation model] are a consequence of pathological persistence of the evolutionarily conserved CDR, [cell danger response] and that the CDR is maintained by dysregulated purine metabolism and secondary abnormalities in purinergic signaling. We found that a single dose of the antipurinergic drug suramin given to adult animals about 6 months of age (21–27 weeks) produced the concerted correction of over 90% of the metabolic pathway disturbances, and all of the behavioral abnormalities that we tested in the MIA model. Six-month-old mice are the human biological age equivalents of about 30 years.[5]

As I read it, Naviaux and his team have strong evidence that suramin will correct the vast majority (90 percent) of metabolic disturbances in these adult mice with autism-like features and "all of the behavioral abnormalities." This is much more than prevention. This is the rescuing of individuals trapped in a body, which will not let them have the control over their lives that the rest of us take for granted.

But still, I have never had a conversation with my daughter. She still wears diapers. She has no friends. She does not call up or text other girls and talk about boys, or clothes, or the latest fashions and music. She is not simply expressing a lifestyle preference, such as wearing Goth outfits, or liking girls more than boys, or wanting to get her entire body tattooed. If she could verbally express any of these preferences, I would consider myself the luckiest father in the world. When I look at my daughter, I see a person trapped in her body like somebody who is paralyzed, or has fallen into a coma. Even somebody who is paralyzed usually has the ability to speak. I have no doubt that she can see and hear the world, but it is as if she were locked behind a thick glass wall.

What are her dreams? I believe that in her brain she is what we would call "intact" in that she has what we would think of as a personality and an inner dialogue. Does she want to simply take a walk outside and feel the wind blowing through her hair? Does she want to go to a department store and pick out a pretty blouse? Does she dream of a trip to Hawaii, or maybe Christmas in Rome? Or maybe she'd simply like to have the physical strength to dress herself in the morning. Or the ability to go into a bathroom, shut the door, pull down her pants, and use the toilet.

Like many autism parents, I am familiar with the death of dreams for what our children will accomplish in their lives. But when I read Naviaux's work, I feel the stirring of a resurrection for such dreams. His mice were the human age equivalent of thirty years old. And even that age did not seem to be a barrier. What about those who are even older? Maybe there is no limit. Does this drug hold the same promise for those who are suffering from Alzheimer's disease or other dementias? Wouldn't it be amazing if this medication, or one like it, could truly unlock the doors to the bodily jails in which so many find themselves imprisoned?

I have been strong in condemning what I believe to be the corruption of science because I believe it has led to enormous suffering, and not just regarding the autism epidemic. But if science can regain its soul and support efforts like those of Dr. Naviaux, it can fulfill its promise to be an exceptional servant of humanity.

***

The so-called "genetic autism" disorders, like Fragile X syndrome, present little controversy over environmental factors. There is a clear genetic abnormality, and usually a consistent pattern of presentation. But if you were to compare those individuals with Fragile X syndrome to those with autism with no known genetic abnormalities, you would be shocked that these two groups are thought to be similar.

However, there is no denying a condition called Fragile X syndrome exists and it has a defined genetic cause. There would be little controversy over curing such a disease, and it's easy to understand why Naviaux might pick this group as his next subject of study. Additionally, it's easy to see how government officials and scientists who still cling to a "genetic cause" for autism (although there has been no significant pattern uncovered after the spending of hundreds of millions of dollars) could still maintain the fiction of a genetic underpinning to the disease. It allows them to skip over the messy part regarding any environmental contributions and push this approach forward as a potential treatment or even a cure.

It wouldn't be the first time science has not been able to explain why something works, yet at the same time advocate for its use. For more than a hundred years scientists did not understand why aspirin was effective against headaches, but that didn't stop just about every person in the industrialized world from using it at some point in their life. As reported in *Science Daily* on January 15, 2015:

> Weekly treatment with suramin in the Fragile X genetic model was started at nine weeks of age, roughly equivalent to 18 years in humans. Metabolite analysis identified 20 biochemical pathways associated with symptom improvements, 17 of which have been identified in human ASD. The findings of the six-month study also support the hypothesis that disturbances in purinergic signaling—a regulator of cellular functions, and mitochondria (prime regulators of the CDR)—play a significant role in ASD.[6]

One of the important pieces of information that Naviaux has been able to add to the discussion of autism relates to how we even measure this disease. In other words, which biological markers can be tested to determine whether somebody has autism, as opposed to somebody who does not? There are behavioral markers, but that's a little like saying somebody feels hot, and we think they have a fever. Having the ability to measure their temperature tells us whether they have a fever or not, and if they do, whether it is a low, mild, or high one, and we should immediately take them to the hospital.

Naviaux has optimized his mass spectrometer to identify seventeen different metabolites that are altered in people who have autism. These are indications that the cell danger response is highly activated and the cells are not communicating with one another in a normal fashion. This is true for mice whose mothers were subjected to an immune challenge during pregnancy, as well as those mice that have the human equivalent of Fragile X syndrome. Whether it is an environmental insult or a genetic deficiency, science has the potential to remedy it:

> "Correcting abnormalities in a mouse is a long way from a cure in humans," cautioned Naviaux, who is also co-director of the Mitochondrial and Metabolomic Center at UC San Diego, "but our study adds momentum to discoveries at the crossroads of genetics, metabolism, innate immunity, and the environment for several childhood chronic disorders. These crossroads represent new leads in our efforts to understand the origins of autism and to develop treatments for children and adults with ASD."[7]

The lines of inquiry start to converge on a possible solution, as all complex questions in science eventually do. Could the answer to autism, and possibly so many other chronic diseases, really be so close at hand? In a newsletter about suramin, Naviaux writes:

> Suramin is unique in all of medicine. It is the oldest man-made drug still in active medical use. It was first synthesized by Bayer scientists in 1916, and has been used for nearly 100 years for the treatment of African sleeping sickness in both children and adults. Because of this long history, we have extensive information about its risks and how to use the drug safely. In addition to its long-known anti-parasitic properties, in 1988, suramin was discovered to bind to cellular receptors that sense and respond to danger. Working in this way, suramin calms the cell danger response (CDR) and reverses the metabolic syndrome that is ultimately caused by this special kind of mitochondrial dysfunction seen in autism.[8]

It makes one interested in knowing how the drug might work in humans, doesn't it? Surely you suspected that I'd save the best for last?

* * *

In May 2015, the UCSD Suramin Autism Treatment Trial was opened, and the clinical studies were completed in March 2016. The first study was small, composed of ten subjects, divided into five pairs. Half of the children received a single infusion of 20 mg/kg of suramin, while the other half received a placebo. The study has since been published, and in a confidential memorandum provided to me in 2016, they were able to make some guesses as to which children received the suramin and which received the placebo.

Naviaux wrote in a clinical trial update from January 18, 2016:

> Suramin produced improvements in all the core symptoms of autism. All 5 of the 5 children we think received the treatment showed significant improvements in language, social interaction, and expression of interests. Suramin appears to be working at a fundamental level. By removing the developmental barriers caused by the cell danger response, suramin permitted children with autism to start moving through developmental stages they had not completed before. Physiological abnormalities controlled by the brainstem and related to low parasympathetic tone are well-known in autism. Many of these were corrected within hours of a single dose. We observed the normalization of "belladonna" pupils (large pupils compared to pupil size in neurotypical individuals in the same room lighting), and correction of diastolic blood pressure elevations. We observed the normalization of the nasal quality of speech-the voice timbre controlled by vagal motor fibers to the soft palate-in one boy of the three who had noticeable nasal speech. Increased attention and eye contact are also observable within a few hours of the infusion.
>
> All 5 boys who received suramin [who he believed had received suramin] started to initiate social interactions with their parents and siblings that they did not do previously. Three of the four also started asking to try new foods from their parents or siblings' plate or from the dinner table for the first time. The interest in novelty was increased. On the behavioral side, a sense of calmness and cheerfulness was seen within hours of the infusion. One 5-year-old child, who had some language before, had a spontaneous bout of the giggle while walking hand in hand with his mom back to the car in the parking lot outside the clinic after his infusion. He looked up at his mom and said, "I just don't know why I'm so happy." The 10-year-old looked at his mother after dinner on the evening after the infusion and said, "I finished my dinner." This was the longest complete sentence he had ever spoken. At the 2-week post-infusion time point, the parents of the 13-year-old boy who was essentially non-verbal, with only word fragments and single-words before the infusion sent the following email: "Overall he has improved. Language is better. Very good eye contact. Seeking his brother. They played hide and seek yesterday. We went shopping and he was very well behaved. Very calm sometimes."
>
> We measured expressive one word picture vocabulary (EOWPV) before, and two days after the infusion. This did not change. However the length

> of the sentences did change. All four children we think received active drug more than doubled their sentence lengths. The 10-year-old spoke in 1–3 word phrases before the infusion, and 4–8 word complete sentences after the infusion. On day 2 after the infusion, he turned to the nurses at the CTRI and said, "I want to go to the bathroom again." The 5-year-old spoke in 4–8 word phrases and sentences before, and 10–20 word sentences after the infusion. The 13-year old spoke only in short sounds, word fragments, and 1–2 word phrases, and had significant oral motor dyspraxia before the infusion. By 8-days after the infusion he started experimenting with new sounds and began speaking in simple 3–5 word sentences like, "I want to eat chips.", and "I want to go outside". Behaviors that required physical, developmental, or social practice did not change after a single dose. The 5-year-old boy began trying to join in enthusiastically with other children on the playground to play tag and chase for the first time, but he did not know the rules of the game and tried to join in a socially awkward and socially disinhibited way. Parents and therapists reported accelerated progress in response to usual behavioral and speech therapies the children have been doing regularly for months.[9]

"Let me tell you what this means for somebody like your daughter, Jacqueline," Naviaux told me after I'd learned about the results. "I think we'd simply need to do an infusion for her every six to eight weeks for probably about two years, and she'd be caught up. For the younger kids, I expect it would take less time, probably a year. And you know how much total medicine we're talking about?"

"How much?" I asked him.

"The equivalent of two and a half aspirin. We are using such low dosages, like one percent of the cancer dosages that caused toxicity. We are using this drug to turn communication back on, not to kill something. We've been monitoring for toxicity, and there's absolutely none. We can use suramin safely. I think what we're going to be doing when we get this published is teaching the world how to use suramin."

Suramin could be the magic bullet to end the autism epidemic and possibly many other diseases that are initiated by the cell danger response.

***

Dr. Eric Gordon is a practicing physician with a thriving practice in Santa Rosa, California, and has partnered with Naviaux to determine whether he can take Naviaux's promising research findings and turn them into effective treatments for patients, particularly among those suffering from chronic fatigue syndrome/ME. Joining them in this effort is Dr. Paul Cheney, one of the country's leading experts on the disease, having been on the scene of what is generally considered to be the first modern emergence, the 1984–1985 Lake Tahoe Incline Village outbreak. I had interviewed Dr. Cheney for my previous book, *PLAGUE*, which argued that a retrovirus, XMRV, was likely behind the chronic fatigue syndrome/ME and autism epidemics. Perhaps this retrovirus was setting off the "cell danger response." In May 2016, Gordon's office began soliciting funds to test a large number of Cheney's chronic fatigue syndrome/ME patients with their new metabolomics test. If a contribution was made in a large enough amount, the donor would get a free metabolomics test for the person of their choice. I quickly contributed in order to get a test for my daughter.

When I interviewed Dr. Gordon in early June 2016, he expressed caution about what physicians may be able to accomplish in the next few years, saying:

> How much we're going to be able to help the individual person, I think is going to take a little time. That's the bummer. Some of the patients we're going to help because we're going to see patterns that we recognize. And some of them we're not going to be able to help that much. That's my bias. I think Bob is more optimistic than I am. But I've been doing this longer and I've been disappointed by a lot more tests. I've been doing this since 1992 with chronic fatigue illnesses and I have watched Kenny De Merlier [a well-known chronic fatigue syndrome/ME researcher] "cure" it many times. And not just to pick on Kenny, but it's a moving target.[10]

While Gordon may have a more conservative view of what may be accomplished with metabolomics and suramin, he is also a strong critic of how the current scientific community has abandoned promising areas of research. Specifically, he thinks the community has ignored the area of biochemistry in favor of genetics:

> About twenty years ago or so people stopped thinking biochemically because things got way too complicated. When we started learning about all the different receptors, well, things were simple for a while. You had one chemical and it did what you thought it did. Then we started to learn things. Like with serotonin. You've got about eight different serotonin receptors and they sometimes do opposite things. It depends where you are in the body. It made

> this whole effort to understand the world biochemically a lot harder to do. And at the same time, genetics were coming on board. And we were going to have all the answers because they'd find the gene that caused the problem and we were going to fix it. Sounded great. Except genetics is the wrong level, if you talk to Bob. And it's something I've felt. It's too complicated and it only explains a few diseases.[11]

Even with his reservations, Gordon is intrigued by what metabolomics may reveal about disease and health in the human body:

> Metabolomics is exciting because we are now going to be looking at the patterns of chemicals in the body and we can begin to see the richness of the interactions. And it's patterns that are going to help us, rather than individual chemicals. And we still can't help ourselves. We're still hooked on looking for the individual chemicals. It's the silver bullet theory of medicine. Ever since we cured people with penicillin, we all want the silver bullet. We can't help ourselves. That's how human beings are wired, I guess. But metabolomics is really going to be more about pattern recognition. Seeing the ebb and flow of chemicals and families of chemicals and what your body is doing. And we're going to learn about what's normal for your body.[12]

Gordon believes metabolomics has the potential to determine what is abnormal in many diseases and the early results with chronic fatigue syndrome/ME are promising:

> We looked at 450 chemicals and there was a group of forty to seventy chemicals that were abnormal in the majority of people with chronic fatigue. But really, the thing that defined chronic fatigue was only a group of nine to thirteen chemicals. My main point is that the 20–25 percent of the abnormal chemicals that these people with chronic fatigue had, defined them. They still had another group of 75 percent abnormal chemicals, but they were individual. They weren't part of a class.[13]

Gordon is hopeful that their efforts will be successful in developing a fairly clean marker for chronic fatigue syndrome/ME patients, providing a simple test that can be administered and interpreted by any physician.

The work that Gordon and Naviaux have done to date backs up Naviaux's theory that diseases like chronic fatigue syndrome/ME and autism are the result of abnormal cellular communication:

When cells don't communicate effectively, you definitely have illness. The thing is, even when cells are sick, they're often communicating very effectively. It's the message they're giving that's the problem. There are other times, when part of the system is down, and the wires are cut. That does happen. With a lot of the chronic fatigue, and maybe autism, the cell might be responding to a danger signal or toxin. It could be anything. I like to call it neurotic. It's jumping up and down over something that's long gone. Or it can still be there, like heavy metal toxicity. You could have had the heavy metal toxicity and maybe you've cleaned it up. But your body is now in a pattern of response that it doesn't know how to change.

Or the heavy metals could still be there. Or it's Lyme. It's just so hard to say.

But what happens, and this is Bob's contribution, is seeing the centrality of the mitochondria. The cell membranes are critical, because that's what we measure in metabolomics. A lot of the things we measure that are abnormal are components of the cell membrane. At least in the chronic fatigue people. I can't speak to the autism because I haven't looked at the data. In the chronic fatigue people, the strongest markers are cell membrane markers. These are components that make up your cell membrane.

And because cell membranes communicate cell to cell. And they change when the cell thinks it's under attack. And they change when they're growing. That's how cells talk to each other, using the chemicals they display on the surface. Just like immune cells. How do immune cells say they're sick? They put up a flag with the proteins that say "I'm sick here." And it can be specific, I'm sick with this bug.

The mitochondria, when they sense invaders, be they biological or toxic chemicals, they brown out. They lower energy production. The analogy I like to use is in the 1400s when you had marauders who would come to raid, and everybody would go into the castle. They'd burn the fields and lock themselves in the castle. And unless the marauders had a lot of supplies, they'd eventually have to leave. That's what your cell does. That's what chronic fatigue is. It's a way of shutting down energy production, so you can't feed whatever else is trying to get nutrients from you. What we do in metabolomics is we can see how this is happening in the body. We can see the changes in these chemicals. And then we're hoping that we're going to be able to see what kind of problems they're causing.[14]

Naviaux and Gordon's theory gives a simple and elegant explanation for many of the observations in chronic fatigue syndrome/ME and autism. The immune system responds to a perceived threat by shutting down the energy production of the mitochondria. As a result, chemicals on the surface of the cell membrane change, affecting communication between the cells. The rest of the body then receives incorrect signals or no signal at all. Further research is needed to determine whether or not this theory is true.

Gordon believes this theory of low energy production also explains one of the great mysteries of autism, the "stimming" or repetitive behaviors often engaged in by children with autism: in the right places of the brain they're not producing enough energy. The body reacts in a way to protect itself from danger. So with autism, on one level, these kids have low energy in neural pathways, and so they can't modulate input. So everything is overwhelming. In what may seem to be a counterintuitive observation, Gordon believes that one needs energy in order to relax: "You have to remember that frenetic energy isn't real energy. Just like muscle spasms. Muscle spasms are an energy slow state. When your muscles have enough energy, they can relax. Because it actually takes ATP [a molecule that produces energy] to let the muscle relax."[15]

Gordon finds Naviaux's preliminary findings with suramin in autism exciting because they suggest that this cell-to-cell communication is capable of being reestablished:

> The suramin trial was profoundly successful. That's the reason he stopped it. [**Author's Note:** He had planned to test twenty children but ended up testing only ten. Five got the placebo and five received the suramin.] It was because the results were so overwhelming and he realized he had missed some of the most important things to measure. And the problem with the study is that you can't change what you're doing in the middle. So he had to stop and rewrite the protocol and get a new IRB [Institutional Review Board] to do a new study so he could capture the important information. So I think that's what's really exciting. The thing I don't know, and I don't think Bob knows, is whether the improvements will sustain. He thinks they will because the kids seemed to be able to learn. They're new behaviors.[16]

Could it be as simple as restoring cell-to-cell communication? It sounds like a silver bullet, and Gordon is on solid ground in suggesting that we should not get our hopes up too high. But it is difficult not to imagine a fundamentally different avenue of hope opening up because of this research. Gordon is cautiously optimistic, but even if the findings confirm his suspicions, other dangers await:

> The problem you have to understand about science is you might lay the groundwork in a way that doesn't fit the paradigm of what other scientists are doing and there will be resistance. The resistance isn't because you have a new idea. The resistance is that they don't know what to do. All the other researchers are looking under a different couch. And you tell them it's on the other side of the room. They have to learn how to get to the other side of the room.[17]

I asked Gordon if those scientists were interested in learning how to get to the other side of the room. His answer was expected, but still disappointing:

> They're not. That's the problem. We would like them to be. Young ones are. But if you've spent your life learning how to measure something like brain scans, and somebody comes to you and says they can do it this way, well, that's nice, but I don't know anything about how to do it your way. That's the problem. If you spent twenty years learning genetics, and I tell you I have a better answer in metabolomics, it doesn't inspire you to learn metabolomics. You don't know biochemistry at that level any more. Say you're an expert in a deep area, even in genetics, but you know only a tiny part of genetics. So that's the problem with research. It's so hard when you have an idea that's truly radical. So many radical ideas just die with the guy who developed them.
>
> It doesn't take a conspiracy to shut down a new idea. It just takes people who ask, why should I do that? They don't know he's right. They're not sure he has the right answers. Somebody else has to replicate it. And if nobody else is bothering to replicate it, nothing changes. It's a lot of work to get a study done. So that's the problem."[18]

***

I try to envision a future that is better than the twenty-two years of my daughter's life that have come before it.

I look two to three years into the future, when my son Ben has graduated from college and is starting out in business. I often say to him that his father is trying to save the world and he wants to own it. This always makes him laugh. Each of us has our own individual path. I imagine that Jacqueline's metabolomics test has revealed a pattern of abnormal cell-to-cell communication, and we have used something like suramin and it has succeeded in restarting normal development. She is starting to become the person she was meant to be.

I imagine Ben coming home on a long weekend, and his older sister, Jacqueline, is in the midst of her recovery. She can talk now, and to the casual eye she might even appear normal. There are still some social deficits, some things that she is learning about the world, but it's easy to see that a time will come when she can function independently in it. Jacqueline asks Ben about college, and he is eager to share the information. The two of them text often, and it surprises me sometimes how close they have become, since for practically their entire childhood period I felt that they were two separate people growing up in the same house. Considering the fact that she could not speak or communicate in any meaningful way, this outcome was not surprising.

Maybe the two of them will decide to go out to a movie, or go visit some of his friends, and she will come along. As they drive off together, my wife and I look at each other in amazement. This is a future neither of us ever believed would come to pass. At one point during Jacqueline's youth, a friend suggested I contact a psychic who specialized in talking to dead people. "Maybe she can talk to living people who can't speak," she suggested.

I contacted this woman and found that while she had never worked with an autistic child before, she had done it with people who were in comas. I figured, what could it hurt? The psychic told me that my daughter was mentally intact inside, that she knew how much her disability affected the family and was sorry about it. "She wants to be a nurse when she grows up," the psychic said. "Because she has been impressed by the kindness of the nurses who took care of her when she has been in the hospital to get her seizures under control." And the psychic told me something else. "She wants a new bed. A big girl bed."

I went out and bought her a new bed, and when she saw it, she giggled like nothing I had ever heard from her before. Maybe there was something to what the psychic had said.

And so maybe in that future where my daughter gets suramin and starts to recover, she will pursue a career in nursing. I imagine her doing other things. "Mom, let's go for a run," she says, and my wife, who has run six marathons and innumerable races, will be shocked by the unbelievable miracle of a child who could never run becoming an adult with whom she goes jogging. They will get their exercise clothes on and head out the door. And while passing motorists may pause for a moment and think how nice it is to see a mother and daughter going for a run, they will never fully appreciate the wonder of that moment.

I have come to believe that the greatest truths are both scientific and spiritual. They satisfy the mind as well as the soul. And maybe in some future time the public will link the quickly disappearing condition of autism to a greater societal issue.

Autism may begin in the body when new chemicals and combinations of viruses or pathogens are encountered and the body responds by shutting down communication, preserving the individual cells, but dangerously unbalancing the whole. Just as cells decided it was not good to communicate with one another, segments of our society have decided to do the same thing. Whether those divisions arose because of differences in skin color, ethnic background, religious beliefs, or political opinions, the result was always the same. Our entire world endured its own type of chronic disease. Few groups have been shunned more by the mainstream media during this time than the autism parents. I often say I'd stand a better chance of getting interviewed if I said I was a member of ISIS, rather than one of those vaccine-autism parents. Maybe we can start the conversation going again. Maybe whenever we have division, there will be sickness in our society. Perhaps that is the greatest lesson of our suffering.

And as our children, infused with suramin, show the way back to health, maybe it will provoke further questions. Might many of the diseases of aging, such as dementia and Alzheimer's diseases, be a similar manifestation of the breakdown of the cell danger response? Might the children of autism lead their elders out of dementia and raise the possibility of a vigorous and healthy old age? What kind of health would we enjoy if we made sure the lines of communication remained open in our own bodies? And what kind of society would we live in if these same lines remained open to those with whom we share this planet?

I believe that would be a much healthier world for us all.

CHAPTER THIRTEEN

Justice

On Saturday, October 24, 2015, a group of protestors numbering around a hundred people gathered at Grant Park in Atlanta, Georgia, just across from the headquarters of the CDC. The organizers hoped to make it a yearly event to bring attention to the allegations of Dr. William Thompson and other acts related to vaccine safety, such as the Simpsonwood Conference of 2000. Speakers included Minister Tony Muhammad, Dr. Andrew Wakefield, Dr. Judy Mikovits, Brandy Vaughn, and Robert F. Kennedy Jr.

Near the end of his speech, Robert Kennedy made a direct challenge:

> Dr. Frank DeStefano, who runs the Vaccine Division, orchestrated this corruption. Dr. DeStefano is a criminal and he committed scientific research fraud and he is guilty of injuring all these people. I am saying that and I am using his name. If what I'm saying is untrue, it is an act of slander and I want him to sue me. If he didn't do it, he ought to sue me. He ought to file a suit this afternoon and enjoin me from ever saying that again. If someone said that about me I would sue them immediately.
>
> I'm saying to you, Frank DeStefano, if you didn't poison these children, you need to sue me right now and shut me up because what I'm saying to you is damaging to your career.

Kennedy paused for a moment to look at the crowd and then said, "Let's see what he does on Monday."[1]

As of this writing, Dr. DeStefano has not brought any legal actions against Robert Kennedy Jr. One gets the impression the CDC is hiding from these allegations and simply hoping that the press does not print these allegations, and that those making them simply get tired and go home.

***

Minister Louis Farrakhan has suggested tribunals for those political officials, African American and otherwise, who were told about this information and did not act upon it. Congressman Elijah Cummings has since died, but it would likely be a long list. Perhaps a few previous occupants of the Oval Office would find themselves in that tribunal, as well.

When President Ronald Reagan signed the National Childhood Vaccine Injury Act of 1986 and set into motion this tragic chain of events, he said some truly amazing things. Reagan said he signed the bill with "mixed feelings" and had "serious reservations" about the program. He thought the Vaccine Court was an "unprecedented arrangement" that was "inconsistent with the constitutional arrangement for separation of powers among the branches of the Federal Government."[2] In retrospect, it's easy to see how that legislation set the table for a remarkable litany of crimes and human depravity. The Constitution is an amazing document because it looks at humanity as it is, not necessarily as we would like it to be. Any powerful group that is not subjected to rigorous oversight will eventually become corrupt. One wishes they could go back in time and tell Reagan to follow his instincts and veto the bill.

But I do not have a time machine, and I cannot travel back in history to try and warn our fortieth president. Even if I could, I don't think I would be met with a warm reception when I arrived in November 1986 as he prepared to sign the National Childhood Vaccine Injury Act, which would usher in this reign of horror. I am certain that even if I time-traveled back to 1986 and tried to get my twenty-three-year-old self to take up this cause, I would pass up that middle-aged version of myself like some crazy homeless person shouting that the end was near.

But even at that stage, my twenty-three-year-old self would be concerned with justice. This was the time in my life when I would work for the US Attorney's Office and compile wiretap evidence for the case against Oakland drug lord Rudy Henderson. I had a plan for my life. Become an assistant district attorney, or maybe even an assistant US attorney, prosecute the bad guys for a few years, and then make the move into politics. My first move would probably be to run for Congress, then after a few terms make the move to run for senator, or maybe even governor. I had lived in that world. I'd worked for a US senator by that time, helped run a campaign for another US Senate candidate, and even driven a car in the motorcade for First Lady Nancy Reagan when she visited the Bay Area.

My future looked bright. Consider what the bestselling historical novelist James Michener wrote about me when our paths crossed at the 1984 Republican Convention in Dallas, Texas. I was a Youth Delegate for Reagan, and he was covering the event for *U.S. News and World Report*:

> **Young enthusiasts.** Kent Heckenlively is one of those bright young college seniors who in 1950 would have been working for the Democrats. Now he is a staunch Republican.
>
> "It's the party of the future," he declared. "I'm a member of the official youth delegation and I'm sure I'll work for the party for the rest of my life."
>
> Heckenlively, who had written his first novel, was excited to meet me and eager to talk literature and politics. I had the feeling that he could not understand how I could be a Democrat, for him the future was so clear: "Reagan inspires the country," he maintained.[3]

It had been a thrill for me to meet Michener, as I also wanted to be one of those writing politicians, just like Winston Churchill. I'd read several of Michener's historical novels and loved his way of telling a story. I wasn't alone. Michener was one of the most popular writers of his day. Still, it's a bit of a shock as a middle-aged man to consider how much I'd ever believed in a single politician or party. It is embarrassing to also recall that at the Convention I'd even worn an Abraham Lincoln black stovepipe hat, and during Reagan's address the television cameras for the network panned to me in my Lincoln regalia, thrilling my family back at home. I'd grown up in a Republican household. My parents had actually met at the 1956 Republican Convention in San Francisco, where my mom was the head of ushers and my dad was cruising the event with a good friend.

But all of my plans changed two years later when I spent a summer assembling wiretap evidence at the US Attorney's Office in San Francisco for the federal case against Oakland drug lord Rudy Henderson. You must understand that at that time there were no greater enemies in America than "drug lords." We were fighting a "drug war," and the "drug lords" were the ultimate villains. In order to get to my office on the fifteenth floor of the Federal Building in San Francisco, I had to pass through blast and fireproof steel doors before going to the large table in the agents' area, putting on my headphones, and pulling out one of the cassette tapes. But I had an epiphany as I listened to the conversations among Rudy Henderson, the members of his gang, and one particular Colombian who often found himself lonely in the United States. Were they doing the wrong thing by selling drugs? Absolutely. But the question for me became whether I feared or pitied them. How different from me were they? They talked about their families, their lives, relationships, even the local sports teams. Their concerns were so similar to mine, but the circumstances of our lives were vastly different.

I had grown up near the fourteenth hole of Roundhill Country Club, a long par-five dogleg with a pond about 120 yards from the green that was often filled with ducks. When I was a boy, my mother would take me out to the pond with bread, and we'd feed them. When I was older, the great green expanse of the golf course was our football field, and we'd pause our nightly games to let those playing twilight golf finish their rounds. By contrast, the members of the Henderson gang had grown up on the mean streets of Oakland, or in the poverty of Columbia. What right did I have to judge them?

I lost my taste for criminal law that summer. It seemed to me like a system in which the top 10 percent of society prosecuted the bottom ten percent of society so that the 80 percent in the middle could live in relative peace. I can't say that I had a better system. I just didn't want to be a part of that one.

The word *criminal* had a specific meaning for me, describing a person who had a clear choice between right and wrong, deciding on the wrong path because of some evil in his or her heart. Yes, I listened to the politicians and even some of the attorneys who talked about how these people were destroying their own neighborhoods with their activities, and I couldn't really disagree. But it seemed to me there were mostly victims.

I'd also lost my taste for politics, as there was so much nastiness on both sides. I liked debating ideas, not making personal attacks. Maybe it never was my destiny. My college roommate saw me differently from so many of my other friends. "Everybody thinks you're going to go into politics, Kent, but they've got it wrong," he said. "You like to be around it, and you like to talk about ideas, but you just don't have the lust for power. You might become a political columnist, but you'll never become a politician."

And so my life went in a different direction. I'd always been good at standing up and talking to people, so it was natural that my elders wanted to guide me into a career that would highlight those skills. After leaving criminal law, I thought I'd venture into some other form of law, but again everything looked like a rigged system. For a time I even went into sales, but that was also unsatisfying. Teaching saved me because I saw it as working on the positive side of humanity, rather than punishing people for what they had done wrong. I teach students the way they should act, how to think, and, most important, how they should treat one another. I love the enthusiasm of students, their strong convictions, their desires to make their marks in the world, and the natural kindness they show to one another. Maybe their kindness to one another is the thing that touches me the most. It makes me believe there is hope for us as a species.

And if Michener worried I might not be able to understand his different take on politics, I'd like to believe that if we were to meet again, he would realize I had developed a broader understanding of the world. That college senior would never have imagined that he would one day find wisdom in the ideals of President Reagan, Robert Kennedy Jr., and Minister Louis Farrakhan. There are strengths and weaknesses in each of these men, just as in each one of us. I live in peace with both the good and the bad. Every morning my students recite the Pledge of Allegiance, and I stand proudly with them and say, "I pledge allegiance to the Flag of the United States of America, and to the Republic for which it stands, one nation, under God, indivisible, with liberty and justice for all." I sometimes hear my fellow teachers complain they don't like to say the Pledge or that they do something else while the students are reciting it, and it makes me a little mad. I understand that in the eyes of many, the issues I have raised in this book make me look like a renegade, a rebel, and maybe even a dangerous person. But I like to believe I am following a hallowed American tradition of principled and well-considered dissent. The country I live in was founded by rebels.

And one of those rebellious American traditions is to demand justice, even from the powerful. It's right there in the Pledge, that good old "with liberty and justice for all." It is not a given. Sometimes it must be fought for. Sometimes those who demand it will have to walk alone. I have always been willing to be a warrior for justice if there are criminals worthy of the name. Drug dealers in poor neighborhoods? I'm not so sure. Top scientists at the CDC or those who work for other governmental agencies or universities who conceal the truth and let children be harmed? It's not even a question for me. This book has been a catalog of the crimes of individuals who should have done better. Individuals who had the benefit of education and learning but out of weakness, greed, or cowardice made terrible choices that affected the lives of millions. Every few days in the papers there are stories of children with autism who wander away from their exhausted caregivers and are found floating in lakes or rivers or dead of exposure in some field or forest. That blood is on their hands.

***

I struggle with the question of what justice would look like for many of the scientists discussed in this book, for their failure to honestly address the question of vaccines and autism. When I asked my wife the question, her response was swift and brutal: "Why don't you give them ten thousand vaccines at the same time and see how they like it?" She was referencing the assertion by prominent vaccine defender and head of infectious diseases at the Children's Hospital of Pittsburgh, Dr. Paul Offit, that the human immune system is strong enough to handle ten thousand vaccinations at the same time. Maybe that will be Offit's unique punishment, but it was his own suggestion. That is one science experiment I would pay money to see.

I look for historical precedents, and the closest analogy I can find is South Africa under apartheid. Prior to apartheid being imposed, there was a de facto separation of the races and a culture that allowed it to happen. Much of the same can be said in the United States between those who support vaccinations and those who question their safety. We have been made the "other." SB 277, the law that mandates children must follow the CDC schedule on vaccines in order to attend school in California, as well as attempts in other states and at the federal level, are trying to legislate us out of existence. What was once allowed, even if frowned upon, will soon become illegal.

But even as South Africa was the setting for decades of injustice, in the person of Nelson Mandela it showed that justice can be delivered without vengeance. I am not saying it is easy. If I were the parent of a child killed or injured while in police custody in South Africa, I would find it difficult to control my rage. As the parent of a vaccine-injured child in America, I am similarly conflicted.

However, South Africa instituted something unique in world history, its Truth and Reconciliation Commission, which was empowered to take testimony from those members of the security services and others who had committed crimes during the reign of apartheid. When I researched these tribunals, I found something truly interesting. Those who appeared before it were not required to apologize for their crimes, as it was considered to be a shortcut for the painful process of getting to the truth. Instead, those who had perpetrated their crimes were required to testify in detail about their actions. Then the surviving family members were allowed to give testimony as to what they had endured because of the injuries or death of their loved ones. It was the goal of the commission to create a complete picture of the tragedy that apartheid had visited upon South Africa. If, after the conclusion of testimony, the judges decided that the accused had testified truthfully about his participation and that of others, his sentence might be reduced or waived.

I believe something similar should happen in the United States in regard to the vaccine/autism question. Many who have participated in this injustice were following the law as it was laid down, like the special masters in the Vaccine Court. Others, particularly those in the public health hierarchy of our government, have in all likelihood committed great crimes.

The eminent scientist, Albert Einstein once said, "Those who have the privilege to know have a duty to act." I believe the following scientists have failed Einstein's directive, and if such a Truth and Reconciliation Commission is ever formed in this country, they should be called to testify:

1. Dr. Walter Orenstein, former head of the National Immunization Program and assistant surgeon general, as well as the deputy director of Immunization for the Bill and Melinda Gates Foundation. Now teaching at Emory University.
2. Dr. Julie Gerberding, former head of the Centers for Disease Control and Prevention (CDC). Currently the head of vaccines at Merck Pharmaceuticals.
3. Dr. Frank DeStefano, current CDC scientist.
4. Dr. Thomas Verstraeten, former CDC scientist.
5. Dr. Coleen Boyle, current CDC scientist.
6. Dr. William Thompson, current CDC scientist.
7. Dr. Melinda Wharton, currently serves as the acting director for the National Center for Immunization and Respiratory Diseases at the CDC.
8. Dr. Poul Thorsen, former CDC scientist and now federal fugitive on charges of embezzling more than a million dollars from autism research, but living openly in Denmark.
9. Dr. Tanya Karapurkar Bhasin, former CDC scientist.
10. Dr. Marshalyn Yeargin-Allsopp, current chief of the Developmental Disabilities Branch at the CDC.
11. Dr. Ian Lipkin, current Columbia University scientist who needs to explain the failings in his XMRV/chronic fatigue syndrome study, as well as his MMR vaccine/autism research.

12. Dr. Mady Hornig, current Columbia University scientist who needs to explain the failings in her XMRV/chronic fatigue syndrome study, as well as her MMR vaccine/autism research.

I'm not sure how long I will have to wait for questions to be put to these individuals. But I am in my early fifties, work out several times a week, and try to watch what I eat. I plan to be around for a long time. And every day I will dedicate some part of my labors to justice and recovery for those who have been harmed by vaccines. I'm not tired and I'm not going away.

I've often said that I want to win this war, but I don't want to become the war. I've learned how to fight for justice and at the same time not be consumed by anger. I find great wisdom in the approach taken by Nelson Mandela in South Africa. He believed there were two pillars upon which he built his advocacy, and they came from the great Indian leader, Mahatma Gandhi, who peacefully freed India from British rule.

The first principle is *satyagraha*, which roughly translates as truth force, or power. It means that you dedicate your life to telling the truth and that such an act creates an enormous opportunity for change. It is only in truth that we can move forward. Even those who oppose you realize that you speak your truth. The second principle is called *ahimsa*, and it means that you intend no harm to your enemies. I freely confess this is often a difficult task for me, and although I struggle with it, I believe I am making progress. To call for justice is not to hate those who have committed injustice. I do not simply refrain from hating my enemies but do my best to demonstrate compassion for them. The burden of having participated in injustice can often be greater than having suffered from it. Justice is not achieved by throwing the wicked into a dark hole, but assisting in their redemption. Although our children with autism and their families have suffered more, science has also suffered. Injustice harms everybody in society.

If I were in charge of ending the autism epidemic, I like to think I'd do it in the way Nelson Mandela ended apartheid. When the South African government realized that they could not move forward unless things changed, they invited Mandela from his incarceration to have a talk. They offered to release him from prison if he promised to refrain from any political activities. He politely turned them down and returned to prison. They invited him again and this time offered that he could participate in political life, and the blacks would have some sort of proportional representation, but it would not be the same as that of the white South Africans. Again, Mandela turned them down, insisting that all South Africans, black or white, had to have the same rights. Finally, the authorities had no choice but to release Mandela and give equal voting rights to all South Africans. Through it all, Mandela was polite, charming, and firm.

Regardless of who is eventually in charge of ending the autism epidemic, I hope they will show similar strength. I imagine the first step of the authorities will be to slightly alter the vaccine schedule. I suggest that be rejected. I expect there to be other proposed half measures. I believe those should be rejected, as well.

Like Nelson Mandela, we have a few simple demands, and they are nonnegotiable. They come from the documentary *VAXXED*, and I believe they are clear and just:

1. That Congress subpoena Dr. William Thompson and investigate the CDC fraud.
2. That Congress repeal the 1986 National Childhood Vaccine Injury Act and hold manufacturers liable for injury caused by their vaccines.
3. That the single measles, mumps, and rubella vaccines be made available immediately.
4. That all vaccines be classified as pharmaceutical drugs and tested accordingly.

Anything less is not worthy of the lives that have been lost and those that have been forever changed as a result of the dereliction of duty and disregard of good science that has been practiced by our governmental health authorities. For certain, there are many of us who await the government's response. And in that process, we hope that science will regain its soul.

EPILOGUE

Trump Hope & Reality

I was one of those rare people who thought Trump would win the election in November 2016. In fact, I even made a bet with my two best friends from high school, an attorney and a Stanford economics professor who's been short-listed to someday win the Nobel Prize. Just to make it fun, I told them I'd take three-to-one odds. I'd owe each one of them five bucks if Trump lost, but they'd owe me fifteen if he won. They were mad about the results but grudgingly gave me the nickname "the Sage of San Ramon," after the small town in which I live.

But I was concealing a secret, fearful that if it were revealed it might change the results of the election.

The secret was not only that Trump had met with autism advocates, including Dr. Andrew Wakefield, but that he'd enjoyed a long association with members of our community, specifically, Dr. Gary Kompothecras, the well-known chiropractor from Florida. There'd been other rumors swirling around Trump and why he was supposed to be so passionate about this issue. Stories that Barron had suffered some type of vaccine reaction (more like cerebral palsy than autism, I was told), and only Melania's intervention in taking him to some unspecified European doctors had healed his injuries. I was never able to confirm these stories, and despite the passions of the topic, most advocates felt strongly about not intruding on the Trump family's privacy.

In my interview with Dr. Wakefield, I asked him about the Trump meeting, and he confirmed it but asked me not to discuss it publicly until after the election. I agreed but requested a favor of Dr. Wakefield. At the meeting, Wakefield had his picture taken with Trump. "If Trump wins," I asked Dr. Wakefield, "can I publish that picture of you with Trump?"

I think Wakefield appreciated my audacity, as he gave a little chuckle and said, "Sure."

In my mind, I imagined the outraged screams of the pharma whores in the media.

I stayed up late election night to watch Trump give his victory speech with a very tired-looking Barron right beside him. I went to bed that night, certain I'd be causing a lot of trouble in the Trump years on the vaccine issue.

Shortly after the election, I published an article in *Bolen Report* titled "Trump Meets Dr. Wakefield."[1] Above the headline was a big picture of Trump standing next to Dr. Wakefield, a flag of the state of Florida at their left. Wakefield has an impish smile on his face, holding his hands in front of him like a child pretending to be well behaved, while Trump seems to be caught in midconversation. Here's how I opened the piece:

> Could the new President-Elect of our country, Donald J. Trump, have really met with Dr. Andrew Wakefield? You've probably heard the rumors, maybe you've even seen the video of Wakefield saying that Trump understands our concerns and will never allow mandatory vaccinations. He goes on to say that he believes Trump will investigate the problems at the CDC, including the allegations of Dr. William Thompson. It's all true.[2]

When the first edition of *Inoculated* was published, I included the start of the chapter from this book about the Trump/Wakefield meeting to hook people to order the book, which was published a few days later, on November 13, 2016. But I was also careful to note that my support for Trump was based on his willingness to meet with members of our community and support their efforts. Here's how I ended the article:

> The election is over, but this is not the time to rest. We need to train and prepare, because when Trump takes office, I want our army of hope to be ready. Maybe he will follow through and be the man I believe him to be. Maybe he will waver. I want to prevent that. If you share this information as widely as possible, and drive this book onto the bestseller list, when Trump takes office, he will see our mighty army in the field, he will have to make a decision. Will he welcome or fight us? I want us to be prepared for either outcome.[3]

The article was widely viewed on Facebook, with more than seven thousand shares. And with Trump's election, others felt comfortable sharing what they knew, including several of the participants.

Science magazine seemed terrified when they learned of Trump's contact with the antivaccine community. This is from an article in *Science* from November 18, 2016, with the title, "Trump Met with Prominent Anti-Vaccine Activists during Campaign":

> "There was a concentrated opportunity to discuss autism," with Trump, says Mark Blaxill, one of the participants. Blaxill is executive director of XLP Capital, a technology investment firm with offices in New York City and Boston, and editor-at-large of the Age of Autism website, which says it gives "voice to those who believe autism is an environmentally induced illness, that is treatable, and that children can recover."
>
> Gary Kompothecras, a chiropractor and Trump doctor from Sarasota, Florida, and Jennifer Larson, a Minnesota-based technology entrepreneur, confirmed to Science Insider that they were also at the event.
>
> > Earlier this week, on Age of Autism, Larson wrote: "Now that Trump won, we can all feel safe in sharing that Mr. Trump met with autism advocates in August. He gave us 45 minutes and was extremely educated on our issues. Mark stated 'You can't make America great with all these sick children and more are coming.' Trump shook his head and agreed. He heard my son's vaccine injury story. Andy told him about Thompson and gave him *Vaxxed*. Dr. Gary ended the meeting by saying 'Donald, you are the only one who can fix this.' He said 'I will.' We left hopeful. Lots of work left to do."[4]

Is it any wonder that those of us who'd fought for so long on the vaccine issue viewed the Trump presidency with great hope? I urged my readers to be optimistic but suggested we needed to stay on the field of battle. Trump might be the president, but he was just one man.

Dr. Wakefield continued to apply public pressure on the new administration. He scandalized the mainstream media by going to Trump's inauguration on January 20, 2017, as well as one of the inaugural balls. The medical magazine *STAT* almost flat-lined when they discovered the news, posting in an article, "Controversial vaccine skeptic Andrew Wakefield attended one of President Trump's inaugural balls late Friday, prompting a flood of mostly negative reaction on social media, with many commentators raising concerns that his discredited ideas will gain traction in the new administration."[5]

To the delight of his followers and the annoyance of his critics, Wakefield even live-streamed a Periscope video from the Inaugural ball.

* * *

While it was unlikely Dr. Wakefield would get a position in the new administration, one who came close was Robert Kennedy Jr., son and namesake of the former US senator and likely Democratic nominee for president in 1968 whose life was cut short by an assassin's bullet. Kennedy received a call from the Trump transition team to talk about his well-known concerns about vaccinations. Kennedy accepted. He made the trip to New York City, entering Trump Tower and ascending the golden elevators on January 5, 2016, to speak with the president-elect. Kennedy later gave an interview to *Science Insider*, describing how he met with the Trump transition team for about an hour. Trump and senior aide Kellyanne Conway were present for most of it, and Vice-President-Elect Mike Pence came in for the last fifteen minutes.

> **Q: Did the president-elect request the meeting or did you?**
>
> **A:** He called me a week ago to request it.
>
> **Q: Why?**
>
> **A:** He wants to make sure that we have the best vaccine science and safest vaccine supply that we can have.
>
> **Q: Did the president-elect indicate that he doesn't believe that to be the case at the moment?**
>
> **A:** He is troubled by questions of the links between certain vaccines and the epidemic of neurodevelopmental disorders including autism. And he has a number—he told me five—friends, he talked about each one of them, who has the same story of a child, a perfectly healthy child who went into a wellness visit around age 2, got a battery of vaccines, spiked a fever, and then developed a suite of deficits in the 3 months following the vaccine.
>
> He said that he understood that anecdote was not science but said that if there's enough anecdotal evidence . . . that we'd be arrogant to dismiss it. Those were his words.[6]

Imagine the potential impact of those words and compare it to your own experiences. Let's say you didn't believe in UFOs and five of your friends described the exact same encounter that happened when they were out camping in the woods. Might it shake your skepticism a little? Or if you heard an allegation your boss was sexually harassing women in your office, and then five of your female friends told similar stories of his misbehavior? I think you'd become a believer. It all seemed to be setting up so well for those activists like myself who'd raised the alarm about vaccines for so many years. We'd finally have a friend in the White House.

Then came Bill Gates.

***

Bill Gates is normally ranked as the world's second-richest person, with a net worth according to *Forbes* magazine of $106.8 billion dollars as of November 6, 2019.[7] Gates is usually neck and neck with European luxury goods kingpin Bernard Arnault, who at the time of the 2019 *Forbes* ranking trailed him by a measly $800 million dollars. Leading both, by a few percentage points, is Amazon titan Jeff Bezos, with a net worth of $111.9 billion, according to the same *Forbes* article. In 2020, Facebook's Mark Zuckerberg joined the ranks of the centibillionaires. However, Gates is the one most strongly associated with public health measures, with his Gates Foundation and its endowment of $46.8 billion dollars as of November 2018.[8]

Some applaud Gates for his business acumen and the creation of Microsoft. Others hail him as a savior of mankind because of his philanthropic efforts, particularly his embrace of Generation IV nuclear energy, which has the potential to drastically cut carbon emissions. A smaller group considers him a villain, obsessed with dreams of shrinking the human population because of what they believe to be his support of eugenics.

But one thing is for certain.

Bill Gates loves vaccines. And he doesn't like people like Robert Kennedy Jr. calling for a closer investigation of them:

> In a recent interview with STAT, the philanthropist talked about how he made the case for global health—and vaccines—to Trump. Though the meeting took place several weeks ago, Gates didn't speak publicly about the most recent meeting after it took place.
>
> "Absolutely!" Gates answered emphatically, when asked if he raised the topic of vaccines with the new president. Vaccines "are miracles and have done great things, and when we get new ones we can do a lot. That definitely came up."
>
> Trump has raised concerns about vaccines, repeating discredited claims that vaccination can trigger autism. Prominent vaccine skeptic Robert

> Kennedy Jr. reported late last year that Trump had asked him to chair a commission exploring vaccine safety. So far, Trump has been publicly silent on whether he intends to proceed with that panel.
>
> Gates said he tried to impress upon the president the danger of confusing the message around vaccines. He knows his comments registered with Trump, because the president repeated them later when he met with a delegation from the pharmaceutical industry.[9]

Bill Gates is many things. But there are several things he is not. He is not a college graduate. He does not have a law degree, an MBA, an engineering degree, or any training in the biological sciences.

Gates has an enormous amount of money and is attempting to spend a great deal of it on vaccines, which he believes should escape the scrutiny given to every other consumer product. In any other context, the meeting Gates had with Trump would be laughable. Gates is worth more than a hundred billion dollars. But when he meets with the president of the United States, what does he want to talk about first?

The Middle East?

Race relations?

American standing in the world?

Climate change?

No. He wanted to denigrate the parents of vaccine-injured children who have had their finances, their hopes and dreams, and often their marriages devastated by their child's affliction. Autism is not good for a family's life, sanity, or pocketbook.

The power of a parent must truly be awesome if it terrifies one of the four centibillionaires on the planet. I guess it's because we're not in awe of all those zeroes in his checking account. And Gates wasn't convinced he'd permanently changed Trump's intention to take action on autism:

> Still, Gates isn't certain that means Trump has been persuaded to drop the idea of a vaccine safety inquiry. "There's a rumor that he is going to do something in that area," Gates said. "But maybe I and others will convince him that's not worthwhile."[10]

It seemed Kennedy and the autism parents he represented were going to have to continue to wait for the government to take any effective action on their behalf. Many wondered why Trump was choosing to listen to Gates, who hated him, rather than the five friends who told him identical stories of what happened to their children.

* * *

There's an old expression that "there's no honor among thieves."

I can think of no better example than the article that broke about the Columbia University researchers Ian Lipkin and Mady Hornig, in May 2017. As you recall, this dubious pair was responsible, with their shoddy science, for the destruction of the reputations of both Dr. Andrew Wakefield and Dr. Judy Mikovits. I devoted an entire chapter in this book to their collaboration, even visiting with the two on September 19, 2013, at a small event at the Soho Grand Hotel in New York City. I believe in knowing my enemies, meeting with them face-to-face, and seeing if there's any way we can deescalate a conflict. Despite my repeated attempts to contact them after that meeting, there was no response from either.

The article was titled "Lawsuit at Columbia University Roils Prominent Chronic Fatigue Syndrome Research Lab" and painted a dark picture of Ian Lipkin and how he treated his research partner, Mady Hornig, with whom he'd also had a romantic relationship of several years. The romantic relationship had ended, but the professional collaboration continued. The article said:

> In the lawsuit, filed on 15 May in the U.S. District Court for the Southern District of New York, Hornig alleges that Lipkin for years has discriminated against her on the basis of her sex and created a hostile work environment, violating U.S. and New York civil rights laws. In particular, it alleges that Lipkin took credit for Hornig's work; diverted or misused funds, thus delaying the publication of Hornig's research results; and improperly added himself as principal investigator to grants.[11]

Let's review what Lipkin's longtime collaborator says about him and compare it to the accusations I have made in this book.

First, Hornig claims Lipkin created a hostile work environment and was discriminating against her because she's a woman, violations of both federal and state law.

Second, he diverted funds, a criminal offense.

Third, he misused funds, another criminal offense.

Fourth, he acted to delay Hornig's research results. That may not be a crime, but perhaps it should be.

Fifth, he improperly added himself as a principal investigator to federal grants, another criminal offense.

The lawsuit went into great detail about Lipkin's pattern of behavior toward Hornig, as reported in the *Science* article:

> The lawsuit alleges that since 2013, Lipkin has refused to allow Hornig to post about her own work on the center's website unless the postings include him; required her to get his permission before giving invited talks; routinely presented Hornig's work as his own in meetings with collaborators; blocked her from meetings with potential donors; and silenced her at meetings, "sometimes kicking Plantiff on the shins, under the table . . . or saying 'shut up, Mady' or 'shut the f**k up, Mady'" at meetings attended by both Columbia and non-Columbia colleagues. He also, she alleges, has repeatedly refused to support her for promotion to full professor, even while supporting a male colleague.
>
> Among the claims of misuse of funds, Hornig alleges Lipkin paid the salary of a researcher studying CFS/ME [chronic fatigue syndrome/myalgic encephalomyelitis] with money from the Simmons Autism Research Initiative, which was supposed to be dedicated to an autism study. The suit also claims Columbia had to return more than $53,000 to the National Institutes of Health (NIH) because Hornig refused to sign off on improper use of autism grants.[12]

In the books I've written with Dr. Judy Mikovits, *PLAGUE* and the *New York Times* bestselling *PLAGUE OF CORRUPTION*, we've made the argument for a strong connection between chronic fatigue syndrome (CFS/ME) and autism. As you can see, our enemies have done much the same thing, although they don't like to publicize it. The Lipkin/Hornig team has been like a one-stop shop for Big Pharma to discredit promising research that might point to corporate and government liability for these conditions.

Lipkin's alleged actions point to something deeply wrong in the character of the man, suggesting flaws both deceitful and pathological in both personal and professional relationships. As alleged by Hornig:

> Hornig, also a physician, also alleges that in 2014 and 2015 Lipkin summoned her to his office, dropped his pants and asked her to examine lesions on his buttocks; the lawsuit states that she complied for fear of retribution. For the

> same reason, according to the lawsuit, Hornig kept quiet about Lipkin's overall pattern of behavior toward her.
>
> When Columbia's human resources department was alerted to the situation in 2015, through an intermediary, the lawsuit alleges Lipkin stripped her of her title as medical director of the center and "severely curtailed" her access to technicians and staff. About this time, according to the lawsuit, a colleague told Hornig that Lipkin was out to 'demolish" her and that his determination to sully her reputation had made him "manic."[13]

I wish I could feel some sympathy for Hornig, as she's clearly suffered. However, I find it difficult as she's been part of the team that has concealed critical research concerning the causes of autism and chronic fatigue syndrome (ME/CFS).

I'd like to tell you that Hornig received several millions of dollars in a high-profile trial that resulted in Lipkin being fired by the university in disgrace, his academic reputation in tatters. The truth is I can't find any more information. I'm assuming the university quietly settled the claim. Both Lipkin and Hornig continue to work at Columbia as of 2020.

And you don't have to worry about Lipkin being a pariah among his colleagues. The allegations by his closest collaborator of misogyny, tyrannical behavior, diversion and misuse of grant money, and being a glory hog don't seem to have hurt him in the slightest. His latest effort has been telling the world not to worry that the SARS-CoV-2 outbreak was a result of the escape of an engineered or modified virus from a laboratory in Wuhan, China.[14] He's certain it was a natural occurrence, even though the bats that are thought to be the natural reservoir of the virus live several hundred miles away.

And the fact that those very types of bats were being studied at the Wuhan Institute of Virology?

It's all just a coincidence.

Just like vaccines and autism.

Or like chronic fatigue syndrome (ME/CFS) and a pesky little mouse retrovirus called XMRV.

That's what Ian Lipkin wants you to believe.

* * *

In January 2019, reporter Sharyl Attkisson broke a story in *The Hill*, which many believed would finally expose the vaccine-autism corruption. The article was titled "How a Pro-Vaccine Doctor Reopened Debate about Link to Autism"[15] and told a shocking tale.

According to Attkisson, the government's own medical expert, Dr. Andrew Zimmerman of Johns Hopkins University, was ready to testify in 2007 that at least a third of autism cases were the result of vaccine injury. Zimmerman was testifying in what was known as the Autism Omnibus Proceeding (OAP), in which five thousand claims were consolidated into a few test cases, which would stand in for the other cases. The first was for a child named Hannah Poling (which settled prior to trial with a reported twenty-million-dollar payout), the second was Michelle Cedillo (her claim was denied), and the third case, in which Zimmerman was scheduled to testify, was for a boy named Rolf Hazlehurst. From Attkisson's article:

> Dr. Zimmerman has now signed a bombshell sworn affidavit. He says that, during a group of 5,000 vaccine-autism cases being heard in court on June 15, 2007, he took aside the Department of Justice (DOJ) lawyers he worked for defending vaccines and told them he'd discovered "exceptions in which vaccines could cause autism."
>
> "I explained that in a subset of children, vaccine-induced fever and immune stimulation did cause regressive brain disease with features of autism spectrum disorder," Dr. Zimmerman now states. He said his opinion was based on "scientific advances" as well as his own experience with patients.
>
> For the government and vaccine industry's own pro-vaccine expert to have this scientific opinion stood to change everything about the vaccine-autism debate—if people were to find out.
>
> But they didn't.
>
> Dr. Zimmerman goes onto say that once the DOJ lawyers learned of his position, they quickly fired him as an expert witness and kept his opinion secret from other parents and the rest of the public.[16]

Who is Dr. Andrew Zimmerman? He graduated from Princeton in 1966 and received his MD from Columbia in 1970. From 1994 to 2010, he was an associate professor of neurology at Johns Hopkins University School of Medicine, from 2010 to 2013 he was an associate professor of neurology and pediatrics at Harvard Medical School, and from 2013 to 2019 he was a clinical professor of neurology and pediatrics at the University of Massachusetts Medical School.[17] (An excellent account of the entire Zimmerman story can be found in J.B. Handley's fine book *How to End the Autism Epidemic*.)[18]

It's difficult to be much more mainstream than Dr. Andrew Zimmerman. This is from Zimmerman's own affidavit, sections 6 through 13:

6. On Friday, June 15, 2007, I was present during a portion of the O.A.P to hear the testimony of the Petitioner's expert in the field of pediatric neurology, Dr. Marcel Kinsborne. During a break in the proceedings, I spoke with DOJ attorneys and specifically the lead attorney, Vincent Matanoski in order to clarify my written expert opinion.
7. I clarified that my written expert opinion regarding Michelle Cedillo was a case specific opinion as to Michelle Cedillo. My written expert opinion regarding Michelle Cedillo was not intended to be a blanket statement as to all children and all medical science.
8. I explained that I was of the opinion that there were exceptions in which vaccinations could cause autism.
9. More specifically, I explained that in a subset of children with an underlying mitochondrial dysfunction, vaccine induced fever and immune stimulation that exceeded metabolic energy reserves could, and in at least one of my patients, did cause regressive encephalopathy with features of autism spectrum disorder.
10. I explained that my opinion regarding exceptions in which vaccines could cause autism was based upon advances in science, medicine, and clinical research of one of my patients in particular.
11. For confidentiality reasons, I did not state the name of my patient. However, I specifically referenced and discussed with Mr. Matanoski and the other DOJ attorneys that were present, the medical paper, **Developmental Regression and Mitochondrial Dysfunction in a Child with Autism**, which was published in the **Journal of Child Neurology** and co-authored by Jon Poling, M.D., Ph.D, Richard Frye, M.D., Ph.D, John Shoffner, M.D. and Andrew Zimmerman, M.D., a copy of which is attached as exhibit C.
12. Shortly after I clarified my opinions with the DOJ attorneys, I was contacted by one of the junior DOJ attorneys and informed that I would no longer be needed as an expert witness on behalf of H.H.S. [Health and Human Services] The telephone call in which I was informed that the DOJ would no longer need me as a witness on behalf of H.H.S. occurred after the above-referenced conversation on Friday, June 15, 2007, and before Monday, June 18, 2007.
13. To the best of my recollection, I was scheduled to testify on behalf of H.H.S. on Monday, June 18, 2007.[19]

When I read this account by Sharyl Attkisson, I was stunned. I had covered each day of the Autism Omnibus Proceeding for the *Age of Autism* website and remembered the curious disappearance of Dr. Zimmerman. However, I never imagined it was because he had gone rogue.

How much of a liability was the United States government looking at if one-third of all autism cases were due to vaccines? At the time, it was estimated there were a million kids with autism, and the lifetime cost of such a child was calculated to be around three million dollars. That's a total liability to the government of $900 billion dollars.

Vincent Matanoski saved the American taxpayers $900 billion dollars by his firing of Dr. Zimmerman. He also condemned another generation of American children to develop autism and other neurological disorders.

In 2007, it was clear to experts like Dr. Andrew Zimmerman that at least one-third of autism cases were due to vaccines. But that knowledge was never given to the public or the scientific community. Maybe further testing would have shown more than a third of the autism cases were due to their vaccines. The number of children with autism in the United States stands at 1.8 million, which means that nearly 600,000 of them are vaccine injuries, if Dr. Zimmerman's belief is correct.

We have lost at least thirteen years of scientific innovation and accountability due to the actions of Vincent Matanoski. He still works for the federal government.

I wonder if they gave him a medal.

* * *

In the wake of the COVID-19 pandemic, Bill Gates has announced his intention to fund a vaccine and get it to the public in record time. If you doubt that he is likely the greatest driving force behind the vaccine push, you need only read what he has written in *Gates Notes: The Blog of Bill Gates:*

> Humankind has never had a more urgent task than creating broad immunity for coronavirus. Realistically, if we're going to return to normal, we need to develop a safe, effective vaccine. We need to make billions of doses, we need to get them out to every part of the world, and we need all of this to happen as quickly as possible.
>
> That sounds daunting, because it is. Our foundation is the biggest funder of vaccines in the world, and this effort dwarfs anything we've ever worked on before. It's going to require a global cooperative effort like the world has never seen. But I know it'll get done. There's simply no alternative.[20]

Are you understanding the Gates criminal enterprise? Prevent any investigation of vaccines in general, but then when an outbreak hits, rush to create a new vaccine, also without liability protection.

This is a partial transcript of Gates explaining vaccine science, the potential harm, and the need for complete liability protection for his proposed COVID-19 vaccine to Becky Quick of CNBC on April 9, 2020:

> The efficacy of vaccines in older people is always a huge challenge. It turns out the flu vaccine isn't that effective in elderly people. Most of the benefit comes from younger people not spreading it, because they're vaccinated. And that benefits on a community basis, the elderly. Here, we clearly need a vaccine that works in the upper age range because they're most at risk.
>
> You amp it up so it works in older people and yet you don't have side effects. If we have one in ten thousand side effects, that's way more, seven hundred thousand people who will suffer from that. Really understanding the safety at gigantic scale, across all age ranges, pregnant, male, female, undernourished, existing comorbidities. It's very, very hard. That actual decision of, okay, let's go and give this vaccine to the entire world; governments will have to be involved because there will have to be some risk and indemnification needed before that can be decided on.[21]

The difficult thing about Bill Gates is he's definitely a smart man. But that doesn't necessarily make him a wise one. What he says about the immune system of older people is absolutely correct. Their immune systems do not respond as robustly as those of younger people. But then it seems he shades the truth, as when he starts to smile a bit when he suggests the serious side effects might be one in ten thousand. And if it's somewhere around that number, you might have seven hundred thousand injured. That's from Bill Gates himself, the guy with more than a hundred billion dollars. But don't come asking for money from Bill Gates if that coronavirus vaccine injures you. You're going to have to look to the government. In other words, you, the taxpayer.

Might there be something uniquely dangerous about any vaccine against a coronavirus? That's what Robert Kennedy Jr. and many scientists appear to believe. This is what Kennedy wrote on his Children's Health Defense blog:

> Scientists first attempted to develop coronavirus vaccines after China's 2002 SARS-CoV outbreak. Teams of US & foreign scientists vaccinated animals

> with the four most promising vaccines. At first, the experiment seemed successful as all the animals developed a robust antibody response to coronavirus. However, when the scientists exposed the vaccinated animals to the wild virus, the results were horrifying. Vaccinated animals suffered hyper-immune responses including inflammation throughout their bodies, especially in their lungs. Researchers had seen this same "enhanced immune response" during human testing of the failed RSV vaccine tests in the 1960s. Two children died.[22]

Let's talk about what's meant by a hyperimmune response. It basically means that your body attacks itself, with death a likely outcome. That's what happened to those two children.

In 2012, a critical paper was released with the frightening title "Immunization with SARS Coronavirus Vaccines Leads to Pulmonary Immunopathology on Challenge with the SARS Virus," authored by scientists from the University of Texas and Baylor University. It's somewhat disturbing to read how prescient the opening is, in light of recent events:

> Severe acute respiratory syndrome (SARS) emerged in China in 2002 and spread to other countries before being brought under control. Because of a concern for reemergence ***or a deliberate release of the coronavirus*** [bold and italics added], vaccine development was initiated. Evaluations of an inactivated whole virus vaccine in ferrets and nonhuman primates and a virus-like-particle in mice induced protection against infection but challenged animals exhibited an immunopathologic-type lung disease.[23]

The researchers were putting forth a nightmare scenario out of a James Bond movie. An evil supervillain could spike vaccines with weakened coronaviruses and then wait a few years to cause a "deliberate release of the coronavirus." Those infected would go on to develop an inflammatory condition in their lungs, which would probably necessitate the use of a respirator as a last-ditch effort to save their lives.

I have no information, either way, that convinces me this is an explanation for the COVID-19 outbreak of 2020 and the subsequent deaths.

The researchers were simply warning that such a terrible plan is biologically possible. The main concern of the researchers was that any coronavirus vaccine carried with it the frightening possibility that reexposure to a coronavirus would cause an immune overreaction, particularly in the lungs. Because of what they saw in the mice given the coronavirus vaccine, and earlier experiments on children, the researchers wrote:

> This combined experience provides concern for trials with SARS-CoV vaccines in humans. Clinical trials with SARS coronavirus vaccines have been conducted and reported to induce antibody responses and to be "safe." However, the evidence for safety is for a short period of observation. The concern arising from the present report is for an immunopathologic reaction occurring among vaccinated individuals on exposure to infectious SARS-CoV, the basis for developing a vaccine against SARS. Additional concerns relate to effectiveness and safety of antigenic variants of SARS-CoV and for the safety of vaccinated persons exposed to other coronaviruses, particularly of the type 2 group.[24]

The vaccinated subjects would appear completely healthy. They would show they have antibodies to the coronavirus. Normally, this would mean the person is "protected" from the virus. But when exposed to a new coronavirus, their immune system would overreact, creating an inflammatory storm in their body, much the way the immune system of a person with a nut allergy might overreact from eating a peanut butter and jelly sandwich.

The main argument in this book is that vaccines have been causing chronic conditions like autism and other developmental problems. Whether that's because of the presence of heavy metals, chemicals, or other pathogens contained in the animal and human biological tissues necessary to produce a vaccine is unclear to me, even at this time.

The 1986 National Childhood Vaccine Injury Act and subsequent mischief by governmental regulators and corporate interests have prevented a full investigation of these complaints. However, this research suggests that any coronavirus vaccine developed has the potential to quickly kill large numbers of the human race whenever another coronavirus makes its inevitable way through the human population.

We will have unwittingly killed ourselves, lining up like sheep for the slaughter.

***

I cannot tell you the inner thoughts of Bill Gates.

I don't know what he thinks. However, I can tell you about his actions. Perhaps they are a window into the man's soul. Maybe he's just as confused as the rest of us. But he's not acting like a person who's confused. He's acting as if he were certain. Here's what I know for certain about Bill Gates:

1. By one estimation, Gates is the second-wealthiest man in the world, with a net worth of approximately $106.8 billion dollars compared to Jeff Bezos of Amazon, whose worth about $111.9 billion dollars.
2. However, that doesn't take into account the Bill and Melinda Gates Foundation with an endowment of $46.8 billion dollars. Combined, that means Bill Gates is effectively in charge of $153.6 billion dollars, easily making him the richest and most powerful man in the world.
3. By his own admission, the Gates Foundation is the world's leading funder of vaccines.
4. Gates claims credit for scuttling the Kennedy Vaccine Safety Commission.
5. Gates wants to give the world "billions of doses" of a coronavirus vaccine.
6. An effective coronavirus vaccine has never been developed because of the devastating side effects upon reexposure to a coronavirus, and these side effects include inflammation in the lungs and death.
7. Although he is arguably the richest man in the world, he admits that such a vaccine may injure at least 700,000 individuals and wants to be free from any legal liability from the injuries caused by his vaccine.

I cannot tell you what's in the heart of Bill Gates. I can only speak of his actions. If anybody else proposed such a reckless course of action, they'd be quickly dismissed by the general public and governmental agencies.

I don't know if Bill Gates is evil or ignorant. I don't really care. Regardless, I believe his plans have the ability to injure tens of millions of people and cause untold damage.

With his vast wealth and influence, I consider Bill Gates to be the greatest threat to humanity on the planet.

* * *

The only possible counter to the plans of Bill Gates was President Donald Trump. I've discussed this at length with many autism advocates who say, "People need to get organized." I understand the sentiment, but it just won't work.

Dr. William Thompson confessed in 2014 to participating in a cover-up of data showing earlier administration of the MMR (measles, mumps, and rubella) vaccine was leading to increased rates of autism in African American boys and others. This is arguably the greatest medical crime against the African American community since the Tuskegee experiments, in which black men with syphilis were left untreated for decades. Congress has not called on Dr. Thompson to testify about these claims.

Dr. Andrew Zimmerman revealed in 2019 that in 2007 he was prepared to testify that at least one-third of autism cases were due to vaccine injuries. Congress has not requested any testimony from Dr. Zimmerman.

Dr. Zimmerman also testified that, in 2007, he was prevented from testifying about the connection between vaccines and autism by Department of Justice attorney Vincent Matanoski, prolonging the vaccine-autism debate for more than a decade. Matanoski continues to work for the federal government. The US Congress has also not requested any testimony from Vincent Matanoski regarding these claims.

Besides the reporting of a few brave journalists, like Sharyl Attkisson, the mainstream media has not reported in depth on any of these allegations.

We need help from the commander in chief, or else our fight will continue for many years to come, and many more children will be injured.

Trump's instincts on this question were the best of any major figure in America today. But he didn't follow through. Even now we're hearing about a "warp speed" plan to bring us a COVID-19 vaccine.

Dr. Judy Mikovits, with whom I've written two books with a third on the way, met with longtime Trump friend Roger Stone in March 2019 and questioned him about Trump's plan to take on the vaccine question. Stone assured her it was a "second-term priority."

That was a slender reed of hope, one that is unfortunately now moot with Trump losing his 2020 reelection bid, and I sympathize with those who say Trump has abandoned us, just as Obama did during his first election for president, signaling his support for investigating the issue and then when elected doing nothing.

Some have made the argument that Trump wishes to first weaken Big Pharma, telling them they need to move their facilities back to the United States from China (a significant blow to their income) and making sure Americans are paying the same price for prescription drugs as others around the world. (This is known as the "favored nations" clause, and it is claimed this will cause a significant drop in prices in America, drastically cutting into Big Pharma's revenues.) Perhaps that was Trump's plan: after Big Pharma is financially weakened, he planned to attack them on the vaccine issue.

However, if that's the case, I believe it has been an enormous mistake.

The health freedom advocates are among Trump's most loyal followers, and he has given them nothing, aside from not enforcing federal vaccine mandates. He has allowed Democratic majorities in some of our largest states, such as New York and California, to pass draconian vaccine laws. And he has said nothing about them.

It would have been a simple matter for a master persuader like Trump to change the debate in our country. Allowing Kennedy's Vaccine Safety Commission to go forward would be a great start. Who could reasonably object in public to a safety review?

Or how about something like this? The provaccine advocates say vaccines are completely safe. Therefore, we don't need the 1986 National Childhood Vaccine Injury Act. Trump could make that argument. How could either side in the debate possibly object? Vaccines can simply return to the traditional civil justice system and products liability laws and let things take their proper course. If vaccines are truly safe, nobody has anything to fear. We, the autism parents, and health freedom advocates have made our voices heard.

Endnotes

Chapter 1: The Call

1. Telephone Interview with Dr. Brian Hooker by Kent Heckenlively, November 20, 2015.
2. Robert Pear, "Reagan Signs Bill on Drug Exports and Payment for Vaccine Injuries," *New York Times*, November 15, 1986, www.nytimes.com/1986/11/15/us/reagan-signs-bill-on-drug-exports-and-payment-forvaccine-injuries.html.
3. Ibid.
4. Email from Dr. Brian Hooker to Dr. William Thompson, November 7, 2013.
5. Email from Dr. William Thompson to Dr. Brian Hooker, November 8, 2013.
6. Email from Dr. Brian Hooker to Dr. William Thompson, November 9, 2013.
7. Email from Dr. William Thompson to Dr. Brian Hooker, November 9, 2013.
8. Email from Dr. Brian Hooker to Dr. William Thompson, November 10, 2013.
9. Email from Dr. William Thompson to Dr. Brian Hooker, November 10, 2013.
10. William H. Thompson, Cristofer Price, Barbara Goodson, et al., "Early Thimerosal Exposure and Neuropsychological Outcomes at 7 to 10 Years," *New England Journal of Medicine*, September 27, 2007, 357: 1281–1292, doi: 10.1056/NEJMoa071434.
11. U.S Department of Health and Human Services website, "NVAC Chair: Walter A. Orenstein, MD," http://www.hhs.gov/nvpo/nvac/roster/orenstein-bio.html, accessed February 6, 2016.
12. Email from Dr. William Thompson to Dr. Brian Hooker, November 11, 2013.
13. Brian Hooker Notes on Conversation with Dr. William Thompson, November 13, 2013.
14. Telephone Interview with Dr. Brian Hooker by Kent Heckenlively, November 20, 2015.
15. Ibid.
16. Ibid.
17. Telephone Interview with Dr. Brian Hooker by Kent Heckenlively, December 9, 2015.

18. Thomas Verstraeten, Robert Davis, Frank DeStefano, et al., "Safety of Thimerosal-Containing Vaccines: A Two-Phased Study of Computerized Health Maintenance Organization Databases," *Pediatrics*, Volume 112, issue 5, November 2003.
19. Thomas Verstraeten, Robert Davis, Frank DeStefano, et al., "Safety of a Two-Phased Study of Computerized Health Maintenance Organizations," *Pediatrics*, November 2003, 112(5): 103901048, www.pediatrics.aappublications.org/content/112/5/1039.
20. "A Brief Review of Verstraeten's 'Generation Zero' VSD Study Results," Safe Minds, April, 2013, accessed October 19, 2020, http://www.safeminds.org/wp-content/uploads/2013/04/GenerationZeroNotes.pdf.
21. Frank DeStefano, Tanya Karapurkar Bhasin, William Thompson, Marshalyn Yeargin-Allsopp, and Coleen Boyle, "Age at First Measles-Mumps-Rubella Vaccination in Children with Autism and School-Matched Control Subjects: A Population-Based Study in Metropolitan Atlanta," *Pediatrics*, Vol. 113, No. 2, 258–266, February 2, 2004, doi: 10.1542/peds.113.2.259.
22. Ibid.
23. Ibid.
24. Telephone Interview with Dr. Brian Hooker by Kent Heckenlively, December 9, 2015.
25. Ibid.
26. Ibid.
27. Ibid.
28. Email from Dr. William Thompson to Dr. Walter Orenstein, October 16, 2002.
29. Email from Dr. Walter Orenstein to Dr. William Thompson, October 18, 2002.
30. Telephone Interview with Dr. Brian Hooker by Kent Heckenlively, December 14, 2015.
31. Email from Dr. William Thompson to Dr. Julie Gerberding, February 2, 2004.
32. Letter from Congressman Dave Weldon to Dr. Julie Gerberding, October 31, 2003, http://www.safeminds.org/wp-content/uploads/2014/08/Weldon-letter-to-Gerberding10-31-03.pdf, accessed February 6, 2015.
33. Biography of Dr. Julie Gerberding, www.merck.com/leadership/julie-l-gerberding-m-d-m-p-h/, accessed November 13, 2020.
34. Memorandum of Being Placed on "Paid Administrative Leave." To William W. Thompson, PhD, Epidemiologist, From Robert T. Chen, MD, MS, Chief of the Immunization Safety Branch, March 9, 2004. Signed by both William Thompson and Robert Chen.
35. Memorandum of "Counseling" to William W. Thompson, PhD, Epidemiologist, from Robert T. Chen, MD, MS, Chief of the Immunization Safety Branch, March 9, 2004. Signed by both William Thompson and Robert Chen.
36. Annex to Memorandum of "Counseling" to William W. Thompson, PhD,

Epidemiologist, from Robert T. Chen, MD, MS, Chief of the Immunization Safety Branch, March 9, 2004.

37. Text Messages between Dr. William Thompson and Dr. Brian Hooker, November 21, 2013.
38. Telephone Interview with Dr. Brian Hooker by Kent Heckenlively, December 14, 2015.
39. Brian Hooker, "Measles-Mumps-Rubella Vaccination Timing and Autism Among Young African-American Boys: A Reanalysis of CDC Data," *Translational Neurodegeneration*, 20143: 16, August 27, 2014, doi: 10.1186/2047-9158-316.
40. Telephone Interview with Dr. Brian Hooker by Kent Heckenlively, December 14, 2015.

Chapter 2: The Insanely Good Soul of Dr. Andrew J. Wakefield

1. Telephone Interview with Dr. Andrew Wakefield by Kent Heckenlively, February 25, 2016.
2. Sophie Borland, "'Dishonest and Irresponsible': Doctor who Triggered MMR Vaccine Scare Is Struck Off," *Daily Mail*, May 24, 2010, www.dailymail.co.uk/news/article -1280840/Andrew-Wakefield-Doctor-heart-MMR-vaccinerow-struck-off.html.
3. Ibid.
4. England and Wales High Court (Administrative Court) Decisions, Professor John Walker-Smith vs. General Medical Council, February 13–17, 2012, Case No: CO/7039/2010, www.bailii.org/EWHC/Admin/2012/503.html.
5. Ibid., 59.
6. "MMR Doctor Wins High Court Appeal," BBC News, March 7, 2012, www.bbc .com/news/health-17283751.
7. Mary Holland, "Co-Author of Lancet MMR-Autism Study Exonerated on All Charges of Professional Misconduct," Elizabeth Birt Center for Autism Law and Advocacy, March 2012, www.ebcala.org/areas-of-law/vaccine-law/coauthor-of-lancet -mmr-autism-study-exonerated-on-all-charges-of-professional-misconduct.
8. Telephone Interview with Dr. Andrew Wakefield by Kent Heckenlively, February 25, 2016.
9. Ibid.
10. Ibid.
11. David Lewis, *Science for Sale: How the US Government Uses Powerful Corporations and Leading Universities to Support Government Policies, Silence Top Scientists, Jeopardize Our Health, and Protect Corporate Profits*, Skyhorse Publishing, New York (2014), 134.
12. Ibid., 147.

13. Telephone Interview with Dr. Andrew Wakefield by Kent Heckenlively, February 25, 2016.
14. Ibid.
15. Ibid.
16. Ibid.
17. Ibid.
18. Ibid.
19. Ibid.
20. Ibid.
21. "Statement of William W. Thompson, Ph.D., Regarding the 2004 Article Examining the Possibility of a Relationship Between MMR Vaccine and Autism," August 27, 2014, www.morganverkamp.com/statement-ofwilliam-w-thompson-ph-d-regarding-the-2004-article-examining-the-possibility-of-a-relationship-between-mmrvaccine-and-autism/.
22. Hooker and Wakefield Complaint to CDC and Office of Research Integrity at Health and Human Services, October 14, 2014, p. 2.
23. Ibid., 1–2.
24. Ibid., 16.
25. Ibid., 18.
26. Ibid., 18.
27. Ibid., 19.
28. Ibid., 20–21.
29. Ibid., 23.
30. Ibid., 24–25.
31. Ibid., 27.
32. Ibid., 28.
33. Ibid., 30–31.

Chapter 3: The Lipkin-Hornig Team

1. Mady Hornig, D Chian, and W. Ian Lipkin, "Neurotoxic Effects of Postnatal Thimerosal are Mouse Strain Dependent," *Molecular Psychiatry*, June 8, 2004, doi: 10.1038.1038/sj.mp.4001529, www.mailman.columbia.edu/sites/default/files/legacy/neurotoxiceffects-ofpostnatalthimerosal.pdf.
2. Ibid.
3. Ibid.
4. Congressional Testimony of Dr. Mady Hornig, U.S. House of Representatives, The Subcommittee on Human Rights and Wellness, Committee on Government Reform, September 8, 2004, www.gpo.gov/fdsys/pkg/CHRG108hhrg98046/html/CHRG-108-hhrg98046.htm.

5. Telephone Interview with Dr. Ian Lipkin by Kent Heckenlively, December 7, 2012.
6. Ibid.
7. Prometheus, "Dr. Hornig's Autistic Mice," *A Photon in the Darkness*, July 29, 2005, www.photoninthedarkness.blogspot.com/2005/07/dr-hornigs-autistic-mice_29.html.
8. Kathryn Shattuck, "Trying to Corner the Market on Charity," *New York Times*, April 22, 2010, www.nytimes.com/2010/04/23/travel/23away.html?_r=0.
9. Douglas Martin, "Donald Hornig, Last to See First A-Bomb, Dies at 92," *New York Times*, January 26, 2013, www.nytimes.com/2013/01/27/science/donald-hornig-a-bomb-scientist-and-brown-president-dies-at92.html?_r=0 .
10. Ibid.
11. Columbia Cocktail Party, September 19, 2013—Soho Grand Hotel—Featuring Mady Hornig, Brent Williams, and William Karesh (recorded with permission).
12. Ibid.
13. Ibid.
14. Ibid.
15. Ibid.
16. Ibid.
17. Vincent C. Lombardi, Francis W. Ruscetti, Jaydip Das Gupta, Max Pfost, Kathryn C. Hagen, Daniel L. Peterson, Sandra K. Ruscetti, Cari Petro-Sandowski, Bert Gold, Michael Dean, Robert H. Silverman, and Judy Mikovits, "Detection of an Infectious Retrovirus, XMRV, in Blood Cells of Patients with Chronic Fatigue Syndrome," *Science*, Vol. 334 (October 14, 2009), 334.
18. Sam Shad, "Interview with Anette Whittemore and Judy Mikovits," *Nevada Newsmakers*, October 14, 2009.
19. Frederick Hecht and Annie Lutekmeyer, "Immunizations and HIV," University of California, San Francisco HIV web-site, Accessed May 3, 2013, www.hivinsite.ucsf.edu/InSite?page=kb-00&doc=kb-03-01-08 .
20. Maurice Brodie and William Park, "Active Immunization against Poliomyelitis," *American Journal of Public Health* (February 1936), 119–125.
21. John F. Kessel, Anson S. Hoyt, and Roy T. Fisk, "Use of Serum and the Routine and Experimental Laboratory Findings in the 1934 Poliomyelitis Epidemic," *American Journal of Public Health*, December 1934, 1215–1223, www.ncbi.nlm.nih.gov/pmc/articles/PMC1558945/.
22. Antoinette Cornelia Van Der Kuyl, Marion Comelissen, and Ben Berkhout, "Of Mice and Men: On the Origin of XMRV," *Frontiers in Microbiology*, Vol. 1, Article 147, January 17, 2011, www.ncbi.nlm.nih.gov/pmc/articles/PMC3109487/.
23. Harvey J. Alter, Judy A. Mikovits, William M Switzer, Francis W. Ruscetti, Shyh-Ching Lo, Nancy Klimas, Anthony L. Komaroff, Jose C. Montoya, Lucinda Bateman,

Susan Levine, Daniel Peterson, Bruce Levin, Maureen R. Hanson, Afia Genfi, Meera Bhat, HaoQiang Zheng, Ricahrd Wang, Bingjie Li, Guo-Chiuan Hung, Li Ling Lee, Stephanie Sameroff, Walid Heine, John Coffin, and W. Ian Lipkin, "A Multicenter Blinded Analysis Indicates No Association between Chronic Fatigue Syndrome /Myalgic Encephalomyelitis and either Xenotropic Murine Leukemia VirusRelated Virus or Polytropic Murine Leukemia Virus," *MBio*, Vol. 3, no. 5, September 18, 2012, doi: 10.1128/mBio.00266–12, www.mbio.asm.org/content/3/5/e00266-12.full.

24. Paul Cheney, Telephone Interview with Kent Heckenlively, July 25, 2013.
25. Ewen Calloway, "The Man Who Put the Nail in XMRV's Coffin," *Nature*, September 18, 2012.
26. Ian Lipkin, Public Conference Call with the Centers for Disease Control and Prevention, September 10, 2013. Transcript by ME/CFS Forums.com/wiki/Lipkin.
27. Ibid.
28. Mady Hornig, Thomas Briese, Timothy Buie, Margaret L. Bauman, Gregory Lauwers, Ulrike Siemetski, Kimberly Hummel, Paul A. Rota, William J. Bellini, John J. O'Leary, Orla Sheils, Errol Alden, Larry Pickering, and W. Ian Lipkin, "Lack of Association between Measles Virus Vaccine and Autism with Enteropathy: A Case Control Study," *PLOS One*, September 4, 2008, www.dx.doi.org/10.1371/journal .pone.0003140.
29. Ibid.
30. Ibid.
31. Ibid.
32. Ibid.
33. Kevin Barry, *Vaccine Whistleblower: Exposing Autism Research Fraud at the CDC*, Skyhorse Publishing, New York (2015), 102–104.
34. Grant Delin, "Discover Interview: The World's Most Celebrated Virus Hunter, Ian Lipkin," *Discover*, April 2012.

Chapter 4: The Documents

1. Draft Analysis Plan, "Autism and Childhood MMR Vaccine," Centers for Disease Control and Prevention, National Immunization Program, April 3, 2001, 1.
2. Ibid., 1.
3. Ibid., 1.
4. Ibid., 2.
5. Ibid., 3.
6. "Events Surrounding the DeStefano et al. (2004) MMR-Autism Study," Prepared by Dr. William Thompson for Congressman William Posey, September 9, 2014, 1–2.
7. Ibid., 5–6.
8. Ibid., 2–5.

Chapter 5: The CDC Runs Away to Simpsonwood to Defend Mercury in Vaccines

1. "Scientific Review of Vaccine Safety Datalink Information, June 7–8, 2000, Simpsonwood Retreat Center, Norcross, Georgia," Centers for Disease Control and Prevention, Accessed February 7, 2016, National Immunization Program, 1–2. http://thinktwice.com/simpsonwood.pdf.
2. Ibid., 1.
3. Ibid., 1.
4. Ibid., 2.
5. Ibid., 2.
6. Ibid., 2.
7. Ibid., 3.
8. Ibid., 3.
9. Ibid., 3.
10. Ibid., 3.
11. Boyd E. Haley, "Mercury Toxicity: Genetic Susceptibility and Synergistic Effects," *Medical Veritas* 2 (2005), 535–542, doi: 10.1588/medver.2005.02.00.67.
12. "Scientific Review of Vaccine Safety Datalink Information, June 7–8, 2000, Simpsonwood Retreat Center, Norcross, Georgia," Centers for Disease Control and Prevention, Accessed February 7, 2016, National Immunization Program, 4. http://thinktwice.com/simpsonwood.pdf.
13. Ibid., 5.
14. Ibid., 7.
15. Ibid., 26.
16. Ibid., 28.
17. Ibid., 28.
18. Ibid., 28–29.
19. Ibid., 29.
20. Ibid., 32.
21. Ibid., 32.
22. Ibid., 36.
23. Ibid., 36.
24. Ibid., 37.
25. Ibid., 42.
26. Ibid., 42.
27. Ibid., 46.
28. Ibid., 46.
29. Ibid., 46.
30. Ibid., 46.

31. Ibid., 48.
32. Email from Robert Kennedy Jr. to Kent Heckenlively et al., September 22, 2015.
33. Email from Thomas Verstraeten to Philippe Grandjean, Robert Chen, Frank DeStefano, et al., July 14, 2000, http://www.putchildrenfirst.org/media/2.20.pdf.
34. Telephone Interview with Beth Clay by Kent Heckenlively, January 16, 2016.
35. Ibid.

Chapter 6: The View from Congress

1. Telephone Interview with Beth Clay by Kent Heckenlively, January 16, 2016.
2. "Profile of Harold Varmus. MD," Memorial Sloan Kettering Cancer Center, Accessed February 8, 2016, www.mskcc.org/harold-varmus-md.
3. Telephone Interview with Beth Clay by Kent Heckenlively, January 16, 2016.
4. Ibid.
5. Ibid.
6. Brian Deer, "How the Case Against the MMR Vaccine was Fixed," *British Medical Journal,* January 6, 2011, doi: http://dx.doi.org/10.1136/bmj.c5347.
7. Lawrence Solomon, "Merck Has Some Explaining to Do over Its MMR Vaccine Claims," *Huffington Post,* September 25, 2014.
8. Telephone Interview with Beth Clay by Kent Heckenlively, January 16, 2016.
9. Ibid.
10. Ibid.
11. "Mercury in Medicine Report," House Committee on Government Reform, Congressional Record, Vol. 149, Number 76, p. E1012, May 21, 2003, www.gpo.gov/fdsys/pkg/CREC-2003-05-21/html/CREC-2003-05-21-pt1PgE1011-3.htm.
12. Ibid.
13. Ibid., E1012-E1013.
14. Ibid., E1013.
15. Telephone Interview with Beth Clay by Kent Heckenlively, January 16, 2016.
16. Congressman Dave Weldon, "Something Is Rotten, but Not Just in Denmark," Speech at Autism One Conference, Chicago, Illinois, May 29, 2004, www.vce.org/mercury/Weldon.pdf.
17. Ibid.
18. Ibid.
19. Ibid.
20. Ibid.
21. Ibid.
22. Ibid.
23. Ibid.
24. Ibid.

25. Ibid.
26. Ibid.
27. "The Week in Science," *Nature,* 472, 264–265, April 20, 2011, doi:10.1038/472264a.
28. "Autism Researcher Indicted for Stealing Grant Money," United States Attorney's Office, Northern District of Georgia, April 13, 2011, www.justice.gov/archive/usao/gan/press/2011/04-13-11.html.
29. Ibid.
30. "Three Years and Counting on Failure to Prosecute Poul Thorsen," Safe Minds, April 14, 2014, www.safeminds.org/blog/2014/04/14/three-years-counting-failure-prosecute-poul-thorsen/.
31. Sharyl Attkisson, "Researcher who Dispelled Vaccine-Autism Link: 'Most-Wanted Fugitive,'" www.sharylyattkisson.com, August 16, 2014, www.sharylattkisson.com/researcher-who-dispelled-vaccine-autismlink-most-wanted-fugitive/.

Chapter 7: The Legal View

1. Clifton Parker, "Federal Program for Vaccine-Injured Children is Failing, Stanford Scholar Says," *Stanford News,* July 6, 2015, www.news.stanford.edu/news/2015/july/vaccine-court-engstrom-070615.html.
2. Telephone Interview with Special Master X by Kent Heckenlively, January 29, 2016.
3. Ibid.
4. Ibid.
5. Ibid.
6. Ibid.
7. Ibid.
8. Ibid.
9. Ibid.
10. Ibid.
11. Ibid.
12. Ibid.
13. Ibid.
14. Ibid.
15. Telephone Interview with Special Master X by Kent Heckenlively, February 1, 2016.
16. Ibid.
17. Ibid.
18. Ibid.
19. Health and Human Services, "Data and Statistics," Accessed February 14, 2016, www.hrsa.gov/vaccinecompensation/vicpmonthlyreport02032016.pdf.
20. David Kessler, "Introducing MedWatch: A New Approach to Reporting Medication and Device Adverse Effects and Product Problems," *Journal of the American Medical*

Association, Vol. 269, No. 21, June 2, 1993, http://www.fda.gov/downloads/Safety/MedWatch/UCM201419.pdf.

21. Nora Freeman Engstrom, "A Dose of Reality for Specialized Courts: Lessons from the VICP," *University of Pennsylvania Law Review*, Vol. 163, 1631–1717, (2015); Nora Freeman Engstrom, "Exit, Adversarialism, and the Stubborn Persistence of Tort," *Journal of Tort Law* (2015), doi: 10.1515/jtl-2015-0002.
22. Nora Freeman Engstrom, "Heeding Vaccine Court's Failures," *National Law Journal*, June 29, 2015, http://law.stanford.edu/wp-content/uploads/2015/07/Nora-Engstrom-Heeding-Vaccine-Court-Failures-NationalLaw-Journal.pdf.
23. Nora Freeman Engstrom, "A Dose of Reality for Specialized Courts: Lessons from the VICP," *University of Pennsylvania Law Review*, Vol. 163, 1659–1660 (2015).
24. Ibid., 1655.
25. Ibid., 1656.
26. Ibid., 1656.
27. Ibid., 1657.
28. Ibid., 1659.
29. Ibid., 1659.
30. Ibid., 1636.
31. Nora Freeman Engstrom, "Heeding Vaccine Court's Failures," *National Law Journal*, June 29, 2015, http://law.stanford.edu/wp-content/uploads/2015/07/Nora-Engstrom-Heeding-Vaccine-Court-Failures-NationalLaw-Journal.pdf.
32. Ibid.
33. Nora Freeman Engstrom, "A Dose of Reality for Specialized Courts: Lessons from the VICP," *University of Pennsylvania Law Review*, Vol. 163, 1659 (2015).
34. Ibid., 1676.
35. Ibid., 1675–1676.
36. Nora Freeman Engstrom, "Exit, Adversarialism, and the Stubborn Persistence of Tort," *Journal of Tort Law* (2015), doi: 10.1515/jtl-2015-0002.
37. Ibid.
38. Telephone Interview with Nora Freeman Engstrom by Kent Heckenlively, February 18, 2016.
39. Ibid.
40. Ibid.
41. Ibid.
42. Ibid.
43. Ibid.
44. Bruesewitz v. Wyeth Laboratories, United States Supreme Court, 562 U.S. (February 22, 2100), Sotomayor dissent, p. 1.

Chapter 8: Those Who Would Oppose Goliath

1. Telephone Interview with Brandy Vaughn by Kent Heckenlively, March 14, 2016.
2. Ibid.
3. Ibid.
4. Ibid.
5. Ibid.
6. Ibid.
7. Ibid.
8. Ibid.
9. Ibid.
10. Ibid.
11. Ibid.
12. David Voreacos and Allen Johnson, "Merck Paid 3,468 Death Claims to Resolve Vioxx Suits," *Bloomberg Business*, July 27, 2010, www.bloomberg.com/news/articles/2010-07-27/merck-paid-3-468-death-claims-toresolve-vioxx-suits.
13. "September 11 Fast Facts," CNN Library, September 7, 2015, accessed March 23, 2016, www.cnn.com/2013/07/27/us/september-11-anniversary-fast-facts/.
14. Alexander Cockburn, "When Half a Million Americans Died and Nobody Noticed," *The Week*, April 27, 2012, www.theweek.co.uk/us/46535/when-half-million-americans-died-and-nobody-noticed.
15. Ibid.
16. Jim Edwards, "Merck Created Hit List to 'Destroy,' 'Neutralize' or 'Discredit' Dissenting Doctors," CBS News, May 6, 2009, www.cbsnews.com/news/merck-created-hit-list-to-destroy-neutralize-or-discredit-dissentingdoctors/.
17. Jim Edwards, "New Merck Allegations: A Fake Journal; Ghostwritten Studies; Vioxx Pop Songs: PR Execs Harass Reporters," CBS News, April 23, 2009, www.cbsnews.com/news/new-merck-allegations-a-fake-journalghostwritten-studies-vioxx-pop-songs-pr-execs-harass-reporters/.
18. Telephone Interview with Brandy Vaughn by Kent Heckenlively, March 14, 2016.
19. "Secret Anitiperspirant/Deodorant Powder Fresh," Drugs.Com, accessed March 24, 2016, www.drugs.com/otc/115982/secret-antiperspirant-deodorant-powder-fresh.html.
20. Telephone Interview with Brandy Vaughn by Kent Heckenlively, March 29, 2016.
21. Telephone Interview with Brandy Vaughn by Kent Heckenlively, March 14, 2016.
22. Ibid.
23. Ibid.
24. Ibid.
25. Nancy Shute, "How a California Law to Encourage Vaccination Could Backfire," National Public Radio, Health Shots, November 9, 2013, www.npr.org/sections/health

-shots/2013/11/09/243937869/how-a-california-law-toencourage-vaccination-could-backfire.

26. Telephone Interview with Brandy Vaughn by Kent Heckenlively, March 14, 2016.
27. Ibid.
28. Ibid.
29. "The Overt and Covert Intimidation of Brandy Vaughn," July 31, 2015, www.youtube.com/watch?v=fuTXlCGjqMc.
30. Telephone Interview with Brandy Vaughn by Kent Heckenlively, March 14, 2016.
31. Ibid.
32. Ibid.
33. Ibid.
34. Council for Vaccine Safety, Accessed March 27, 2016, www.councilforvaccinesafety.org/.
35. Telephone Interview with Brandy Vaughn by Kent Heckenlively, March 14, 2016.
36. Telephone Interview with Brandy Vaughn by Kent Heckenlively, March 29, 2016.
37. Ibid.
38. Ibid.
39. Kent Heckenlively, "Dr. Bradstreet, Nagalese and the Viral Issue in Autism," *Age of Autism*, October 20, 2011, www.ageofautism.com/2011/10/dr-bradstreet-nagalase-and-the-viral-issue-in-autism.html.
40. United States District Court, Northern District of Georgia, Search Warrant Case Number: 1:15-MC674, June 16, 2015, www.naturalnews.com/files/GCMAF-Bradstreet-Search-Warrant.pdf.
41. Scott Creighton, "The Strange Story of Dr. Bradstreet, SB 277, the FDA and the Struggle against Mandatory Vacinations," *American Everyman*, June 27, 2015, www.willyloman.wordpress.com/2015/06/27/the-strange-storyof-dr-bradstreet-sb277-the-fda-and-the-struggle-against-mandatory-vaccinations/.
42. CBS Staff, "Authorities: Anti-Vaccine Doctor Dead in Apparent Suicide," CBS News, June 27, 2015, www.cbsnews.com/news/authorities-anti-vaccine-doctor-dead-in-apparent-suicide/.
43. Erin Elizabeth, "Recap on my Unintended Series: The Holistic Doctor Deaths," *Health Nut News*, March 12, 2016, www.healthnutnews.com/recap-on-my-unintended-series-the-holistic-doctor-deaths/.
44. Erin Elizabeth, "6 Myths Debunked about Dr. Bradstreet, GCMAF, Autism, and his Research," *The Examiner*, March 27, 2016, www.examiner.com/article/6-myths-debunked-about-dr-bradstreet-gmcaf-autism-and-hisresearch.
45. Telephone Interview with Thomas and Candace Bradstreet by Kent Heckenlively, May 25, 2016.
46. Ibid.

47. Ibid.
48. Ibid.
49. United States District Court, Northern District of Georgia, Search Warrant Case Number: 1:15-MC674, June 16, 2015, www.naturalnews.com/files/GCMAF-Bradstreet-Search-Warrant.pdf.
50. Telephone Interview with Thomas Bradstreet by Kent Heckenlively, June 1, 2016.
51. Ibid.
52. Ibid.
53. Telephone Interview with Thomas and Candace Bradstreet by Kent Heckenlively, May 25, 2016.
54. Ibid.
55. Ibid.
56. Telephone Interview with Thomas Bradstreet by Kent Heckenlively, June 1, 2016.
57. Telephone Interview with Brandy Vaughn by Kent Heckenlively, March 14, 2016.

Chapter 9: Curious Alliances

1. Telephone Interview with Minister Tony Muhammad by Kent Heckenlively, April 26, 2016.
2. Conversation with Minister Tony Muhammad and Brian Hooker.
3. Ibid.
4. Ibid.
5. Ibid.
6. Ibid.
7. Ibid.
8. Ibid.
9. Telephone Interview with Minister Tony Muhammad by Kent Heckenlively, April 26, 2016.
10. Ibid.
11. Jim Miller, "Drug Companies Donated Millions to California Lawmakers before Vaccine Debate," *Sacramento Bee*, June 15, 2015, www.sacbee.com/news/politics-government/capitol-alert/article24913978.html.
12. Ibid.
13. Conversation with Minister Tony Muhammad and Brian Hooker.
14. "Bill Posey on CDC Whistleblower," C-Span, July 29, 2015, www.c-span.org/video/?c4554834/user-clip-rep-bill-posey-cdc-whistleblower.

Chapter 10: DeNiro, Tribeca, and the Real *Goodfellas*

1. Pam Belluck and Melena Ryzick, "Robert De Niro Defends Screening of Anti-Vaccine Film at the Tribeca Film Festival," *New York Times*, March 25, 2016, www.nytimes

.com/2016/03/26/health/vaccines-autism-robert-deniro-tribeca-film-festival-andrew-wakefield-vaxxed.html?_r=0.

2. Maggie Mallon, "Tribeca Film Festival Will Screen Highly Questionable Anti-Vaccine Film," *Glamour*, March 25, 2016, www.glamour.com/inspired/blogs/the-conversation/2016/03/tribeca-film-festival-anti-vaccinedocumentary.

3. Ibid.

4. Email from Robert Kennedy Jr. to Kent Heckenlively= et al., March 25, 2016.

5. Pam Belluck and Melena Ryzick, "Robert De Niro Defends Screening of Anti-Vaccine Film at the Tribeca Film Festival," *New York Times*, March 25, 2016, www.nytimes.com/2016/03/26/health/vaccines-autism-robert-deniro-tribeca-film-festival-andrew-wakefield-vaxxed.html?_r=0.

6. Stephanie Goodman, "Robert De Niro Pulls Anti-Vaccine Documentary From Tribeca Film Festival," *New York Times*, March 26, 2016, www.nytimes.com/2016/03/27/movies/robert-de-niro-pulls-anti-vaccinedocumentary-from-tribeca-film-festival.html?_r=0.

7. Ibid.

8. Jeremy Gerard, "'Vaxxed' Filmmakers Accuse Robert De Niro and Tribeca of 'Censorship' in Wake of Cancellation," *Deadline*, March 26, 2016, www.deadline.com/2016/03/robert-de-niro-vaxxed-tribeca-film-festivalstatement-1201726799/.

9. Mike Adams, "Pressure to Censor VAXXED Documentary at Tribeca Film Festival Came from the Nazi-Linked Sloan Foundation, Headed by CFR Member Paul Joskow," *Natural News*, March 28, 2016, www.naturalnews.com/053449_Tribeca_Film_Festival_Sloan_Foundation_eugenics_and_depopulation.html.

10. Email from Robert Kennedy Jr. to Kent Heckenlively et al., March 27, 2016.

11. Emily Willingham, "Why was Rep. Bill Posey Involved in Tribeca-De Niro-Wakefield Kerfuffle?" *Forbes*, March 27, 2016, www.forbes.com/sites/emilywillingham/2016/03/27/why-was-rep-bill-posey-involved-in-tribeca-deniro-wakefield-kerfuffle/#6a09f27-902b4.

12. Jon Rappaport, "The Vaccine Film Robert De Niro Won't Let His Audience See," *Infowars*, March 27, 2016, www.infowars.com/the-vaccine-film-robert-de-niro-wont-let-his-audience-see/

13. Ed Pilkington, "How the Scientific Community United Against Tribeca's Anti-Vaccination Film," *The Guardian*, March 29, 2016, www.theguardian.com/society/2016/mar/29/tribeca-de-niro-anti-vaccination-film-scientistsresponse.

14. Ibid.

15. Immunization Action Coalition, "IAC Funding 2015," accessed April 7, 2016, www.immunize.org/aboutus/funding.asp.

16. Ed Pilkington, "How the Scientific Community United Against Tribeca's

Anti-Vaccination Film," *The Guardian*, March 29, 2016, www.theguardian.com/society/2016/mar/29/tribeca-de-niro-anti-vaccination-film-scientistsresponse.

17. Statement of Del Bigtree, April 3, 2016, Facebook post, www.facebook.com/photo.php?fbid=10154088256654025&set=p.10154088256654025&type=3&theater.
18. Dave McNary, "Controversial Anti-Vaccination Documentary Gets Release from Cinema Libre (EXCLUSIVE)," *Variety*, March 29, 2016, www.variety.com/2016/film/news/vaxxed-anti-vaccine-documentary-cinema-libre1201741603/.
19. Ibid.
20. W. Ian Lipkin, "Anti-Vaccination Lunacy Won't Stop: Robert De Niro Made the Right Call in Pulling 'Vaxxed' from His Film Festival. But the Bogus Message Rolls On," *Wall Street Journal*, April 3, 2016, www.wsj.com/articles/anti-vaccination-lunacy-wont-stop-1459721652.
21. Ibid.
22. Ibid.
23. Alisyn Camerota, "Interview with Bob Wright on Donald Trump, Autism Speaks," *New Day*, April 1, 2016, www.transcripts.cnn.com/TRANSCRIPTS/1604/01/nday.06.html.
24. Sharyl Attkisson, "Former NBC Chief Bob Wright on Vaccines and Autism," *Sharyl Attkisson*, April 5, 2016, www.sharylattkisson.com/former-nbc-chief-bob-wright-on-vaccines-and-autism/.
25. Ibid.
26. Fandango, "Fan Reviews of VAXXED: From Cover-up to Catastrophe," Fandango Movie Reviews, Accessed April 8, 2016, www.fandango.com/vaxxed:fromcoveruptocatastrophe_191505/moviereviews.
27. Box Office Mojo, "Daily Box Office for April 1, 2, & 3, 2016," Box Office Mojo Daily Box Office, Accessed April 8, 2016, www.boxofficemojo.com/daily/chart/?sortdate=2016-04-03&p=.htm.
28. J.B. Handley, "Backfire: How a Pharma-Funded 'Listserv' and Censorship are Turning the Movie Vaxxed into a Worldwide Phenomenon," *Medium*, April 7, 2016, www.medium.com/@jbhandley/backfire-how-a-pharmafunded-listserv-and-censorship-are-turning-the-movie-vaxxed-into-a-53b7c849a13f#.z6h1nuz9q
29. Ibid.
30. Ibid.
31. Celia Farber, "What They Said: Andrew and Carmel Wakefield, on Houston-Gate," *Truth Barrier*, April 7, 2016, www.truthbarrier.com/2016/04/07/what-they-said-andrew-and-carmel-wakefield-on-houston-gate/.
32. "Robert De Niro on Anti-Vaccine Film Controversy: Let's Find Out the Truth," *Today* show, April 13, 2016, www.today.com/video/robert-de-niro-on-anti-vaccine-film-controversy-let-s-find-out-the-truth-665031235642.

33. Email from Leslie Manookian to Kent Heckenlively et al., April 13, 2016, 3:02 PM.
34. Richard Ford Burley, "Damnit De Niro, I Thought We Were Good," *This Week in Tomorrow*, April 14, 2016, www.thisweekintomorrow.com/damnit-de-niro-i-thought-we-were-good-vol-3-no-23-4/.
35. Jeffey Jaxen, "Sacramento Senator Pan Runs Away from Vaxxed Movie Producers," May 9, 2016, www.jeffereyjaxen.com/blog/sacramento-senator-pan-runs-away-from-vaxxed-movie-producers.
36. Francesca Alesse, "Interview with Del Bigtree at State Capitol after Visit to Pan Office," May 10, 2016, www.vimeo.com/166083184.
37. Jeffrey Jaxen, "Compton Mayors Office Offers Free Premiere of VAXXED, Community Revolts against Big Pharma Control," May 19, 2016, www.jefferey-jaxen.com/blog/video-compton-mayor-offers-free-premiere-of-vaxxedcommunity-revolts-against-big-pharma-control.
38. Ibid.
39. Ibid.
40. Jada Yuan, "Robert De Niro on Making His Own Vaccines Documentary and Returning to Boxing and Cannes with Hands of Stone," *Vulture*, May 20, 2016, www.vulture.com/2016/05/robert-de-niro-vaccines-cannes-handsof-stone.html.
41. Ibid.
42. Email from Robert Kennedy Jr. to Kent Heckenlively et al., May 21, 2016, 11:55 p.m.
43. Andy Wakefield, "Statement of March 2016," VAXXED: From Cover-Up to Catastrophe website, accessed May 8, 2016, www.vaxxedthemovie.com/directors-statement/.

Chapter 11: The Battle for California, America, and My Hometown

1. Tim Bolen, "The Lawsuit Against SB 277 Has Been Filed," *The Bolen Report*, July 1, 2016, www.bolenreport.com/lawsuit-sb-277-filed/.
2. Kent Heckenlively, "Battle for California—Stopping the Pharma Juggernaut," *Age of Autism*, July 22, 2016, www.ageofautism.com/2016/07/battle-for-california-stopping-the-pharma-juggernaut.html.
3. Tim Bolen, "Federal Court Lawsuit to be Filed Against SB 277 Tomorrow Morning," *The Bolen Report*, June 30, 2016, www.bolenreport.com/federal-court-lawsuit-filed-sb-277-tomorrow-morning/.
4. Ibid.
5. Ibid.
6. Ibid.
7. Ibid.
8. "Chaffetz Office with Tami Canal and Del Bigtree," June 16, 2016, *YouTube*, www.youtube.com/watch?v=lApsnJkndrk.

9. Ibid.
10. "VAXXED: News From Del Bigtree after Meeting Rep. Chaffetz in D.C.," June 17, 2016, *YouTube*, www.youtube.com/watch?v=8UP11XeSN2s.
11. Ibid.
12. Levi Quackenboss, "Quackenboss Goes to Washington," Levi Quackenboss, June 16, 2016, www.leviquackenboss.wordpress.com/2016/06/16/quackenboss-goes-to-washington/.
13. Ibid.
14. Ibid.
15. Ibid.
16. VAXXED: From Cover-Up to Catastrophe, Rotten Tomatoes, accessed July 29, 2016, www.rottentomatoes.com/m/vaxxed_from_cover_up_to_catastrophe_2016/?search=VAXXED.
17. Email from Peter Klenow to Kent Heckenlively, July 22, 2016.
18. Kent Heckenlively, "The Coming War and its Aftermath," *Bolen Report*, July 26, 2016, www.bolenreport.com/coming-war-aftermath/.
19. Ibid.
20. Ibid.
21. Ibid.

Chapter 12: Can Science End the Autism Epidemic?

1. Robert Naviaux, "Breakthroughs in the Diagnosis and Treatment of Autism," UCSD Mitochondrial and Metabolic Center, November 10, 2013.
2. Robert Naviaux, Zarazuela Zolkipli, Lin Wang, Tomohiro Nakayama, Jane Naviaux, Thuy Le, Michael Schucbacher, Mihael Rogac, Qinbo Tang, Laura Dugan, and Susan Powell, "Antipurinergic Therapy Corrects Autism-Like Features in the Poly(IC) Mouse Model," PLOS One, March 13, 2013, doi: 10.137/journal.pone.0057380.
3. "Drug Treatment Corrects Autism Symptoms in Mouse Model," *Science Daily*, March 13, 2013, www.sciencedaily.com/releases/2013/03/130313182019.htm.
4. Ibid.
5. Jane Naviaux, Michael Schuchbacher, K. Li, L. Wang, V. B. Risborough, Susan Powell, and Robert Naviaux, "Reversal of Autism-Like Behaviors and Metabolism in Adult Mice with Single-Dose Antipurinergic Therapy," *Translation Psychiatry*, June 17, 2014, doi: 10.1038/tp.2014.33.
6. "Century-Old Drug Reverses Autism-Like Symptoms in Fragile X Mouse Model," *Science Daily*, January 15, 2015, www.sciencedaily.com/releases/2015/01/150115163535.htm.
7. Ibid.
8. Robert Naviaux, "Newsletter—The UCSD Suramin Autism Study," UCSD Mitochondrial and Metabolic Disease Center, Newsletter #2, April 15, 2016.

9. Robert Naviaux, "Suramin Treatment of Autism-Clinical Trial Update," University of California, San Diego, School of Medicine, January 18, 2016.
10. Telephone Interview with Dr. Eric Gordon by Kent Heckenlively, June 3, 2016.
11. Ibid.
12. Ibid.
13. Ibid.
14. Ibid.
15. Ibid.
16. Ibid.
17. Ibid.
18. Ibid.

Chapter 13: Justice

1. Barbara Loe Fisher, "RFK Issues Ultimatum to CDC's Dr. Frank Destefano," *Age of Autism*, October 26, 2015, www.ageofautism.com/2015/10/rfk-destefano-cdc.html.
2. Robert Pear, "Reagan Signs Bill on Drug Exports and Payment for Vaccine Injuries," *New York Times*, November 15, 1986, www.nytimes.com/1986/11/15/us/reagan-signs-bill-on-drug-exports-and-payment-forvaccine-injuries.html.
3. James Michener, "A Buoyant, Optimistic Convention," *U.S. News and World Report*, September 3, 1984.

Epilogue: Trump Hope & Reality

1. Kent Heckenlively, "Trump Meets Dr. Wakefield," *Bolen Report*, November 10, 2016, www.bolenreport.com/trump-meets-dr-wakefield/.
2. Ibid.
3. Ibid.
4. Zack Kopplin, "Trump Met with Prominent Anti-Vaccine Activists during Campaign," *Science*, November 18, 2016, www.sciencemag.org/news/2016/11/trump-met-prominent-anti-vaccine-activists-during-campaign.
5. Casey Ross, "Andrew Wakefield Appearance at Trump Inaugural Ball Triggers Social Media Backlash," STAT, January 21, 2017, www.statnews.com/2017/01/21/andrew-wakefield-trump-inaugural-ball/.
6. Meredith Wadman, "Exclusive Q&A: Robert F. Kennedy Jr. on Trump's Proposed Vaccine Commission," *Science*, January 10, 2016, www.sciencemag.org/news/2017/01/exclusive-qa-robert-f-kennedy-jr-trumps-proposed-vaccine-commission.
7. Carter Coudriet, "Bill Gates Again World's Second Richest Person after One Day Behind Arnault," *Forbes*, November 6, 2019, www.forbes.com/sites/cartercoudriet/2019/11/06/bill-gates-second-richest-arnault-bezos/#77d624784e7c.
8. Bill and Melinda Gates Foundation, "Foundation Fact Sheet" (Accessed August 31,

2020), www.gatesfoundation.org/Who-We-Are/General-Information/Foundation; Factsheet Helen Branswell, "The Education of President Trump by Bill Gates, Global Health Advocate," STAT News, April 20, 2017, www.statnews.com/2017/04/20/bill-gates-donald-trump/.

9. Ibid.
10. Ibid.
11. Meredith Waldman, "Lawsuit at Columbia University Roils Prominent Chronic Fatigue Syndrome Research Lab," *Science*, May 23, 2017, www.sciencemag.org/news/2017/05/lawsuit-columbia-university-roils-prominent-chronic-fatigue-syndrome-research-lab.
12. Ibid.
13. Ibid.
14. Kristian G. Andersen, Andrew Rambault, W. Ian Lipkin, Edward Holmes, and Robert Garry, "The Proximal Origins of SARS-CoV-2," *Nature Medicine*, March 17, 2020: DOI: 10.1038/s41591-020-0820-9.
15. Sharyl Attkisson, "How a Pro-Vaccine Doctor Reopened Debate about Link to Autism," *The Hill*, January 13, 2019, www.thehill.com/opinion/healthcare/425061-how-a-pro-vaccine-doctor-reopened-debate-about-link-to-autism.
16. Ibid.
17. Sharyl Attkisson, "Dr. Andrew Zimmerman's Full Affidavit on Alleged Link Between Vaccines and Autism that U.S. Government Covered Up," Sharyl Attkisson, January 6, 2019, www.sharylattkisson.com/2019/01/dr-andrew-zimmermans-full-affidavit-on-alleged-link-between-vaccines-and-autism-that-u-s-govt-covered-up/.
18. J.B. Handley, *How to End the Autism Epidemic*, Chelsea Green Publishing, White River Junction, Vermont (2018).
19. Sharyl Attkisson, "Dr. Andrew Zimmerman's Full Affidavit on Alleged Link between Vaccines and Autism that U.S. Government Covered Up," Sharyl Attkisson, January 6, 2019, www.sharylattkisson.com/2019/01/dr-andrew-zimmermans-full-affidavit-on-alleged-link-between-vaccines-and-autism-that-u-s-govt-covered-up/.
20. Bill Gates, "What You Need to Know About the Covid-19 Vaccine," April 30, 2020, *Gates Notes: The Blog of Bill Gates*, www.gatesnotes.com/Health/What-you-need-to-know-about-the-COVID-19-vaccine?WT.mc_id=20200430165003_COVID-19-vaccine_BG-TW&WT.tsrc=BGTW&linkId=87665522.
21. Becky Quick, "Watch CNBC's Full Interview with Microsoft Co-Founder Bill Gates on the Coronavirus Pandemic and His Work Toward a Vaccine," CNBC, April 9, 2020, www.cnbc.com/video/2020/04/09/watch-cnbcs-full-interview-with-microsoft-co-founder-bill-gates-on-past-pandemic-warnings.html?__source=iosappshare%7C-com.apple.UIKit.activity.Mail&fbclid=IwAR0RG79OtbdXUpY_ylULT6sY_Xo5D-cBQ0awSo6vGS19VnLGqB9z.

22. Robert Kennedy Jr., "Here's Why Bill Gates Wants Indemnity . . . Are You Willing to Take the Risk?", Children's Health Defense, April 11, 2020, www.childrenshealthdefense .org/news/heres-why-bill-gates-wants-indemnity-are-you-willing-to-take-the-risk/.
23. ChienTe Tseng, Elena Sbrana, et al., "Immunization with SARS Coronavirus Vaccines Leads to Pulmonary Immunopathology on Challenge with SARS Virus," *PLOS One*, April 20, 2012: doi: 10.1371/journal.pone.0035421.
24. Ibid.